AAP Advances in Nutraceuticals

IMMUNE-BOOSTING NUTRACEUTICALS FOR BETTER HUMAN HEALTH

Novel Applications

Edited by

Urmila Jarouliya, PhD

Raj K. Keservani, PhD, MPharm

Rajesh K. Kesharwani, PhD

Virendra K. Patel, PhD

Adi D. Bharti, MPharm

AAP | APPLE ACADEMIC PRESS

First edition published 2024

Apple Academic Press Inc.
1265 Goldenrod Circle, NE,
Palm Bay, FL 32905 USA

760 Laurentian Drive, Unit 19,
Burlington, ON L7N 0A4, CANADA

CRC Press
2385 NW Executive Center Drive,
Suite 320, Boca Raton FL 33431

4 Park Square, Milton Park,
Abingdon, Oxon, OX14 4RN UK

Library and Archives Canada Cataloguing in Publication

Title: Immune-boosting nutraceuticals for better human health : novel applications / edited by Urmila Jarouliya, PhD, Raj K. Keservani, MPharm, Rajesh K. Kesharwani, PhD, Virendra K. Patel, PhD, Adi D. Bharti, MPharm.
Names: Jarouliya, Urmila, editor. | Keservani, Raj K., 1981- editor. | Kesharwani, Rajesh Kumar, 1978- editor. | Patel, Virendra K., editor. | Bharti, Adi D., editor.
Series: AAP advances in nutraceuticals.
Description: First edition. | Series statement: AAP advances in nutraceuticals | Includes bibliographical references and index.
Identifiers: Canadiana (print) 20230506518 | Canadiana (ebook) 20230506542 | ISBN 9781774913574 (hardcover) | ISBN 9781774913581 (softcover) | ISBN 9781003371069 (ebook)
Subjects: LCSH: Functional foods. | LCSH: Immune system—Nutritional aspects.
Classification: LCC QP144.F85 I46 2024 | DDC 613.2—dc23

Library of Congress Cataloging-in-Publication Data

CIP data on file with US Library of Congress

ISBN: 978-1-77491-357-4 (hbk)
ISBN: 978-1-77491-358-1 (pbk)
ISBN: 978-1-00337-106-9 (ebk)

ABOUT THE BOOK SERIES: AAP ADVANCES IN NUTRACEUTICALS

In the modern era, mankind has witnessed a paradigm shift with respect to fundamental eating behavior. The lack of physical workouts and busy schedules at offices and in households have promoted the consumption of junk foods, which eventually results in numerous diseases and disorders since the nutritional content from fast food is inadequate. Plumpness and obesity have become a global health threat. The leading causes of death from most developing nations are noncommunicable diseases such as cardiovascular diseases, cancer, arthritis, osteoporosis, and liver toxicity. Patients suffering from such lifestyle ailments are a bit apprehensive towards prolonged use of costly modern therapeutics, which encourages instead the use of alternative approaches for management of such diseases and disorders.

The emerging sector of the nutraceutical industry encompasses products derived from nature, dietary supplements, and functional foods. Nutraceuticals are used for treatment and prevention of a broad range of diseases, such as the common cold, arthritis, sleep-related disorders, cancers, cardiovascular complications, metabolic disorders, and others.

The research on nutraceuticals is increasing day by day, considering the beneficial effects of food or food supplements in the management of diverse diseases. The issue of paramount concern is standardization and establishment of clinical efficacy of nutraceuticals, which is in fact a challenge for researchers around the globe.

This book series aims at realizing the significance of the variety of nutraceuticals for human well-being. The books in the series will also emphasize the role of dieticians and nutritionists for the prescription of judicious eating habits. The key food components such as carbohydrates, proteins, and lipids as well as micronutrients (vitamins, minerals) are demonstrated to maintain good health, obviating the need for medicines. Thus, nutraceuticals that are indeed derived from food ingredients are believed to be potential alternative therapeutics.

Asymptomatic diseases/disorders necessitate proper diagnosis. Stressed circumstances have been reported to cause weight loss. Workplace stress is a key variable in the genesis and propagation of metabolic disorders.

The nuances of the underlying mechanisms are still to be deciphered. Poor lifestyle trends have been attributed along with the consumption of junk foods in several such instances. Sugar-rich carbonated beverages, the use of contraband drugs, and the consumption of liquor in excess are assigned as risk factors. This book series will cover advances and applications in addition to providing the basics of nutraceuticals.

The books in this series may be valuable resources for industry professionals to design and develop quality products for end users. In addition, the book series will be fortified by ideas of innovation, concept building, manufacturing aspects, quality control, and regulatory status of nutraceuticals.

Ostensibly, business has seen a rise in the nutraceutical consumption in recent years. Pure, safe, efficacious products are the need of the hour globally. There are a number of books on nutraceuticals in the market, yet these are only a few books presenting certain aspects of nutraceuticals therein. This book series aims to offer a holistic view of this promising strategy, which is picking up pace with time. This book series will be a unique endeavor to bring manufacturing, research and development, and marketing strategies under a single umbrella.

CURRENT BOOKS IN THE SERIES

- **Micronutrients and Macronutrients as Nutraceuticals**
  Editors: Prakash Chandra Gupta, PhD, Sayan Bhattacharyya, MBBS, MD, Nisha Sharma, PhD, MPharm, Rajesh K. Kesharwani, PhD, and Raj K. Keservani, MPharm
- **Formulations, Regulations, and Challenges of Nutraceuticals**
  Editors: Tingirikari Jagan Mohan Rao, PhD, Rajesh K. Kesharwani, PhD, Raj. K. Keservani, MPharm, and Anil K. Sharma, PhD
- **Food Supplements and Dietary Fibers in Health and Disease**
  Editors: Bhushan R. Rane, PhD, MPharm, Raj K. Keservani, MPharm, Durgesh Singh, DPhil, Nayan A. Gujarathi, PhD, and Ashish S. Jain, PhD
- **Nutraceuticals in Cancer Prevention, Management, and Treatment**
  Editors: Raj K. Keservani, MPharm, Bui Thanh Tung, PhD, Sippy Singh, DPhil, and Rajesh K. Kesharwani, PhD
- **Advances in Flavonoids for Human Health and Prevention of Diseases**
  Editors: Nisha Sharma, PhD, Deepika Saini, PhD, Rajesh K. Kesharwani, PhD, Prakash Chandra Gupta, and Raj. K. Keservani, MPharm

- **The Flavonoids: Extraction and Applications**
  Editors: Deepika Saini, PhD, Rajesh K. Kesharwani, PhD, and
  Raj K. Keservani, MPharm
- **Immune-Boosting Nutraceuticals for Better Human Health: Novel Applications**
  Editors: Urmila Jarouliya, PhD, Raj K. Keservani, MPharm,
  Rajesh K. Kesharwani, PhD, Virendra K. Patel, PhD, and
  Adi D. Bharti, MPharm
- **Applications of Functional Foods in Disease Prevention**
  Editors: Raj K. Keservani, MPharm, and Eknath D. Ahire, MSPharm
- **Plant Metabolites and Vegetables as Nutraceuticals**
  Editors: Raj K. Keservani, MPharm, Bui Thanh Tung, PhD,
  Rajesh K. Kesharwani, PhD, MPharm, and Eknath D. Ahire
- **Preventive and Therapeutic Role of Vitamins as Nutraceuticals**
  Editors: Khemchand R. Surana, Eknath D. Ahire,
  Raj K. Keservani, MPharm, and Rajesh K. Kesharwani, PhD
- **Nutraceuticals and Bone Health**
  Editors: Deepak Sharma, PhD, Madan Mohan Gupta, PhD,
  Anil K. Sharma, PhD, MPharm, Raj K. Keservani, MPharm, and
  Rajesh K. Kesharwani, PhD
- **Flavonoids as Nutraceuticals**
  Editors: Rajesh K. Kesharwani, PhD, MTech, Deepika Saini, PhD,
  Raj K. Keservani, MPharm, and Anil Kumar Sharma, PhD, MPharm
- **Herbals as Nutraceuticals: Their Role in Healthcare**
  Editors: Raj K. Keservani, MPharm,
  Rajeshwar Kamal Kant Arya, PhD, MPharm, and
  Rajesh K. Kesharwani, PhD, MTech
- **Nutraceutical Fruits: Overview and Disease Prevention**
  Editor: Raj K. Keservani, MPharm
- **Nutrigenomics and Nutraceuticals**
  Editors: Raj K. Keservani, MPharm, Eknath D. Ahire,
  Shubham J. Khainar, PhD, Sanjay J. Kshirsagar, PhD, and
  Rajesh K. Kesharwani, PhD
- **The Nature of Nutraceuticals: History, Properties, Sources, and Nanotechnology**
  Editors: Rajesh K. Kesharwani, PhD, Prashant Kumar, PhD, and
  Raj K. Keservani, MPharm

- **Nutraceuticals for the Treatment and Prevention of Sexual Disorders**
  Editors: Raj K. Keservani, MPharm, Sharangouda J. Patil, PhD, and Ivan Aranha, PhD
- **Antioxidants as Nutraceuticals**
  Editors: Nayan A. Gujaratih, PhD, Raj K. Keservani, PhD, MPharm, Rajesh K. Kesharwani, PhD, Bhushan R. Rane, PhD, and Yogeeta Sameer Goyal, PhD
- **Nutraceuticals in Cardiac Health Management**
  Editors: Raj K. Keservani, PhD, Rajesh K. Kesharwani, PhD, Praveen Kumar Jain, PhD, Bhushan R. Rane, PhD, and Nayan A. Gujarathi, PhD
- **Nutraceuticals in Obesity Management and Control**
  Editors: Raj K. Keservani, PhD, and Alka Lohani, PhD
- **Nutraceuticals in Arthritis and Psoriasis**
  Editors: Meenakshi Jaiswal, PhD, Raj K. Keservani, PhD, Rajesh K. Kesharwani, PhD, and Swati G. Talele, PhD

ABOUT THE EDITORS

Urmila Jarouliya, PhD
School of Studies in Biochemistry and Biotechnology, Jiwaji University, Gwalior, Madhya Pradesh, India

Urmila Jarouliya, PhD, was a Project Associate in the University Grants Commission-funded research project (2005–2008) at the School of Studies in Biochemistry and Biotechnology, Jiwaji University, Gwalior, Madhya Pradesh, India. In the year 2011, she earned her PhD at Jiwaji University, Gwalior, Madhya Pradesh, India. Dr. Jarouliya's research interest includes clinical biochemistry (mainly diabetes mellitus and basic complications related to it) and algal biotechnology (*Spirulina* Sp.). She has published 11 research papers in national and international journals and 12 book chapters with international publishers and has presented research papers as well as poster papers at several scientific conferences.

Raj K. Keservani, PhD, MPharm
Associate Professor, Faculty of BPharmacy, CSM Group of Institutions, Prayagraj, Uttar Pradesh, India

Raj K. Keservani, PhD, MPharm, is currently working as an Associate Professor, Faculty of BPharmacy, CSM Group of Institutions, Prayagraj, India. He has more than 12 years of academic (teaching) experience from various institutes of India in pharmaceutical education. He has supervised more than 25 UG students for their research work, has published 31 peer-reviewed papers in the field of pharmaceutical sciences in national and international journals, one patent, 43 book chapters, three co-authored books, and 19 edited books. He is also active as a reviewer for several international scientific journals. Mr. Keservani graduated with a pharmacy degree from the Department of Pharmacy, Kumaun University, Nainital (Uttarakhand), India. He received his Master of Pharmacy (MPharm) (specialization in pharmaceutics) from the School of Pharmaceutical Sciences, Rajiv Gandhi Proudyogiki Vishwavidyalaya, Bhopal, India. Mr. Keservani is a life member of the Society of Pharmaceutical Education and Research (SPER). His research interests include nutraceutical and functional foods, novel drug delivery systems (NDDS), transdermal drug delivery/drug delivery, health science, cancer biology, and neurobiology.

Rajesh K. Kesharwani, PhD

Associate Professor, Department of Computer Application, Nehru Gram Bharati (Deemed to be University), Prayagraj, Uttar Pradesh, India

Rajesh K. Kesharwani, PhD, MTech, is working as an Associate Professor in the Department of Computer Application at Nehru Gram Bharati (Deemed to be University), Prayagraj, India. He has more than 11 years of research and nine years of teaching experience in various institutes of India, imparting bioinformatics and biotechnology education. He has received several awards, including the NASI-Swarna Jayanti Puruskar by The National Academy of Sciences of India. He has supervised one PhD, and more than 20 UG, PG students for their research work, has authored over 49 peer-reviewed articles, 20 book chapters, and 14 edited books with international publishers (e.g., Springer Nature, IGI Global, Taylor & Francis, etc.). He has been a member of many scientific communities as well as a reviewer for many international journals. He has presented many papers at various national and international conferences. Dr. Kesharwani received his PhD from the Indian Institute of Information Technology, Allahabad, and worked at NIT Warangal for two-semester. He has been a recipient of the Ministry of Human Resource recipient of Development (India) Fellowship and a Senior Research Fellowship from the Indian Council of Medical Research, India. His research fields of interest are medical informatics, protein structure and function prediction, computer-aided drug designing, structural biology, drug delivery, cancer biology, nano-biotechnology, and biomedical sciences.

Virendra K. Patel, PhD

Professor, Sri Sathya Sai Institute of Pharmaceutical Sciences, RKDF University, Bhopal, Madhya Pradesh, India

Virendra K. Patel, PhD, MPharm, is working as a Professor at Sri Sathya Sai Institute of Pharmaceutical Sciences RKDF University, Bhopal, Madhya Pradesh, India. Dr. Patel has 15 years of academic experience in pharmacy institutes. He did his BPharm in 2004 at RKDF College of B. Pharmacy Bhopal, Madhya Pradesh, India, and MPharm (Pharmaceutical Chemistry) in 2007 at the School of Pharmaceutical Sciences, Rajiv Gandhi Proudyogiki Vishwavidyalaya, Bhopal, Madhya Pradesh, India. Dr. Patel earned PhD in Pharmaceutical Sciences in 2021 from Ram Krishna Dharmarth Foundation University, Bhopal, Madhya Pradesh, India. He has published seven review and research articles in national and international journals of repute and published four Indian patents. Dr. Patel supervised 10 post-graduate students. His research interest includes drug analysis, synthetic chemistry, phytochemical screening, and nutraceuticals.

Adi D. Bharti, MPharm
Deputy Manager, Indian Medicines Pharmaceutical Corporation Limited, Mohan, Uttarakhand, India

Adi D. Bharti, MPharm, is presently working as a Deputy Manager at Indian Medicines Pharmaceutical Corporation Limited, Mohaan, Uttarakhand, India. Mr. Bharti graduated from the Department of Pharmacy, Kumaun University in 2005, Nainital, Uttarakhand, India, and post-graduated from the School of Pharmaceutical Sciences, Rajiv Gandhi Proudyogiki Vishwavidyalaya in 2009, Bhopal, India. Mr. Bharti qualified GATE in 2005. He has published articles in national and international journals of repute. His research interest includes NDDS, drug delivery, and nutraceuticals.

CONTENTS

CONTRIBUTORS

Anshu Agarwal
Department of Zoology, Agra College, Dr. B. R. Ambedkar University, Agra, Uttar Pradesh, India

Nitin Agrawal
Faculty of Pharmacy, Raja Balwant Singh Engineering Technical Campus, Uttar Pradesh, India

Eknath D. Ahire
Department of Pharmaceutics, MET's Institute of Pharmacy, Affiliated to Savitribai Phule Pune University, Maharashtra, India

Ashima Ahuja
Institute of Pharmaceutical Research, GLA University, Mathura, Uttar Pradesh, India

Ehsan Amiri
Department of Phytopharmaceuticals (Traditional Pharmacy), Faculty of Pharmacy, Shiraz University of Medical Sciences, Shiraz, Iran

Suresh B. Arakera
Department of Applied Genetics, Karnatak University, Dharwad, Karnataka, India

Bharti Bairagi
Department of Pharmaceutics, Corporate Institute of Pharmacy, Bhopal, Madhya Pradesh, India

Deepti Dwivedi
Institute of Pharmacy, Dr. Ram Manohar Lohia Avadh University, Ayodhya, Uttar Pradesh, India

Ena Gupta
Department of Family and Community Sciences, University of Allahabad, Uttar Pradesh, India

Swati Gupta
Center of Biotechnology, Nehru Science Center, University of Allahabad, Prayagraj, Uttar Pradesh, India

Meenu Jain
Gajra Raja Medical College, Gwalior, Madhya Pradesh, India

Urmila Jarouliya
SOS in Biochemistry, Jiwaji University, Gwalior, Madhya Pradesh, India

Shubham J. Khairnar
Department of Pharmacology, MET's Institute of Pharmacy, Affiliated to Savitribai Phule Pune University, BKC, Adgaon, Nashik, Maharashtra, India

A. Sameera Khathoon
Post-Graduate and Research Department of Zoology, J.B.A.S. College for Women, Teynampet, Chennai, Tamil Nadu, India

Anurag Samson Lall
Department of Chemistry, Sam Higginbottom University of Agriculture, Technology, and Sciences, Prayagraj, Uttar Pradesh, India

Sunil K. Mahajan
Department of Pharmaceutical Chemistry, MGV's Pharmacy College, Panchavati, Nashik, Maharashtra, India

Jyoti Vandana Mani
Department of Chemistry, Sam Higginbottom University of Agriculture, Technology, and Sciences, Prayagraj, Uttar Pradesh, India

Shanti Bhushan Mishra
Department of Pharmacognosy, United Institute of Pharmacy, Prayagraj, Uttar Pradesh, India

R. Md. Anwar Nawaz
Department of Biomedical Engineering, Karpaga Vinayaga College of Engineering and Technology, Chengalpattu, Tamil Nadu, India

Uroosa Noor
Department of Family and Community Sciences, University of Allahabad, Uttar Pradesh, India

Dhananjay M. Patil
Department of Pharmaceutics, Shreeshakti Shaikshanik Sanstha, Divine College of Pharmacy, Satana, Nashik, Maharashtra, India

R. S. Raghuwanshi
Department of Pharmaceutical Chemistry, Corporate Institute of Pharmacy, Bhopal, Madhya Pradesh, India

Md. Akhlaqur Rahman
Department of Biotechnology, S.S. Khanna Girls' Degree College, Prayagraj, Uttar Pradesh, India.

Supriya Ray
Department of Applied Genetics, Karnatak University, Dharwad, Karnataka, India

Divya Rani Sharma
Department of Pharmacology, United Institute of Pharmacy, Prayagraj, Uttar Pradesh, India

Shiwangi Sharma
Institute of Pharmaceutical Research, GLA University, Mathura, Uttar Pradesh, India

Ajay Kumar Shukla
Institute of Pharmacy, Dr. Ram Manohar Lohia Avadh University, Ayodhya, Uttar Pradesh, India

Amit Kumar Singh
Department of Pharmaceutics, United Institute of Pharmacy, Prayagraj, Uttar Pradesh, India

Kavita Singh
National Institute of Immunology, New Delhi, India

Sandeep Kumar Singh
Institute of Pharmacy, Dr. Ram Manohar Lohia Avadh University, Ayodhya, Uttar Pradesh, India

Satya Prakash Singh
Institute of Pharmacy, Dr. Ram Manohar Lohia Avadh University, Ayodhya, Uttar Pradesh, India

Shradhanjali Singh
Department of Pharmaceutical Chemistry, United Institute of Pharmacy, Prayagraj, Uttar Pradesh, India

Sonia Singh
Institute of Pharmaceutical Research, GLA University, Mathura, Uttar Pradesh, India

K. Sivakumar
Department of Biomedical Engineering, Karpaga Vinayaga College of Engineering and Technology, Chengalpattu, Tamil Nadu, India

Shashi Soni
Department of Family and Community Sciences, University of Allahabad, Uttar Pradesh, India

Ankur Srivastava
Institute of Pharmacy, Dr. Ram Manohar Lohia Avadh University, Ayodhya, Uttar Pradesh, India

Shruti Srivastava
Center of Biotechnology, Nehru Science Center, University of Allahabad, Prayagraj, Uttar Pradesh, India

Shanthy Sundaram
Center of Biotechnology, Nehru Science Center, University of Allahabad, Prayagraj, Uttar Pradesh, India

Khemchand R. Surana
Department of Pharmaceutical Chemistry, Shreeshakti Shaikshanik Sanstha, Divine College of Pharmacy, Satana, Nashik, Maharashtra, India

Aarti Tiwari
Institute of Pharmacy, Dr. Ram Manohar Lohia Avadh University, Ayodhya, Uttar Pradesh, India

Deep Shikhar Varshney
Faculty of Pharmacy, Raja Balwant Singh Engineering Technical Campus, Bichpuri, Agra, Uttar Pradesh, India

ABBREVIATIONS

6-OHDA	6-hydroxydopamine
AAD	antibiotic-associated diarrhea
A-Beta	amyloid-β
AD	Alzheimer's disease
AGEs	advanced glycation end
AIA	adjuvant-induced arthritis
AIDS	acquired immunodeficiency syndrome
AKK	*Akkermansia muciniphila*
Akt	protein kinase B
ALA	α-linolenic acid
AMD	age-related macular degeneration
AMPs	antimicrobial peptides
AP-1	activator protein-1
APCs	antigen-presenting cells
ARA	arachidonic acid
ASD	autism spectrum disorder
ASX	astaxanthin
ATP	adenosine triphosphate
Aβ	amyloid beta
Bad	Bcl-2 associated agonist of cell death
Bax	Bcl-2-associated X protein
Bcl-2	B-cell lymphoma 2
Bcl-w	alias of Bcl-2-like protein
Bcl-Xl	B-cell lymphoma-extra-large
C6 Rat Glioma	cell line derived from a rat glial tumor
CA	*Candida albicans*
Cacna1c	calcium voltage-gated channel subunit alpha1 c genes
Cacna1d	calcium voltage-gated channel subunit alpha1 d genes
CAM	cell adhesion molecules
CAMKII	CAM-based protein kinase II
Caspase	cysteine-aspartic proteases
CAT	catalase
CFA	complete Freund's adjuvant
CLRs	C-type lectin receptors

COPD	chronic obstructive pulmonary disease
CoQ	coenzyme Q
COX-2	cyclooxygenase-2
CRC	colorectal cancer
CRP	C-reactive protein
CTLA-4	cytotoxic T-lymphocyte-associated
Cu	copper
CVD	cardiovascular disease
CWSP	*C. pyrenoidosa* water-soluble polysaccharide
Cyt C	cytochrome C
DC	dendritic cell
DHA	docosahexaenoic acid
DP	dipeptidase
EC	endothelial cell
ECHO	educational concerns for hunger organization
EGCG	epigallocatechin gallate
EOs	essential oils
EPA	eicosapentaenoic acid
EPO	erythropoietin hormone
EPR	epiregulin
EPS	exocellular polysaccharides
ER	endoplasmic reticulum
ERK	extracellular signal-regulated kinase
ETC	electron transport chain
$FADH_2$	flavin adenine dinucleotide
FAO	Food and Agricultural Organization
FDA	Food and Drug Administration
Fe^{2+}	ferrous ion
FFA	free fatty acids
FFAR-2	free fatty acid receptors
FOS	fructooligosaccharides
GALT	gut-associated lymphoid tissue
GCS	γ-glutamyl cysteine synthetase
GGT	γ-glutamyl transferase
GI	gastrointestinal
GIT	gastrointestinal tract
GLA	γ-linoleic acid
GOFA	4′geranyloxyferulic
GOS	galacto-oligosaccharides

GPx	glutathione peroxidase
GR	glutathione reductase
GRAS	generally recognized as safe
GSH	glutathione
GSK-3β	glycogen synthase kinase-3β
GSSG	oxidized glutathione
GST	glutathione sulfotransferase
H_2O_2	hydrogen peroxide
HA	humoral antibody
HAMP	hepcidin antibacterial protein
HCMV	human cytomegalovirus
Hdac2	histone deacetylase
HIV	human immunodeficiency virus
HMGB1	high mobility group 1
HO-1	heme oxygenase-1
HPLC	high pressure liquid chromatography
HPV	human papillomavirus
HSV	herpes simplex virus
IAA	indoleacetic acid
IBD	inflammatory bowel disease
ICAM-1	intercellular adhesion molecule-1
IECs	intestinal epithelial cells
IFN	interferon
IKK	IκB kinase
IKKβ	inhibitor of nuclear factor kappa-B kinase subunit β
IL	interleukin
IL-1β	interleukin-1β
ILFs	isolated lymphoid follicles
iNOS	inducible nitric oxide synthase
IP	isoprenoid
Iqgap2	IQ Motif Containing GTPase Activating Protein 2
IRFs	IFN regulatory factors
IUPAC	International Union of Pure and Applied Chemists
IκBα	nuclear factor kappa-light-chain-enhancer of activated B-cells inhibitor alpha
JNK	c-Jun N-terminal kinase
Keap 1	Kelch-like ECH-associated protein 1
LAG-3	lymphocyte-activation gene 3
LCL	lymphoblastoid cell line

LCOFA	long-chain omega-3 fatty acids
LDL	low-density level
LDL-C	low-density level-cholesterol
LP	lamina propria
LPO	lipid peroxides
LPS	lipopolysaccharide
LRR	leucine-rich repeat
MALT	mucosa-associated lymphoid tissue
MAMPs	microbe- or pathogen-associated molecular patterns
MAO	monoamine oxidase
MAPK	mitogen-activated protein kinase
MCP1	monocyte chemoattractant protein 1
MDA	malondialdehyde
MDA-5	melanoma differentiation-associated 5
ME	metabolic endotoxemia
MMP9	matrix metalloproteinase nine
Mn	manganese
MPP^{+}	1-methyl-4-phenylpyridinium
MPTP	1-methyl-4-phenyl-1,2,3,6-tetrahydropyridine
mTOR	mammalian target of rapamycin
NADH	nicotinamide adenine dinucleotide
NADPH	nicotinamide adenine dinucleotide phosphate
NAFLD	nonalcoholic fatty liver disease
NFAT	nuclear factor of activated T-cells
NF-Kb	nuclear factor kappa B
NGOs	non-governmental organizations
NK	natural killer
NLRs	NOD-like receptors
NO	nitric oxide
NOD	nucleotide-binding oligomerization domain
NOX	NADPH oxidase
Nrf2	nuclear factor erythroid 2-related factor 2
NUS	National University of Singapore
O_2	oxygen
OA	osteoarthritis
OH	hydroxide radical
ONOO	peroxynitrite
OPCs	oligomeric proanthocyanidins
OS	oxidative stress

p70S6K	ribosomal protein S6 kinase
PAMPs	pathogen-associated molecular patterns
PARP-1	poly(ADP-ribose) polymerase-1
PBMC	proliferation of mononuclear cells
PCBs	polychlorinated biphenyls
PD	Parkinson disease
PDPN	painful diabetic polyneuropathies
PEM	protein-energy malnutrition
PHA	phytohemagglutinin
PI	propidium iodide
PI3K	phosphoinositide 3-kinase
p-JNK	phosphorylated JNK
PMFs	polyethoxylated flavones
PMN	polymorphonuclear
PP	Peyer's patches
PPARγ	peroxisome proliferator activator receptor γ
PRRs	pattern recognition receptors
PRX	peroxiredoxin
PS	polysaccharides
PUFA	polyunsaturated fatty acid
RA	retinoic acid
RAR	retinoic acid receptors
ROS	reactive oxygen species
SAR	seasonal allergic rhinitis
SCFA	short-chain fatty acids
Se	selenium
SFAs	saturated fatty acids
SLE	systemic lupus erythematosus
SOC	store-operated calcium
SOD	superoxide dismutase
sPS	sulfated polysaccharides
STZ	streptozotocin
TAG	triacylglycerol
TB	tuberculosis
TLR	toll-like receptor
TNFs	tumor necrosis factors
TNF-α	tumor necrosis factor α
USEPA	United States Environmental Protection Agency
UTI	urinary tract infections

UV	ultraviolet
WBC	white blood cells
WHO	World Health Organization
XOS	xylo-oligosaccharides
Zn	zinc

PREFACE

We all understand that we've been going through a pandemic because of COVID-19 for more than two years, and all of us are concerned on how to build a strong immunity to be able to live a healthy life. Everybody is concerned about building robust immunity. If our immune system is powerful, it is going to play an important role in shielding and defending us from diverse diseases.

Keeping this in mind, we are happy to have edited this book on this thrilling and upcoming new subject of immune-boosting nutraceuticals. We aim for the book to be a useful guide on nutraceuticals or nutrients that should be included in our everyday eating regimen to make our immune system more potent. We hope that it will be well-accepted by dieticians, medical students, and researchers and that it will display newer paths for college students and researchers in this very exciting and upcoming area.

This book focuses on how dietary antioxidants, vitamins, minerals, edible microalgae, herbs, phytonutrients, omega 3-fatty acids, or even probiotics, which are resources of numerous active nutraceutical components, facilitate in strengthening the immune system, which, in consequence, fights diverse infectious diseases and makes our system strong and healthy.

This book, Immune-Boosting Nutraceuticals for Better Human Health: Novel Applications, comprises 15 chapters in four sections that describe using nutraceuticals to boost the immune system for well-being.

PART I: IMMUNOBOOSTING PROPERTIES OF HERBS AND VEGETABLES

Various medicinal species and herbal plants are used in Chapter 1 – "Herbs to Improve Human Health," written by Sameera Khathoon A., to show the importance of traditional medicine in the treatment of various disorders involved in day-to-day human life.

Chapter 2 – "Novel Importance of Herbs and Their Effects in Human Health," written by Ena Gupta et al., focuses on the pharmacological and therapeutic importance of commonly used Indian medicinal herbs on human health.

The immunological and biological actions of the primary carotenoids used to treat various illnesses, as well as their probable mechanism of action, are briefly described in Chapter 3 – "Consumption of Carotenoid-Rich Vegetables and Functions of Human T-Lymphocytes," written by Shanti Bhushan Mishra and his associates.

Chapter 4 – "Role of Vegetables Containing Carotenoids on Human Immune Functions," written by Ajay Kumar Shukla and his associates, report that carotenoids present in vegetables play an important role in human health.

PART II: NUTRITIONAL COMPONENTS TO BOOST IMMUNITY

Probiotics have been accepted as a better preventive and therapeutic option that mitigate the duration/symptoms of the diseases and also serve as an alternative choice to antibiotics, as described by Meenu Jain and her associates in Chapter 5 – "Probiotics and Immune Response in Disease Prevention and Treatment."

Minerals such as magnesium, as well as micronutrients such as selenium, zinc, copper, and others, are required for proper immune function, and those who are lacking in these nutrients may be in danger, as discussed in Chapter 6 – "Minerals as a Gatekeeper of Immune Function" written by R. Md. Anwar Nawaz and K. Sivakumar.

Md. Akhlaqur Rahman et al. discusses phytochemical substances that are present in different citrus fruit peels such as orange (*Citrus sinensis*), kinnow (*Citrus reticulata*) and mosambi (Citrus limetta) grapefruit, which are a rich source of immune-boosting substances in Chapter 7 – "Bioactive Compound of Fruits in Boosting Immunity."

Supriya Ray and Suresh B. Arakera focus on the close association and interaction between nutraceuticals and gut microbiota and the after-effects of these interactions in human and gut health in Chapter 8 – "Role of Nutraceuticals in Gut Health."

Sonia Singh and her co-worker explain the phytochemical classes of some significant nutraceuticals and antioxidants as nutraceuticals with their mode of action in Chapter 9 – "Antioxidants Properties of Nutraceuticals."

Chapter 10 – "Understanding the Role of ω-3 Polyunsaturated Fatty Acids on Autoimmunity," written by Ajay Kumar Shukla et al., summarizes the role of ω-3 PUFAs on autoimmune diseases.

PART III: MARINE-DERIVED SOURCES USED TO BOOST IMMUNITY

Chapter 11 – "Immunomodulatory Effects of Edible Microalgae," written by Md. Akhlaqur Rahman et al., discusses that microalgae-derived compounds could promote better health and alleviate the risk of the development of degenerative diseases.

Chapter 12 – "Antioxidant Action of Glutathione Against Heavy Metal Toxicity in Fishes," written by Jyoti Vandana Mani and her colleagues, provides a comprehensive overview of the mechanism of heavy metal-induced toxicity, generation of reactive oxygen species (ROS), and the antioxidant defense of glutathione (GSH) and associated enzymes in fishes.

Chapter 13 – "Coral Reef: A Source of Value-Added Compounds Beneath the Water," written by Shanti Bhushan Mishra and Shradhanjali Singh, illustrates various compounds obtained from coral reefs and their associated biological activities.

PART IV: AYURVEDIC AND OTHER SOURCES TO BOOST IMMUNITY

Chapter 14 – "Immune-Boosting Property of Ayurvedic Formulations," written by Bharti Bairagi and R. S. Raghuwanshi, presents an overview and introduction and discusses the importance and effects of ayurvedic formulations on the immune system.

Nutrition plays a main role in the control of each acute and persistent illness, in phases of the body's response to pathogenic organisms described in Chapter 15 – "Nutrition, Malnutrition, Infection, and Immunity," by Eknath D. Ahire and his associates.

PART I

Immuno-Boosting Properties of Herbs and Vegetables

CHAPTER 1

HERBS TO IMPROVE HUMAN HEALTH

A. SAMEERA KHATHOON

Post-Graduate and Research Department of Zoology, J.B.A.S. College for Women, Teynampet, Chennai, Tamil Nadu, India

ABSTRACT

Many chronic diseases, such as cancer and cardiovascular disease (CVD), are associated with the inflammatory process, which has been widely focused on the pathogenicity of such diseases and mechanistic studies. To prevent these diseases, various natural and pharmacological methods are used. Various herbal plants are used in this review chapter to show the importance of traditional medicine in the treatment of various disorders involved in day-to-day human life. In addition to this, the main purpose of this study is to show the known action pathways and the potential synergistic effects of the related compounds of various extracts.

1.1 INTRODUCTION

In the prevention of human diseases, Medicinal plants play a vital role and act as a potential source of natural products. Many spices and herbs have been widely used for culinary and medicinal purposes for the past many centuries. Acute and chronic diseases are protected by spices as we know that spices also enhance the flavor, aroma, and color of food and beverages. Excellent sources of antioxidants are present in spices and herbs, including sage, rosemary, clove, oregano, and cinnamon which possess a high content of phenolic compounds. There is some evidence, which shows the consumption of spices

Immune-Boosting Nutraceuticals for Better Human Health: Novel Applications. Urmila Jarouliya, PhD, Raj K. Keservani, MPharm, Rajesh K. Kesharwani, PhD, Virendra K. Patel, PhD, and Adi D. Bharti, MPharm (Eds.)

frequently reduces the risk of ischemic heart and respiratory diseases and also protects against cancer death. However, the actual role of spices and herbs is to maintain good health and also protect against the development of chronic, non-communicable diseases, etc. (Jiang, 2019). The use of herbal extract as a natural alternative medicine in the prevention, treatment, and promotion of various human diseases has grown over time (Sharma et al., 2020; Rane et al., 2020a-c).

Humans used herbal medicines as ancient therapy (Li & Vederas, 2009; Keservani et al., 2010a, b, 2017, 2020). In recent years, due to low complications and lesser side effects compared to synthetic drugs, many people are enthusiastic to use various herbal medicines (Boyd et al., 2013). In the Regard of demand increased by people for daily usage of medicinal plants and other constituent compounds which are involved in the management of various painful neuropathy have been developing all over the world along with the phytopharmaceutical studies (Forouzanfar & Hosseinzadeh, 2018).

1.2 *ARMORACIA RUSTICANA* (HORSERADISH)

As Horseradish has enormous therapeutic properties and has a valuable influence on human health, it has been used by various cultures all over history in treating many human ailments, like bronchitis, sinusitis, urinary bladder infection, paradontosis, rheumatism, anemia, gastritis, pleurisy, and it is also helpful in reducing the pain associated with sciatica and rheumatism (Stillo et al., 2018). The roots of horseradish being a rich source of glucosinolates possess a high amount of other human health constituents, like phenolics, vitamins, and minerals (Stillo et al., 2018; Rivelli et al., 2017). Horseradish is also rich in the following constituents like potassium, calcium, iron, magnesium, phosphorus, and potassium (Velasco et al., 2010).

In recent studies, glucosinolates help in prevention and also inhibit the influences on different types of cancer. There are some other medicinal benefits of horseradish as it is well-known antibacterial properties associated with isothiocyanates, and for the benefit of human health, horseradish possesses a high content of other antioxidants. In the treatment of periodontal disease, horseradish juice is diluted and used as an antibiotic mouthwash.

In addition, horseradish also provides an enzyme that helps in the stimulation of digestion, reduction of constipation, and regulation of bowel movement (Walters, 2021).

Today, Germany has approved horseradish in treating the respiratory tract infections like bronchitis, sinusitis, which also supports the treatment of urinary tract infections (UTI) and urinary stones and minor muscle pains (Shehata et al., 2009).

1.3 *PEGANUM HARMALA* (WILD RUE)

Candidiasis, anti-inflammatory, anti-cholinesterase, anti-bacterial, anti-microbial, anti-tumor, angiogenesis, antiparasitic, antioxidant, prevention of cytotoxicity activities, hepatoprotective, abortifacient potential, and pesticide effects are some of the most important therapeutic effects of *P. harmala* (Majid, 2018). Wang et al. (2016) reported that *P. harmala* has been traditionally used in Chinese and Uygur medicine in the treatment of cancer and also found that *P. harmala* is a promising leading compound in the development of an anti-lung cancer drug.

1.4 *TRIGONELLA FOENUM-GRAECUM* (FENUGREEK)

In the history of India and China, fenugreek has been widely used as a traditional medicine that helps in the treatment of weakness, leg edema, indigestion, baldness, and fever. It also acts as an appetite stimulant and helps in lactation, myalgia, wound treatment and cellulitis. Fenugreek helps in the elevation of blood glucose and lipid levels, which are associated with diabetes and obesity (Jarouliya & Keservani, 2019a) like chronic conditions and it is one of its potential benefits (Keith, 2017).

Human data suggest that higher intakes may be required for a lipid-lowing activity to become significant. Another clinical trial study also showed that serum levels of triglycerides and VLDL-C were decreased significantly (30 and 30.6%, respectively) after taking 10 grams/day of powdered seeds of fenugreek soaked in hot water for 8 weeks in type 2 diabetic patients (Chevassus et al., 2010 a.).

In a recent study, the food which is incorporated with fenugreek could reduce the glycemic index by 21% when it is compared with standard food (Gopalpura et al., 2007). It has been reported that fenugreek seeds at 10 g/day potentially reduces fasting blood glucose level and also decreases HbA1c, serum levels of insulin, a homeostatic model is used to the identify the insulin resistance, and total cholesterol and triglycerides, and in type 2 diabetic patients the incorporation of fenugreek increases the level of serum (Rafraf et al., 2014).

A small-scale, randomized, a double-blind, placebo-controlled human trial showed consumption of 1,200 mg of fenugreek seed extract daily for 14 days significantly decreased fat consumption by 17% compared with a placebo in 12 healthy subjects (Chevassus et al., 2010b).

The extract of fenugreek seeds has been used to demonstrate the activity of modulators hormone and also probably helps in relieving the menopause symptoms. In a randomized, placebo-controlled trial, treatment with fenugreek de-husked seed extract at 600 mg/day resulted in a significant reduction in menopausal symptoms; women aged 40–65 years in the treatment group reported significantly fewer daytime hot flashes and night sweats at 12 weeks (Steels et al., 2017). In addition to improvements in various postmenopausal discomforts and life of women and its quality, plasma estradiol has also been increased in the extract-treated group (Shamshad et al., 2016).

Moreover, Rao et al. (2015) reported that when a placebo is compared with the extract of fenugreek as a supplementary diet resulted in a potential increase of blood-free testosterone and also it helps in increasing the E2 levels, arousal, and sexual desires, etc. The results indicate that fenugreek extract may be a useful treatment for increasing sexual arousal.

Recently, clinical studies also documented the effect of a specialized fenugreek seed extract (e.g., Testofen) on the symptoms of possible androgen deficiency, sexual function, and serum androgen concentrations in healthy aging males (Rao et al., 2016; Maheshwari et al., 2017; Steels et al., 2011; Wilborn et al., 2010). The dose of 600 mg/day for 12 weeks of extract given as supplementation may improve the Aging Male Symptom questionnaire, also helps in measuring the androgen deficiency symptoms, proper function of sexual organs and raises the level of free and total testosterone in healthy middle-aged and older men (Rao et al., 2016). The demonstration of testofen showed a significant positive effect on the physiological aspect of libido and also assist to have a normal healthy level of testosterone (Steels et al., 2011). A protodioscin-enriched fenugreek seed extract (500 mg/day) increased serum-free testosterone levels up to 46% as well as sperm counts, and it improved mental alertness, mood, and libido in the male study population (Maheshwari et al., 2017).

1.5 *CAPSICUM ANNUUM* (CHILI PEPPER)

Red pepper capsaicin has anti-inflammatory effects and also possess the antioxidant capacity in mitigating oxidative stress (OS) in *in-vitro* and as well as in animal model especially in tissue or organ (Anandakumar et al., 2008; Manjunatha & Srinivasan, 2007; Spiller et al., 2008). In the animal study, Capsaicin-inhibits neutrophils (inflammatory cells) which migrate towards the inflammatory focus leading to the reduction of vascular permeability and pro-inflammatory cytokine production (Spiller et al., 2008). In *in-vitro* studies, Capsaicin modulates the messenger molecules by suppressing obesity-induced inflammation which is released by fat cells of obese mice and releases pro-inflammatory mediators by inactivating macrophage molecules (Kang et al., 2007).

Capsaicin has antioxidant and antiplatelet properties and regulation of energy metabolism is one of the important roles of capsaicin. It may also contribute to beneficial effects on the cardiovascular system (Nilius & Appendino, 2013; Yang et al., 2019; Saito & Yoneshiro, 2013; Srinivasan, 2013).

Some of the human trials found that usage of 5 g or more of chili pepper (Capsicum frutescens) helps in a decrease in blood glucose levels and is also associated with the maintenance of normal insulin levels (Chaiyata et al., 2003; Ahuja et al., 2006; Chaiyasit et al., 2009).

In peptic ulcers, capsaicin acts as a gastroprotective agent (Mózsik et al., 2005). Acid secretion is inhibited, and alkali and mucous secretions are stimulated by capsaicin to prevent and heal the ulcer issues (Satyanarayana, 2006). The contractility and motility response of the human esophagus is increased by capsaicin which is a high amount in Red pepper sauce to rectify the swallowing issues. In patients with ineffective esophagus motility, an acute administration of capsaicin helps in the improvement of the motor performance of the esophageal body (Grossi et al., 2006).

1.6 *CINNAMOMUM VERUM* (CINNAMON)

In *in-vitro* studies, pathogens that live in the tract of the gastrointestinal (GI) pathway as well as in the respiratory system are killed by the extracts of cinnamon and its major components like cinnamaldehyde and eugenol (Fabio et al., 2007; Azumi et al., 1997). There are some reports which suggest that in chronic salmonella infection, cinnamon plays a beneficial effect (Rosti & Gastaldi, 2005).

In the study of *in-vitro* gene expression, the extract of cinnamon polyphenol suppresses the inflammation processes by regulating anti- and pro-inflammatory gene expression (Cao et al., 2008; Kim et al., 2007a).

Various species of cinnamon and their biologically active ingredients have shown great interest in the treatment of mild-to-moderate Alzheimer's disease (AD) in patients by inhibiting the tau protein aggregation and preventing the formation and accumulation of amyloid-β (A-beta) peptides into the neurotoxic oligomeric inclusions, which are considered to be the AD trademarks. Cinnamon's aqueous extract inhibits tau protein activity (when a protein accumulates and twists inside the brain's nerve cells and becomes toxic) and A-beta insult of neuronal cells in an *in-vitro* model (Peterson et al., 2009; Kim et al., 2007b). Indeed, cinnamon interferes with multiple OS and pro-inflammatory pathways to possess neuroprotective effects.

1.7 *ZINGIER OFFICINALE* (GINGER)

Clinical trials show that the usage of ginger (1 g/day) during pregnancy may be safe and effective in decreasing nausea and vomiting (Ozgoli et al., 2009) or when induced by chemotherapy (Levine et al., 2008).

Ginger and its extracts, including 6-gingerols and 6-shogaol, have strong anti-inflammatory properties and inhibit the production of inflammatory mediators like prostaglandin E2 and NO. In a recent human clinical trial, patients with osteoarthritis (OA) received capsules containing 500 mg of ginger powder for 3 months which decreases the serum levels of TNF-α and interleukin-1β (IL-1β) (Mozaffari-Khosravi et al., 2016). Ginger also importantly lowered COX-1 protein (Keservani et al., 2015; Singh et al., 2017b; Jarouliya & Keservani, 2019b) expression in patients with colorectal cancer (CRC) increasing the risk at an early stage (Jiang et al., 2013).

Ginger possesses anti-inflammatory, antioxidant (Singh et al., 2017a), antiplatelet, antihypertensive, and hypolipidemic effects (Nicoll & Henein, 2009; Chrubasik et al., 2005; Ghayur et al., 2005; Alizadeh-Navaei et al., 2008; Bordia et al., 1997).

Animal studies suggest that joint swelling, cartilage destruction, and serum levels of inflammatory cytokines associated with rheumatoid arthritis and joint and muscle pain are reduced by the use of ginger (Fouda & Berika, 2009; Funk et al., 2009). In a randomized human trial, before exercise, the oral administration of 2 g raw or heat-treated ginger for 11 days may reduce muscle pain caused by eccentric exercise and also markers of inflammation and muscle function are reduced slightly by the use of ginger (Black et al., 2010). It was also reported that muscle strength can be improved by a 4 g of ginger supplement (Matsumura et al., 2015).

The advanced glycation end (AGEs) product accumulation results in a nonenzymatic reaction between proteins and sugar in association with aging and diabetes have been implicated in unhealthy conditions. Thus, inhibiting AGEs formation is helpful in the prevention of diabetic complications. *In vitro* studies showed the extract of ginger could help prevent and/or inhibit protein glycation (Saraswat et al., 2009; Dearlove et al., 2008). Animal data also indicated a hypoglycemic effect for ginger extract (Nammi et al., 2009). In randomized clinical trials, it was found that ginger supplement (1.6–3.0 g daily) improves insulin sensitivity and some fractions of lipid profile and reduces C-reactive protein (CRP) and prostaglandin E2 in type 2 diabetic patients. Glycemic indexes (lowering blood glucose and HbA1c), the sensitivity of insulin, and lipid profile, as well as total antioxidant capacity, MDA (malondialdehyde), CRP, and paraoxonase-1 activity, were also improved in patients with T2D (Arablou et al., 2014; Shidfar et al., 2015; Mahluji et al., 2013).

There is evidence that oral ingestion of ginger may induce the function of thermoregulatory and oxidation in human fat. Intake of 1.0 g of dried ginger root powder may elevate the serum levels of free fatty acids (FFAs) after 120 min in the morning and afternoon. Intake of ginger in the morning significantly reduces the respiratory exchange ratios and elevates the fat oxidation by 13.5% at 120 min after ingestion. These results show that the administration of single oral ginger helps effectively in facilitating fat use (Miyamoto et al., 2015).

In vitro studies found that the extract of ginger inhibited the expression of a wide range of inflammation-related genes in microglial-like cells (non-neuronal brain cells) and protective in brain cells from A-beta insult (too much A-beta protein in the brain is involved in the development of AD), suggesting that neuroprotective effects are mainly caused by ginger (Grzanna et al., 2004; Kim et al., 2007b).

1.8 *PIPER NIGRUM* (BLACK PEPPER)

In *in-vivo*, there is evidence of lowering lipid peroxidation by the treatment of Black pepper or piperine and cellular antioxidant status is influenced in several experimental situations of OS (Kaleem et al., 2005; Vijayakumar & Nalini, 2006).

Analgesic activities and anti-inflammatory are the remarkable properties have been revealed by Piperine (Tasleem et al., 2014).

The digestion process may be accelerated by Black pepper by influencing the activity of digestive enzymes, gastric acid and secretion of bile acid also increased, and reduces food transit time (Srinivasan, 2007). In animal studies, the activities of pancreatic amylase, lipase, and chymotrypsin by 87%, 37%, and 30%, respectively, are enhanced by piperine, when consumed through the diet (Platel & Srinivasan, 2004).

Zanzer et al. (2018) recently examined the black pepper postprandial effect-based beverage (BPB) on glycemia, appetite, gastro intestine of well-being, hormones of gut (peptide tyrosine-tyrosine and GLP-1), and thyroid hormones of thyroid postprandially after a white wheat bread challenge in healthy adults and concluded.

Antidepressant-like activity has been exhibited by Black pepper and has a cognition-enhancing effect through the regulation of neurotransmitter metabolism in animals (Li et al., 2007; Wattanathorn et al., 2008).

1.9 *CURCUMA LONGA* (TURMERIC)

Various laboratory, animal, and human studies suggest that the function of the cardiac system, vascular health, and lipid profiles are protected by the effect of the curcumin (Srivastava & Mehta, 2009; Fang et al., 2009). In acute coronary syndrome patients, curcumin played a role in reducing the cholesterol levels in a clinical trial when administered in various doses (45, 90, or 180 mg/day).

It is also evidenced that the usage of lower doses shows more effectiveness than the higher doses like 45 mg/day of curcumin decreased LDL (low density level) and total cholesterol levels with increased HDL concentrations (Alwi et al., 2008).

Curcumin has been proven to be effective against the development of hepatic steatosis and it also helps in steatohepatitis progression (Panahi et al., 2016a; Rahmani et al., 2016). In a randomized trial, supplementation of curcumin for an 8-week study was linked with an essential reduction in liver fat content (78.9% improvement in the curcumin versus 27.5% improvement in the placebo group) in patients with nonalcoholic fatty liver disease (NAFLD) (Rahmani et al., 2016). There were results which proved in the reductions in BMI and serum levels of total cholesterol, LDL-C (low density level-cholesterol), triglycerides, liver enzymes, and uric acid when compared with the placebo group (Panahi et al., 2016a; Rahmani et al., 2016). Curcumin possesses an antidepressant effect. In depressed patients,

Chronic supplementation with curcumin (1,000 mg/day) produced a significant antidepressant behavioral response by a reduction of 17 items Hamilton Depression Rating Scale and Montgomery-Asberg Depression Rating Scale scores (Yu et al., 2015).

Curcumin has anti-inflammatory activity properties and helps in maintaining the healthy join function by activating the bioactive compound (Park et al., 2007). In a rheumatoid arthritis animal model, the effect of turmeric extracts was assessed as a primary endpoint in treating the arthritic index, and joint swelling (Funk et al., 2006). In the management of arthritis, various human clinical trials have demonstrated that curcumin acts as an analgesic and anti-inflammatory agent (Panahi et al., 2016b; Amalraj et al., 2017). Kuptniratsaikul et al. (2009) found that primary knee OA patients who consumed turmeric extract for 6 weeks dose of 2 g/day had similar pain relief as the patients who consumed 800 mg/day of ibuprofen. Haroyan et al. (2018) compared CuraMed® (Terry Naturally) 500 mg capsules (333 mg curcuminoids) and Curamin® (Terry Naturally) 500 mg capsules (350 mg curcuminoids and 150 mg boswellic acid) given orally three times a day for 12 weeks in 201 individuals with OA in a three-arm, randomized placebo-controlled study. They found that patients with OA have been relieved from pain-related symptoms by the consumption of curcumin complex and the combination of boswellic acid with this curcumin has become more effective, probably because of the synergistic effects of boswellic acid and curcumin.

Recent studies in both *in-vitro* and *in-vivo* found that the glucose level is lowered by the suppression of glucose in the liver by the usage of curcumin (Kim et al., 2009a), in fat cell cultures reversed insulin resistance (Wang et al., 2009), in the skeletal muscle glucose uptake is increased (Best et al., 2007) and also stimulated the pancreatic beta-cell functions (Cheng et al., 2009). Supplementation of turmeric has been shown to improve glucose indexes (Usharani et al., 2008).

Some patients who are with a 73% mark tumor regression and pancreatic cancer are treated with curcumin where oral consumption of large doses of curcumin has shown some biological active change was observed in one subject (Dhillon et al., 2008).

1.10 *ROSMARINUS OFFICINALIS* (ROSEMARY)

Under OS conditions, Membrane damage by 40–50% is reduced by Carnosic acid and carnosol reduced and lipid peroxidation is inhibited by 88–100 and

38–89%, respectively, in a cell culture testing and also lowered DNA damage induced by a dietary oxidant (Wijeratne & Cuppett, 2007).

A random study using a cognitive assessment battery test and also a self-assessment mood scale for 140 subjects showed that the Inhalation of oil such as lavender and rosemary oil improves cognitive function (Moss et al., 2003). In graduate students, test-taking stress is reduced by the aroma of rosemary oil (Mc Caffrey et al., 2009).

His observation has been done on rosemary as a protective effect on dopaminergic neural cells including behavior and cognition, mood, attention, and learning in several rosemary constituents (Kim et al., 2006; Park et al., 2008). There is emerging evidence that *in vitro* studies, acetylcholinesterase and butyrylcholinesterase are inhibited by rosemary to promote brain health and which may help to facilitate communication among cells in the brain (Adsersen et al., 2006; Orhan et al., 2008). Furthermore, cell viability and blood flow to the brain may be improved by carnosic acid based on *in vitro* experiments (Kim et al., 2006; Satoh et al., 2008).

Antithrombotic activity has been shown by Rosemary and in both *in vitro* and *in vivo* endothelial function has also been increased (Lee et al., 2007; Naemura et al., 2008).

By inhibiting platelet reactivity and arterial blood clot formation, vascular function is significantly improved by rosemary at a 5% concentration, and also in animals, it enhances flow-mediated vasodilation (Naemura et al., 2008).

An activated Rosemary PPAR-γ plays an important role in cellular functions and metabolism regulation, helping in lowering blood glucose levels as well as lowering blood fatty acid levels (Rau et al., 2006). Furthermore, alpha-glucosidase is potentially inhibited by rosemary which helps to reduce sugar absorption (Ercan & El, 2018). In invitro, AGE formation is also inhibited by rosemary (Hsieh et al., 2007), which suggests the development of diabetic complications has been potentially protected.

In *in-vitro* studies, rosemary showed a potential effect on treating and protecting the skin from UV (ultraviolet) photodamage using the water-soluble extract of rosemary (Martin et al., 2008). Further, when human cells are exposed to UVA light, carnosic acid has demonstrated photoprotective action *in vitro*. In both, *in-vitro* and *in-vivo* studies, the extract of rosemary inhibited oxidative damage to skin-surface lipids (Calabrese et al., 2000). *In-vitro* studies suggest that in many human cell lines, extract of rosemary may decrease the effects of carcinogenic or toxic agents (Cheung & Tai, 2007). By reducing the expression of several pro-inflammatory genes, rosemary may exert anticancer effects (Scheckel et al., 2008).

1.11 *ALLIUM SATIVUM* (GARLIC)

The daily supplement of 1 gram of garlic improves flexibility, pain, and other physical function in overweight or obese women (Salimzadeh et al., 2018).

Various *in-vitro* studies show that animal and human research showed that consuming garlic may reduce the atherosclerotic process and its development (arteries harden) (Budoff et al., 2009; Efendy et al., 1997) which leads to heart attacks and strokes, by reducing fatty streak formation in blood vessels and atherosclerotic plaque size if not treated (Ferri et al., 2003; Durak et al., 2002), also inhibits oxidation of LDL-C (as oxidized LDL is what damages the blood vessels) (Lau, 2006; Lei et al., 2008) prevents adhesion of inflammatory cell and endothelial cells (ECs) (Lei et al., 2008), which in turn improves impaired endothelial function (Gorinstein et al., 2007; Gonen et al., 2005). Some human trials showed that increased brachial artery flow-mediated, endothelium-dependent dilation by 44% has been due to the consumption of garlic (Williams et al., 2005). More recently, Szulińska et al. (2018). Has proved that 400 mg of garlic extract supplementation may favorably modify CRP like, endothelial biomarkers and plasminogen activator inhibitor-1, and LDL-C) in association with cardiovascular risk. Furthermore, in patients with coronary artery disease, supplementation of garlic powder tablets (1,200 μg allicin/tab) twice daily could prevent the progression of carotid intima-media thickness (Mahdavi-Roshan et al., 2013).

There is contradictory evidence that garlic usage has a potential effect on cholesterol and triglyceride levels. Blood lipids levels have been reduced by the consumption of garlic via inhibition of HMG-CoA reductase demonstrated in animal and human cell lines studies (Rai et al., 2009). As garlic can reduce blood total cholesterol (7.3 mg/dL) and triglycerides (9.7 mg/dL) found in a meta-analysis but did not exhibit any significant effect on LDL or HDL (Reinhart et al., 2009). A later review of eight meta-analyzes says that supplementation of garlic may reduce total cholesterol by 7.4–29.8 mg/dL (Varshney & Budoff, 2016).

The synthesis of NO is stimulated by Garlic (Lei et al., 2010) and inhibited ACE activity (Hosseini et al., 2007). Furthermore, through the vascular smooth-muscle cell signaling pathway, as well red blood cells convert the garlic-derived organic polysulfides into hydrogen sulfide gas, which leads to vasorelaxation (Benavides et al., 2007).

In both *in-vitro* and *in-vivo* human studies, Antithrombotic activity has been identified in the extract of garlic (Rahman, 2007; Steiner & Li, 2001; Fukao et al., 2007). Human platelet *in vitro* studies activity of COX-1 and formation of A2 thromboxane (a clotting factor) are inhibited by garlic (Rahman, 2007; Fukao et al., 2007). Additionally, fibrinolytic activity, increasing fibrinolysis has been potentially activated by garlic extracts (dissolving small blood clots) (Hiyasat et al., 2009; Pierre et al., 2005).

A clinically controlled study shows that the involvement of 30 patients affected with coronary artery disease has been given garlic extract at a dose equivalent to 4 g of garlic could increase the activity of fibrinolytic (Fukao et al., 2007).

However, it proves that the intake of raw garlic or garlic oil does not impair the functions of platelets or changes in a sore activity as well fibrinolytic potential in healthy volunteers (Scharbert et al., 2007; Wojcikowski et al., 2007; Womack et al., 2015).

In a human trial, it was demonstrated that when time-released garlic product called Allicor was used for treatment to show the better metabolic control as it helps in lowering fasting blood glucose levels and also triglycerides levels (Sobenin et al., 2008). More recently, Wang et al. (2017) systematically evaluated the safer use of garlic supplement and its efficacy helps in the management of T2DM and found a significant reduction in the fasting blood glucose level from 1 to 2 weeks up to 24 weeks, and also a significant decrease in fructosamine and glycated hemoglobin (both in 12 and 24 weeks), was achieved in the garlic group. Thus, current data confirm that garlic supplements play a positive and sustained role in blood glucose and favorable lipoprotein regulation in the management of T2DM.

Experimental evidence has shown that some garlic-derived products have a protective effect against ischemic brain injury, thereby improving learning and memory retention (Aguilera et al., 2010; Saleem et al., 2006; Borek, 2006).

1.12 *HIBISCUS SABDARIFFA* (ROSELLE) AND *OLEA EUROPAEA* (EUROPEAN OLIVE)

Hibiscus sabdariffa (roselle) and *Olea europaea* (European olive) herbs are highly treated as essential herbs in historical periods because of their high nutritious and therapeutic values (Hoda et al., 2021).

Many studies have identified potential beneficial health effects, such as antihypertensive, hypoglycemic, hepatoprotective, antimicrobial, and antioxidant properties (Singh et al., 2015; Ali et al., 2005; El & Karakaya, 2009). Due to the presence of high amounts of bioactive compounds, these effects are mainly attributed.

1.13 *BUTEA MONOSPERMA* (FLAME-OF-FOREST)

For various therapeutic effects, *Butea monosperma* has been used in traditional medicine to treat the following effects, such as anti-diabetic, diuretic, wound healing, antimicrobial, anthelmintic, arthritis, burning sensation of the body (Thiagarajan et al., 2012, 2013). Development of histopathological, biochemical, and behavioral alterations are induced by the pretreatment of *B. monosperma* attenuated CCI (Chronic constriction injury) which is dose-dependent. *B. monosperma* may have significant anti-oxidative effects as well helps in neuroprotection effect and also possess modulators calcium channel effect (Thiagarajan et al., 2012).

1.14 *CITRULLUS COLOCYNTHIS* (WILD GOURD)

Citrullus colocynthis is used in folk medicine, especially as an anti-inflammatory agent and also as an analgesic and anti-inflammatory agent in folk medicine (Marzouk et al., 2010). In a recent study, a two-arm double-blind clinical random trial used 60 painful diabetic polyneuropathies (PDPN) patients, a parallel design, were randomly treated either with *C. colocynthis* topical formulation or placebo, which showed the results in patients after 3 months as topical formulation of *C. colocynthis* fruit extract helps in reducing pain (Heydari et al., 2016).

1.15 *CROCUS SATIVUS* (SAFFRON CROCUS)

Due to its high remedial property, *Crocus sativus* has been widely used as traditional food and acts as a remedial product in treating various disorders like insomnia, menstrual issues, cardiovascular disorders, bronchospasm, etc. (Amin & Hosseinzadeh, 2012). Recent research showed that saffron acts as adjunctive therapy when it combined with amitriptyline leading to improve therapeutic outcomes in the management of neuropathic pain (Amin et al., 2017).

1.16 *ELAEAGNUS ANGUSTIFOLIA* (RUSSIAN OLIVE)

E. angustifolia showed muscle relaxant activity (Hosseinzadeh et al., 2003). When a hot plate test is conducted in nerve lighted mice showed a potential analgesic effect due to the administration of different doses of *E. angustifolia* (Karimi et al., 2010). Flavonoids (Keservani & Sharma, 2014) have been related to inflammatory activities and prescribed as the most important components in *E. angustifolia* (Ramezani et al., 2001). In a recent controlled random trial study, the symptoms of OA are reduced by the consumption of *E. Angustifolia* and possess good efficacy when compared to ibuprofen. It was also found that it is safe and well tolerable and has no adverse effects observed (Panahi et al., 2016c).

1.17 *GINKGO BILOBA* (GINKGO)

G. biloba is a well-known herb that helps in the protection of both transient and permanent focal cerebral dementia and ischemia as it possesses neuroprotective effects (Lee et al., 2002; Ahlemeyer et al., 2003). The extract of *G. biloba* has some useful effects on neuropathic pain because of the following effects anti-inflammatory, antioxidant, and factor which helps in platelet activation and also helps in protecting against neurotoxicity induced by NMDA (Kim et al., 2009b).

1.18 *MOMORDICA CHARANTIA* (BITTER GOURD)

Momordica charantia (Bitter gourd) is a medicinal herb used traditionally for various purposes including asthma, neuropathic pain and anemia (Basch et al., 2003).

1.19 *NIGELLA SATIVA* (BLACK CUMIN)

N. sativa and Thymoquinone essentially help in reducing the increased serum glucose level and increase the decreased serum insulin level. Immunoreactive β-cells of insulin have been elevated by *N. sativa* and Thymoquinone. Some morphological changes also occurred due to the treatment of *N. sativa* and Thymoquinone in the diabetic caused animal tissue and finally result shows that all these physiological changes are due to the direct and indirect antioxidant effect caused by TQ and *N. sativa* (Kanter, 2008).

1.20 *OCIMUM SANCTUM* (BASIL)

Treatment with *O. sanctum* reduced axonal degeneration, decreased nociceptive threshold, and motor incoordination caused by sciatic nerve transection. It also reduced the axotomy-induced rise in TBARS, total calcium, and GSH levels (Muthuraman et al., 2008). In another study, treatment with *O. sanctum* and its saponin rich fraction significantly reduced the vincristine-induced increase in the hind paw withdrawal duration in response to non-noxious cold stimuli and noxious mechanical stimuli, as well as the vincristine-induced increase in OS markers and total calcium levels in vincristine-induced neuropathic pain in rats, possibly due to a reduction in OS and calcium (Kaur et al., 2010).

According to a recent study, *O. sanctum* has a strong effect on neuropathic pain, particularly in CCI-induced peripheral neuropathy, and saponins are an important critical element in neuropathic pain. Although the author indicates that the pain-alleviating activity of *O. sanctum* and saponin rich fraction generates an increase in free radicals and calcium levels owing to the attenuation of the nerve injury triggering agent (Kaur et al., 2015).

1.21 *PHYLLANTHUS AMARUS* (GALE OF WIND)

In various countries, different species from the genus *Phyllanthus* are widely used as a traditional medicine in treating several disorders like dropsy. Bladder calculus, jaundice, flu, etc. (Kassuya et al., 2003).

In mice, after its partial ligation of the sciatic nerve process, the hexane extract of *P. amarus* helps resist the mechanical allodynia, which shows similar efficacy to that obtained with gabapentin. The activity of MPO is decreased by the administration of hexane extract, either by injecting complete Freund's adjuvant (CFA) or by anti-inflammatory action including partial sciatic nerve ligation (PSNL) (Kassuya et al., 2003). *P. amarus* has anti-hyperalgesic and anti-inflammatory properties which are identified as the model of chronic musculoskeletal inflammatory pain which is mediated through spinal or supraspinal neuronal mechanisms, by inhibiting PGE2 (Amin & Hosseinzadeh, 2016).

1.22 *RUBIA CORDIFOLIA* (MADDER)

All the parts of the plant *R. cordifolia* have a wide range of therapeutic properties including analgesic and anti-inflammatory activities. When Patel et al. Identified the therapeutic properties *of R. cordifolia*, it showed that methanolic extract of the root of *R. cordifolia* essentially reduces edema in the paw caused by carrageenan and also in tail-flick test it helps increase the reaction time (Patel et al., 2010).

1.23 *SALVIA OFFICINALIS* (SAGE)

Salvia officinalis herb can be used to treat unilateral migraines as well as headaches caused by a neurological condition (Mansourabadi et al., 2015). In both enzyme-dependent and enzyme-independent lipid peroxidation assays, various extracts of *S. officinalis* displayed anti-inflammatory effects (Schroder et al., 2013). On chemical behavioral models of nociception involving an opioid pathway, the hydroalcoholic extract of *S. officinalis* leaves has considerable anti-inflammatory and antinociceptive effects. Furthermore, the plant's carnosol and ursolic acid/oleanolic acid appear to contribute to the extract's antinociceptive activity, most likely through a modulatory impact on TRPA1-receptors (Rodrigues et al., 2012).

1.24 *ALLIUM NIGRUM* (BLACK GARLIC) GARLIC

Allium nigrum possesses unique bioactivity as compared to fresh garlic, providing several advantages such as anti-inflammatory, anticancer, anti-obesity, antioxidant, and antibacterial characteristics (Kimura et al., 2017; Horita et al., 2016).

1.25 *HARPAGOPHYTUM PROCUMBENS* (DEVIL'S CLAW)

Harpagophytum procumbens have been utilized as anti-inflammatory medications to treat the symptoms of arthritis and rheumatism. It was connected to beneficial qualities such as antimalarial, anticancer, and uterotonic effects (Mncwangi et al., 2012). The glycoside portion of the devil's claw is antimutagenic against injury when exposed to the environment.

1.26 *MYRCIARIA DUBIA* (CAMU CAMU)

Due to the presence of high content of vitamin C and polyphenol content, *Myrciaria dubia* possesses antioxidant, anti-inflammatory, and antibacterial properties (Akter et al., 2011). Camu juice has been shown to have the physiological antioxidant capacity, lowering total reactive oxygen species (ROS), anti-inflammatory activity, and circulating C reactive protein in humans at a dosage of 70 mL/day, due to its vitamin C concentration (Inoue et al., 2008). Camu camu polyphenols such as proanthocyanidins, ellagitannins, and ellagic acid derivatives, on the other hand, have antioxidant and anti-inflammatory properties (Fracassetti et al., 2013).

1.27 CONCLUSION

In recent days, many major issues like cardiovascular diseases (CVDs), and cancer creates a great challenge among society and public health where it causes a serious effect on the environment as it is mainly identified as a very crucial concern. The increase of interest in public health may lead us to do more research to promote advancement in the treatment of acute and chronic illness by providing healthy food and nutraceuticals. This review is mainly focused on the usage of selected extracts of traditional herbs and plant species in treating various disorders based on their dosage as well summarizes the importance of medicinal plants and herbs which possess disease resistance capacity as well a dose-dependent range of plant extracts, including liquid or powder form and some raw usage of species is also helpful to show the extent to which it works in the treatment of disorders.

ACKNOWLEDGMENT

The author of this chapter would like to thank the management of her institution for their support to complete this chapter.

KEYWORDS

- **antioxidants**
- **extract**
- **metabolism**
- **phenol**
- **phytochemicals**

REFERENCES

Adsersen, A., Gauguin, B., Gudiksen, L., & Jäger, A. K., (2006). *J. Ethnopharmacol., 104*, 418–422. doi: 10.1016/j.jep.2005.09.032.

Aguilera, P., Chánez-Cárdenas, M. E., Ortiz-Plata, A., Léon-Aparicio, D., Barrera, D., Espinoza-Rojo, M., Villeda-Hernández, J., et al., (2010). *Phytomedicine, 17*, 241–247. doi: 10.1016/j.phymed.2009.06.004.

Ahlemeyer, B., & Krieglstein, J., (2003). Neuroprotective effects of *Ginkgo biloba* extract. *Cell Mol. Life Sci., 60*, 1779–1792.

Ahuja, K. D., Robertson, I. K., Geraghty, D. P., & Ball, M. J., (2006). *Am. J. Clin. Nutr., 84*, 63–69. doi: 10.1093/ ajcn/84.1.63.

Akter, M., Oh, S., Eun, J., & Ahmed, M., (2011). Nutritional compositions and health promoting phytochemicals of camu-camu (*Myrciaria dubia*) fruit: A review. *Food Res. Int., 44*, 1728–1732.

Ali, B. H., Wabel, N. A., & Blunden, G., (2005). Phytochemical, pharmacological and toxicological aspects of *Hibiscus sabdariffa* L.: A review. *Phytother. Res., 19*, 369–375.

Alizadeh-Navaei, R., Roozbeh, F., Saravi, M., Pouramir, M., Jalali, F., & Moghadamnia, A. A., (2008). *Saudi Med. J., 29*, 1280–1284.

Alwi, I., Santoso, T., Suyono, S., Sutrisna, B., Suyatna, F. D., Kresno, S. B., & Ernie, S., (2008). *Acta Med. Indones., 40*, 201–210.

Amalraj, A., Varma, K., Jacob, J., Divya, C., Kunnumakkara, A. B., Stohs, S. J., & Gopi, S., (2017). *J. Med. Food., 20*, 1022–1030. doi: 10.1089/jmf.2017.3930.

Amin, B., & Hosseinzadeh, H., (2012). Evaluation of aqueous and ethanolic extracts of saffron, *Crocus sativus* L., and its constituents, safranal and crocin in allodynia and hyperalgesia induced by chronic constriction injury model of neuropathic pain in rats. *Fitoterapia, 83*, 888–895.

Amin, B., & Hosseinzadeh, H., (2016). Black cumin (*Nigella sativa*) and its active constituent, thymoquinone: An overview on the analgesic and anti-inflammatory effects. *Planta Med., 82*, 8–16.

Amin, B., Hosseini, S., & Hosseinzadeh, H., (2017). Enhancement of antinociceptive effect by co-administration of amitriptyline and *Crocus sativus* in a rat model of neuropathic pain. *Iran J. Pharm. Res., 16*, 187–200.

Anandakumar, P., Kamaraj, S., Jagan, S., Ramakrishnan, G., Vinodjkumar, R., & Devaki, T., (2008). *Phytother. Res., 22*, 529–533. doi: 10.1002/ptr.2393.

Arablou, T., Aryaeian, N., Valizadeh, M., Sharifi, F., Hosseini, A., & Djalali, M., (2014). *Int. J. Food Sci. Nutr., 65*, 515–520. doi: 10.3109/09637486.2014.880671.

Azumi, S., Tanimura, A., & Tanamoto, K., (1997). *Biochem. Biophys. Res. Commun., 234*, 506–510. doi: 10.1006/bbrc.1997.6668.

Basch, E., Gabardi, S., & Ulbricht, C., (2003). Bitter melon (*Momordica charantia*): A review of efficacy and safety. *Am. J. Health Syst. Pharm., 60*, 356–359.

Benavides, G., Squadrito, G. L., Mills, R. W., Patel, H. D., Isbell, T. S., Patel, R. P., Darley-Usmar, V. M., et al., (2007). *Proc. Natl. Acad. Sci. USA, 104*, 17977–17982. doi: 10.1073/pnas.0705710104.

Best, L., Elliott, A. C., & Brown, P. D., (2007). *Biochem. Pharmacol., 73*, 1768–1775. doi: 10.1016/j.bcp.2007.02.006.

Black, C. D., Herring, M. P., Hurley, D. J., & O'Connor, P., (2010). *J. Pain, 11*, 894–903.

Bordia, A., Verma, S. K., & Srivastava, K. C., (1997). Prostaglandins, leukotrienes Essent. *Fatty Acids, 56*, 379–384. doi: 10.1016/S0952-3278(97)90587-1.
Borek, C., (2006). *J. Nutr., 136*, 810S–812S. doi: 10.1093/jn/136.3.810S.
Boyd, A., Bleakley, C., Gill, C., Mc Donough, S., Hurley, D. A., Bell, P., et al., (2013). Herbal medicinal products or preparations for neuropathic pain and fibromyalgia. *Cochrane Database Syst. Rev.*
Budoff, M. J., Ahmadi, N., Gul, K. M., Liu, S. T., Flores, F. R., Tiano, J., Takasu, J., et al., (2009). *Prev. Med., 49*, 101–107. doi: 10.1016/j.ypmed.2009.06.018.
Calabrese, V., Scapagnini, G., Catalano, C., Dinotta, F., Geraci, D., & Morganti, P., (2000). *Int. J. Tissue React., 22*, 5–1.
Cao, H., Urban, J. F. Jr., & Anderson, R. A., (2008). *J. Nutr., 138*, 833–840. doi: 10.1093/jn/138.5.833.
Chaiyasit, K., Khovidhunkit, W., & Wittayalertpanya, S., (2009). *J. Med. Assoc. Thai., 92*, 108–113.
Chaiyata, P., Puttadechakum, S., & Komindr, S., (2003). *J. Med. Assoc. Thai., 86*, 854–860.
Cheng, T. C., Lin, C. S., Hsu, C. C., Chen, L. J., Cheng, K. C., & Cheng, J. T., (2009). *Neurosci. Lett., 465*, 238–241. doi: 10.1016/j.neulet.2009.09.012.
Cheung, S., & Tai, J., (2007). *Oncol. Rep., 17*(6), 1525–1531.
Chevassus, H., Gaillard, J. B., Farret, A., Costa, F., Gabillaud, I., Mas, E., Dupuy, A. M., et al., (2010a). *Eur. J. Clin. Pharmacol., 66*, 449–455. doi: 10.1007/s00228-009-0770-0.
Chevassus, H., Molinier, N., Costa, F., Gabillaud, I., Mas, E., Dupuy, A. M., Michel, F., et al., (2010b). *Eur. J. Clin. Pharmacol., 66*, 449–455. doi: 10.1007/s00228-009-0770-0.
Chrubasik, S., Pittler, M. H., & Roufogalis, B. D., (2005). *Phytomedicine, 12*, 684–670. doi: 10.1016/j.phymed.2004.07.009.
Dearlove, R. P., Greenspan, P., Hartle, D. K., Swanson, R. B., & Hargrove, J. L., (2008). *J. Med. Food, 11*, 275–281. doi: 10.1089/jmf.2007.536.
Dhillon, N., Aggarwal, B. B., Newman, R. A., Wolff, R. A., Kunnumakkara, A. B., Abbruzzese, J. L., Ng, C. S., Badmaev, V., & Kurzrock, R., (2008). *Clin. Cancer Res., 14*, 4491–4499. doi: 10.1158/1078-0432.CCR-08-0024.
Durak, I., Oztürk, H. S., Olcay, E., & Güven, C., (2002). *J. Herb. Pharmacother., 2*, 19–32.
El, S. N., & Karakaya, S., (2009). Olive tree (*Olea europaea*) Leaves: Potential beneficial effects on human health. *Nutr. Rev., 67*, 632–638.
Ercan, P., & El, S. N., (2018). *Int. J. Biol. Macromol., 115*, 933–939. doi: 10.1016/j.ijbiomac.2018.04.139.
Fabio, A., Cermelli, C., Fabio, G., Nicoletti, P., & Quaglio, P., (2007). *Phytother. Res., 21*, 374–377. doi: 10.1002/ptr.1968.
Fang, X. D., Yang, F., Zhu, L., Shen, Y. L., Wang, L. L., & Cheng, Y. Y., (2009). *Clin. Exp. Pharmacol. Physiol., 36*, 1177–1182.
Ferri, N., Yokoyama, K., Sadilek, M., Paoletti, R., Apitz-Castro, R., Gelb, M. H., & Corsini, A., (2003). *Br. J. Pharmacol., 138*, 811–818. doi: 10.1038/sj.bjp.0705126.
Forouzanfar, F., & Hosseinzadeh, H., (2018). Medicinal herbs in the treatment of neuropathic pain: A review. *Iran J. Basic Med. Sci., 21*, 347–358. doi: 10.22038/IJBMS.2018.24026.6021.
Fouda, A. M., & Berika, M. Y., (2009). *Basic Clin. Pharmacol. Toxicol., 104*, 262–271. doi: 10.1111/j.1742-7843.2008.00363.x.
Fracassetti, D., Costa, C., Moulay, L., & Tomás-Barberán, F., (2013). Ellagic acid derivatives, ellagitannins, proanthocyanidins and other phenolics, vitamin C and antioxidant capacity of two powder products from camu-camu fruit (*Myrciaria dubia*). *Food Chem., 139*, 578–588.

Fukao, H., Yoshida, H., Tazawa, Y., & Hada, T., (2007). *Biosci. Biotechnol. Biochem., 71*, 84–90. doi: 10.1271/bbb.60380.

Funk, J. L., Frye, J. B., Oyarzo, J. N., & Timmermann, B. N., (2009). *J. Nat. Prod., 27*, 403–407. doi: 10.1021/np8006183.

Funk, J. L., Oyarzo, J. N., Frye, J. B., Chen, G., Lantz, R. C., Sólyom, A. M., & Timmermann, B. N., (2006). *J. Nat. Prod., 69*, 351–355. doi: 10.1021/np050327j.

Ghayur, M. N., Gilani, A. H., & Afridi, M. B., (2005). *Vascul. Pharmacol., 43*(4), 234–241.

Gonen, A., Harats, D., Rabinkov, A., Miron, T., Mirelman, D., Wilchek, M., Weiner, L., et al., (2005). *Pathobiolog., 72*, 325–334. doi: 10.1159/000091330.

Gopalpura, P. B., Jayanthi, C., & Dubey, S., (2007). *Int. J. Diabetes Dev. Ctries., 27*, 4145.

Gorinstein, S., Jastrzebski, Z., Namiesnik, J., Leontowicz, H., & Trakhtenberg, S., (2007). *Mol. Nutr. Food Res., 51,* 1365–1381. doi: 10.1002/mnfr.200700064.

Grossi, L., Cappello, G., & Marzio, L., (2006). *Neurogastroenterol Motil., 18*, 632–636. doi: 10.1111/j.1365-2982.2006.00793.x.

Grzanna, R., Phan, P., Polotsky, A., Lindmark, L., & Fondoza, C. G., (2004). *J. Altern. Complement. Med., 10,* 1009–1013. doi: 10.1089/acm.2004.10.1009.

Haroyan, A., Mukuchyan, V., Mkrtchyan, N., Minasyan, N., Gasparyan, S., Sargsyan, A., Narimanyan, M., & Hovhannisyan, A., (2018). *BMC Complement. Altern. Med., 18*, 7.

Heydari, M., Homayouni, K., Hashempur, M. H., & Shams, M., (2016). Topical *Citrullus colocynthis* (bitter apple) extract oil in painful diabetic neuropathy: A double-blind randomized placebo-controlled clinical trial. *J. Diabetes, 8*, 246–252.

Hiyasat, B., Sabha, D., Grotzinger, K., Kempfert, J., Rauwalk, J. W., Mohr, F. W., & Dhein, S., (2009). *Pharmacology, 83*, 197–204. doi: 10.1159/000196811.

Hoda, M., Marzouka, C., Mahmoud, H., Teaimab, C., Ahmed, M., El-Desokyd, & Shanaz, O. M., (2021). A cost-effective HPLC-DAD method for quality and stability assessment of antihypertensive herbal extracts in their recently co-formulated nutraceutical product. *Microchemical Journal, 169*, 106563.

Horita, C., Farías-Campomanes, A., Barbosa, T., Esmerino, E., Da Cruz, A., Bolini, H., Meireles, M., & Pollonio, M., (2016). The antimicrobial, antioxidant and sensory properties of garlic and its derivatives in Brazilian low-sodium frankfurters along shelf-life. *Food Res. Int., 84*, 1–8.

Hosseini, M., Shafiee, S. M., & Baluchnejadmojarad, T., (2007). *Pathophysiology, 14*, 109–112. doi: 10.1016/j. pathophys.2007.07.002.

Hosseinzadeh, H., Ramezani, M., & Namjo, N., (2003). Muscle relaxant activity of *Elaeagnus angustifolia* L. fruit seeds in mice. *J. Ethnopharmacol., 84*, 275–278.

Hsieh, C. L., Peng, C. H., Chyau, C. C., Lin, Y. C., Wang, H. W., & Peng, R. Y., (2007). *J. Agric. Food Chem., 55,* 2884–2891. doi: 10.1021/jf0631833.

Inoue, T., Komoda, H., Uchida, T., & Node, K., (2008). Tropical fruit camu-camu (*Myrciaria dubia*) has anti-oxidative and anti-inflammatory properties. *J. Cardiol., 52*, 127–132.

Jarouliya, U., & Keservani, R. K., (2019a). Pathways leading to child obesity. In: Debasis, B., (ed.), *Global Perspectives on Childhood Obesity: Current Status, Consequences and Prevention* (2nd edn., pp. 137–146). Academic Press, Elsevier, Chapter 12. ISBN: 9780128128404.

Jarouliya, U., & Keservani, R. K., (2019b). Protein function as cell surface and nuclear receptor in human diseases. In: Sanjay, B., & Debarshi, K. M., (eds.), *Medicinal Chemistry with Pharmaceutical Product Development* (pp. 1–32). Apple Academic Press, CRC Press, Taylor & Francis Group. Chapter 1. ISBN: 9781771887106.

Jiang, T. A., (2019). Health benefits of culinary herbs and spices. *Journal of AOAC International, 102*(2), 395.

Jiang, Y., Turgeon, D. K., Wright, B. D., Sidahmed, E., Ruffin, M. T., Brenner, D. E., Sen, A., & Zick, S. M., (2013). *Eur. J. Cancer Prev., 22*, 455–460. doi: 10.1097/CEJ.0b013e32835c829b.

Kaleem, M., Sheema, K., Sarmad, H., & Bano, B., (2005). *Indian J. Physiol. Pharmacol., 49*, 65–71.

Kang, J. H., Kim, C. S., Han, I. S., Kwanda, T., & Yu, R., (2007). *FEBS Lett., 581*, 4389–4396. doi: 10.1016/j.febslet.2007.07.082.

Kanter, M., (2008). Effects of *Nigella sativa* and its major constituent, thymoquinone on sciatic nerves in experimental diabetic neuropathy. *Neurochem. Res., 33*, 87–96.

Karimi, G., Hosseinzadeh, H., Rassoulzadeh, M., Razavi, B. M., & Taghiabadi, E., (2010). Antinociceptive effect of *Elaeagnus angustifolia* fruits on sciatic nerve ligated mice. Iran J *Basic Med. Sci., 13*, 97–101.

Kassuya, C. A., Silvestre, A. A., Rehder, V. L. G., & Calixto, J. B., (2003). Anti-allodynic and anti-oedematogenic properties of the extract and lignans from *Phyllanthus amarus* in models of persistent inflammatory and neuropathic pain. *Eur. J. Pharmaco., 478*, 145–153.

Kaur, G., Bali, A., Singh, N., & Jaggi, A. S., (2015). *Ameliorative Potential of Ocimum Sanctum in Chronic Constriction Injury-Induced Neuropathic Pain in Rats* (Vol. 87, pp. 417–429). Anais da Academia Brasileira de Ciências.

Kaur, G., Jaggi, A. S., & Singh, N., (2010). Exploring the potential effect of *Ocimum sanctum* in vincristine-induced neuropathic pain in rats. *J. Brachial. Plex. Peripher. Nerve, Inj., 5*, 3.

Keith, W. S., (2017). Fenugreek-overview of potential health benefits. *Nutrition Today, Culinary Spices Series, 52*(2).

Keservani, R. K., & Sharma, A. K., (2014). Flavonoids: Emerging trends and potential health benefits. *Journal of Chinese Pharmaceutical Sciences, 23*(12), 815.

Keservani, R. K., Kesharwani, R. K., Sharma, A. K., Vyas, N., & Chadoker, A., (2010b). Nutritional supplements: An overview. *International Journal of Current Pharmaceutical Review and Research, 1*(1), 59–75.

Keservani, R. K., Kesharwani, R. K., Vyas, N., Jain, S., Raghuvanshi, R., & Sharma, A. K., (2010a). Nutraceutical and functional food as future food: A review. *Der Pharmacia Lettre, 2*(1), 106–116.

Keservani, R. K., Sharma, A. K., & Jarouliya, U., (2015). Protein and peptide in drug targeting and its therapeutic approach. *Ars Pharmaceutica (Internet), 56*(3), 165–177.

Keservani, R. K., Sharma, A. K., & Kesharwani, R. K., (2017). An overview and therapeutic applications of nutraceutical and functional foods. *Recent Advances in Drug Delivery Technology*, 160–201. ISBN:9781522507543.

Keservani, R. K., Sharma, A. K., & Kesharwani, R. K., (2020). *Nutraceuticals and Dietary Supplements: Applications in Health Improvement and Disease Management*. CRC Press. ISBN: 9781771888738.

Kim, D. H., Kim, C. H., Kim, M. S., Kim, J. Y., Chung, J. H., Lee, J. W., Yu, B. P., & Chung, H. Y., (2007a). *Biogerontology, 8*, 545–554. doi: 10.1007/s10522-007-9098.

Kim, D. S., Kim, J. Y., & Han, Y. S., (2007b). *J. Altern. Complement. Med., 13*, 333–340. doi: 10.1089/acm.2006.6107.

Kim, S. J., Kim, J. S., Cho, H. S., Lee, H. J., Kim, S. Y., Kim, S., Lee, S. Y., & Chun, H. S., (2006). *Neuroreport., 17*, 1729–1733. doi: 10.1097/01.wnr.0000239951.14954.10.

Kim, T., Davis, J., Zhang, A. J., He, X., & Matthews, S. T., (2009a). *Biochem. Biophys. Res. Commun., 388*, 377–382. doi: 10.1016/j. bbrc.2009.08.018.

Kim, Y. S., Park, H. J., Kim, T. K., Moon, D. E., & Lee, H. J., (2009b). The effects of *Ginkgo biloba* extract EGb 761 on mechanical and cold allodynia in a rat model of neuropathic pain. *Anesth Analg., 108*, 1958–1963.

Kimura, S., Tung, Y., Pan, M., Su, N., Lai, Y., & Cheng, K., (2017). Black garlic: A critical review of its production, bioactivity, and application. *J. Food Drug Anal., 25*, 62–70.

Kuptniratsaikul, V., Thanakhumtorn, S., Chinswangwantanakul, P., Wattanamongkonsil, L., & Thamlikitkul, V., (2009). *J. Altern. Complement. Med., 15*, 891–897. doi: 10.1089/acm.2008.0186.

Lau, B. H. S., (2006). *J. Nutr., 136*, 765S–768S. doi: 10.1093/ jn/136.3.765S.

Lee, E., Chen, H. Y., Wu, T. S., Chen, T. Y., Ayoub, I. A., & Maynard, K. I., (2002). Acute administration of *Ginkgo biloba* extract (EGb 761) affords neuroprotection against permanent and transient focal cerebral ischemia in Sprague-Dawley rats. *J. Neurosci. Res., 68*, 636–645.

Lee, J. J., Jin, Y. R., Lee, J. H., Yu, J. Y., Han, X. H., Oh, K. W., Hong, J. T., Kim, T. J., & Yun, Y. P., (2007). *Planta Med., 73*, 121–127. doi: 10.1055/s-2006-957066.

Lei, Y. P., Chen, H. W., Sheen, L. Y., & Lii, C. K., (2008). *J. Nutr., 138*, 996–1003. doi: 10.1093/jn/138.6.996.

Lei, Y. P., Liu, C. T., Sheen, L. Y., Chen, H. W., & Lii, C. K., (2010). *Mol. Nutr. Food Res., 54*, S42–S52. doi: 10.1002/ mnfr.200900278.

Levine, M. E., Gillis, M. G., Koch, S. Y., Voss, A. C., Stern, R. M., & Koch, K. L., (2008). *J. Altern. Complement. Med., 14*, 545–551. doi: 10.1089/acm.2007.0817.

Li, J. W. H., & Vederas, J. C., (2009). Drug discovery and natural products: End of an era or an endless frontier? *Science, 325*, 161–165.

Li, S., Wan, C., Wang, M., Li, W., Matsumoto, K., & Tang, Y., (2007). *Life Sci., 80*, 1373–1381. doi: 10.1016/j.lfs.2006.12.027.

Mahdavi-Roshan, M., Zahedmehr, A., Mohammad-Zadeh, A., Sanati, H. R., Shakerian, F., Firouzi, A., Kiani, R., & Nasrollahzadeh, J., (2013). *Nutr. Health, 22*, 143–155. doi: 10.1177/0260106014563446.

Maheshwari, A., Verma, N., Swaroop, A., Bagchi, M., Preuss, H. G., Tiwari, K., & Bagchi, D., (2017). *Int. J. Med. Sci., 14*, 58–66. doi: 10.7150/ijms.17256.

Mahluji, S., Attari, V. E., Mobasseri, M., Payahoo, L., Ostadrahimi, A., & Golzari, S. E., (2013). *Int. J. Food Sci. Nutr., 64*, 682–686. doi: 10.3109/09637486.2013.775223.

Majid, A., (2018). A review study of the chemical constituents and therapeutic effects of *Peganum harmala L. Global J. Pure Appl. Chem. Res., 6*(2), 12–19.

Manjunatha, H., & Srinivasan, K., (2007). *Can. J. Physiol. Pharmacol., 85*, 588–596. doi: 10.1139/y07-044.

Mansourabadi, A. H., Sadeghi, H. M., Razavi, N., & Rezvani, E., (2015). Anti-inflammatory and analgesic properties of salvigenin, *Salvia officinalis* flavonoid extracted. *Advanced Herbal Medicine, 1*, 31–41.

Martin, R., Pierrard, C., Lejeune, F., Hilaire, P., Breton, L., & Bererd, F., (2008). *Eur. J. Dermatol., 18*, 128–135. doi: 10.1684/ejd.2008.0349.

Marzouk, B., Marzouk, Z., Haloui, E., Fenina, N., Bouraoui, A., & Aouni, M., (2010). Screening of analgesic and anti-inflammatory activities of *Citrullus colocynthis* from southern Tunisia. *J Ethnopharmacol., 128*, 15–19.

Matsumura, M. D., Zavorsky, G. S., & Smoliga, J. M., (2015). *Phytother. Res., 29*, 887–893. doi: 10.1002/ptr.5328.
Mc Caffrey, R., Thomas, D. J., & Kinzelman, A. O., (2009). *Holist. Nurs. Pract., 23*, 88–93. doi: 10.1097/HNP.0b013e3181a110aa.
Miyamoto, M., Matsuzaki, K., Katakura, M., Hara, T., Tanabe, Y., & Shido, O., (2015). *Int. J. Biometeorol., 59*, 1461–1474. doi: 10.1007/s00484-015-0957-2.
Mncwangi, N., Chen, W., Vermaak, I., Viljoen, A., & Gericke, N., (2012). Devil's claw—A review of the ethnobotany, phytochemistry and biological activity of *Harpagophytum procumbens*. *J. Ethnopharmacol., 143*, 755–771.
Moss, M., Cook, J., Wesnes, K., & Duckett, P., (2003). *Int. J. Neurosci., 113*, 15–38.
Mozaffari-Khosravi, H., Naderi, Z., Dehghan, A., Nadjarzadeh, A., & Fallah, H. H., (2016). *J. Nutr. Gerontol. Geriatr., 35*, 209–218. doi: 10.1080/21551197.2016.1206762.
Mózsik, G., Szolcsányi, J., & Rácz, I., (2005). *World J. Gastroenterol., 11*, 5180–5184. doi: 10.3748/wjg.v11.i33.5180.
Muthuraman, A., Diwan, V., Jaggi, A. S., Singh, N., & Singh, D., (2008). Ameliorative effects of *Ocimum sanctum* in sciatic nerve transection-induced neuropathy in rats. *J. Ethnopharmacol., 120*, 56–62.
Naemura, A., Ura, M., Yamashita, T., Arai, R., & Yamamoto, J., (2008). *Thromb. Res., 122*, 517–522. doi: 10.1016/j. thromres.2008.01.014.
Nammi, S., Sreemantula, S., & Roufogalis, B. D., (2009). *Basic Clin. Pharmacol. Toxicol., 104*, 366–373. doi: 10.1111/j.1742-7843.2008.00362.x.
Nicoll, R., & Henein, M. Y., (2009). *Int. J. Cardiol., 131*, 408–409. doi: 10.1016/j. ijcard.2007.07.107.
Nilius, B., & Appendino, G., (2013). *Rev. Physiol. Biochem. Pharmacol., 164*, 1–76. doi: 10.1007/112_2013_11.
Orhan, I., Aslan, S., Kartal, M., Sener, B., & Hüsnü, C. B. K., (2008). *Food Chem., 108*, 663–668. doi: 10.1016/j.foodchem.2007.11.023.
Ozgoli, G., Goli, M., & Simbar, M., (2009). *J. Altern. Complement. Med., 15*, 243–246. doi: 10.1089/acm.2008.0406.
Panahi, Y., Alishiri, G. H., Bayat, N., Hosseini, S. M., & Sahebkar, A., (2016c). Efficacy of *Elaeagnus angustifolia* extract in the treatment of knee osteoarthritis: A randomized controlled trial. *Excli. J., 15*, 203–210.
Panahi, Y., Alishiri, G. H., Parvin, S., & Sahebkar, A., (2016b). *J. Diet Suppl., 13*, 209–220. doi: 10.3109/19390211.2015.1008611.
Panahi, Y., Kianpour, P., Mohtashami, R., Jafari, R., Simental-Mendía, L. E., & Sahebkar, A., (2016a). *J. Cardiovasc. Pharmacol., 68*, 223–229. doi: 10.1097/FJC.0000000000000406.
Park, C., Moon, D. O., Choi, I. W., Choi, B. T., Nam, T. J., Rhu, C. H., Kwon, T. K., Lee, W. H., Kim, G. Y., & Choi, Y. H., (2007). *Int. J. Mol. Med., 20*, 365–372. doi: 10.3892/ ijmm.20.3.365.
Park, J. A., Kim, S., Lee, S. Y., Kim, C. S., Kim, D. K., Kim, S. J., & Chun, H. S., (2008). *Neuro Report, 19*, 1301–1304. doi: 10.1097/WNR.0b013e32830abc1f.
Patel, A., Patel, T., Macwan, C., Patel, M., Chauhan, K., & Patel, J., (2010). Evaluation of anti-inflammatory and analgesic activity of roots of *Rubia cordifolia* in rats. *Journal of Pharmaceutical Sciences and Research, 2*, 809–813.
Peterson, D. W., George, R. C., & Scaramozzino, F., (2009). *J. Alzheimers Dis., 17*, 585–597. doi: 10.3233/JAD-2009-1083.

Pierre, S., Crosbie, L., & Duttaroy, A. K., (2005). *Platelets, 16*, 469–473. doi: 10.1080/09537100500129540.

Platel, K., & Srinivasan, K., (2004). *Indian J. Med. Res., 119*, 167–179.

Rafraf, M., Malekiyan, M., Asghari-Jafarabadi, M., & Aliasgarzadeh, A., (2014). *Int. J. Vitam. Nutr. Res., 84*, 196–205. doi: 10.1024/0300–9831/a000206.

Rahman, K., (2007). *Mol. Nutr. Food. Res., 51*, 1335–1344. doi: 10.1002/mnfr.200700058.

Rahmani, S., Asgary, S., Askar, G., Keshvari, M., Hataminpour, M., Feizi, A., & Sahebkar, A., (2016). *Phytother. Res., 30*, 1540–1548. doi: 10.1002/ptr.5659.

Rai, S. K., Sharma, M., & Tiwari, M., (2009). *Life Sci., 85*, 211–219. doi: 10.1016/j.lfs.2009.05.020.

Ramezani, M., Hosseinzadeh, H., & Daneshmand, N., (2001). Antinociceptive effect of *Elaeagnus angustifolia* fruit seeds in mice. *Fitoterapia, 72*, 255–262.

Rane, B. R., Bharath, M. S., Patil, R. R., Keservani, R. K., & Jain, A. S., (2020b). Novel approaches in nutraceuticals. In: Rajesh, K. K., Raj, K. K., & Anil, K. S., (eds.), *Enhancing the Therapeutic Efficacy of Herbal Formulations Through Novel Drug Delivery Systems* (pp. 241–266). IGI Global International Publisher, Pennsylvania, USA, Chapter 11. ISBN13: 9781799844532.

Rane, B. R., Patil, A. S., Keservani, R. K., & Jain, A. S., (2020a). Novel approaches in herbal formulation. In: Rajesh, K. K., Raj, K. K., & Anil, K. S., (eds.), *Enhancing the Therapeutic Efficacy of Herbal Formulations Through Novel Drug Delivery Systems* (pp. 43–68). IGI Global International Publisher, Pennsylvania, USA, Chapter 2. ISBN13: 9781799844532.

Rane, B. R., Tadavi, S. A., & Keservani, R. K., (2020c). Naturopathy. In: Anil, K. S., Raj, K. K., & Surya, P. G., (eds.), *Herbal Product Development* (pp. 321–347). Apple Academic Press, CRC Press, Taylor & Francis Group, Chapter 12. ISBN: 9781771888776.

Rao, A., Steels, E., Beccaria, G., Inder, W. J., & Vitetta, L., (2015). *Phytother. Res., 29*, 1123–1130. doi: 10.1002/ptr.5355.

Rao, A., Steels, E., Inder, W. J., Abraham, S., & Vitetta, L., (2016). *Aging Male., 19*, 134–142. doi: 10.3109/13685538.2015.1135323.

Rau, O., Wurglics, M., Paulke, A., Zitzkowski, J., Meindl, N., Bock, A., Dingermann, T., et al., (2006). *Planta Med., 72*, 881–887. doi: 10.1055/s-2006–946680.

Reinhart, K. M., Talati, R., White, C. M., & Coleman, C. I., (2009). *Nutr. Res. Rev., 22*, 39–48. doi: 10.1017/ S0954422409350003.

Rivelli, A. R., Caruso, M. C., De Maria, S., & Galgano, F., (2017). Vitamin C content in leaves and roots of horseradish (*Armoracia rusticana*): Seasonal variation in fresh tissues and retention as affected by storage conditions. *Emir. J. Food Agric., 29*, 799–806.

Rodrigues, M. R. A., Kanazawa, L. K. S., Das Neves, T. L. M., Da Silva, C. F., Horst, H., Pizzolatti, M. G., et al., (2012). Antinociceptive and anti-inflammatory potential of extract and isolated compounds from the leaves of *Salvia officinalis* in mice. *J. Ethnopharmacol., 139*, 519–526.

Rosti, L., & Gastaldi, G., (2005). *Pediatrics, 116*, 1057. doi: 10.1542/peds.2005-1521.

Saito, M., & Yoneshiro, T., (2013). *Curr. Opin. Lipidol., 24*, 71–77. doi: 10.1097/MOL.0b013e32835a4f40.

Saleem, S., Ahmad, M., Ahmad, A. S., Yousuf, S., Asari, M. A., Khan, M. B., Ishrat, T., & Islam, F., (2006). *J. Med. Food, 9*, 537–544. doi: 10.1089/jmf.2006.9.537.

Salimzadeh, A., Alipoor, E., Dehghani, S., Yaseri, M., Hosseini, M., Feinle-Bisset, C., & Hosseinzadeh-Attar, M. J., (2018). *Int. J. Clin. Pract., 23*, e13208. doi: 10.1111/ijcp.13208.

Saraswat, M., Reddy, P. Y., Muthenna, P., & Reddy, G. B., (2009). *Br. J. Nutr., 101*, 1714–1721. doi: 10.1017/S0007114508116270.

Satoh, T., Kosaka, K., Itoh, K., Kobayashi, A., Yamamoto, M., Shimojo, Y., Kitajima, C., et al., (2008). *J. Neurochem., 104*, 1116–1131. doi: 10.1111/j.1471-4159.2007.05039.x.

Satyanarayana, M. N., (2006). *Crit. Rev. Food Sci. Nutr., 46*, 275–328. doi: 10.1080/1040-830491379236.

Scharbert, G., Kalb, M. L., Duris, M., Marschalek, C., & Kozek-Langenecker, S. A., (2007). *Anesth. Analg., 105*, 1214–1218. doi: 10.1213/01.ane.0000287253.92211.06.

Scheckel, K. A., Degner, S. C., & Romagnolo, D. F., (2008). *J. Nutr., 138*, 2098–2105. doi: 10.3945/jn.108.090431.

Schroder, S., Beckmann, K., Franconi, G., Meyer-Hamme, G., Friedemann, T., Greten, H. J., et al., Can medical herbs stimulate regeneration or neuroprotection and treat neuropathic pain in chemotherapy-induced peripheral neuropathy. *Evid. Based Complement Alternat Med., 2013*, 423713.

Shamshad, B. S., Jayalakshmi, H. K., Vidyavathi, H. G., Gopakumar, G., Abin, I., Balu, M., Geetha, K., et al., (2016). *Phytother. Res., 30*, 1775–1784. doi: 10.1002/ptr.5680.

Sharma, A. K., Keservani, R. K., & Gautam, S. P., (2020). *Herbal Product Development.* Apple Academic Press, CRC Press, Taylor & Francis Group. ISBN: 9781771888776.

Shehata, A., Mulwa, R. M. S., Babadoost, M., Uchanski, M., Norton, M. A., Skirvin, R., & Walters, S. A., (2009). Horseradish: Botany, horticulture, breeding. *Hort. Rev., 35*, 221–261.

Shidfar, F., Rajab, A., Rahideh, T., Khandouzi, N., Hosseini, S., & Shidfar, S., (2015). *J. Complement Integr. Med., 12*, 165–170. doi: 10.1515/jcim-2014-0021.

Singh, P., Kesharwani, R. K., & Keservani, R. K., (2017a). Antioxidants and vitamins: Roles in cellular function and metabolism. In: *Sustained Energy for Enhanced Human Functions and Activity* (pp. 385–407). Academic Press. ISBN:9780128054130.

Singh, P., Kesharwani, R. K., & Keservani, R. K., (2017b). Protein, carbohydrates, and fats: Energy metabolism. In: *Sustained Energy for Enhanced Human Functions and Activity* (pp. 103–115). Academic Press. ISBN:9780128054130.

Singh, P., Mishra, A., Singh, P., Goswami, S., Singh, A., & Tiwari, K. D., (2015). Hypertension and herbal plant for its treatment: A review. *Indian J. Biotech. Pharm. Res., 3*, 358–366.

Sobenin, I. A., Nedosugova, L. V., Filatova, L. V., Balabolkin, M. I., Gorchakova, T. V., & Orekhov, A. N., (2008). *Acta Diabetol., 45*, 1–6. doi: 10.1007/s00592-007-0011-x.

Spiller, F., Alves, M. K., Vieira, S. M., Carvalho, T. A., Leite, C. E., Lunardelli, A., Poloni, J. A., et al., (2008). *J. Pharm. Pharmacol., 60*, 473–478. doi: 10.1211/jpp.60.4.0010.

Srinivasan, K., (2007). *Crit. Rev. Food Sci. Nutr., 47*, 735–748. doi: 10.1080/10408390601062054.

Srinivasan, K., (2013). *Food Funct., 4*, 503–521. doi: 10.1039/ c2fo30249g.

Srivastava, G., & Mehta, J. L., (2009). *J. Cardiovasc. Pharmacol. Ther., 14*, 22–27. doi: 10.1177/1074248408329608.

Steels, E., Ra, A., & Vitetta, L., (2011). *Phytother. Res., 25*, 1294–1300. doi: 10.1002/ptr.3360.

Steels, E., Steele, M. L., Harold, M., & Coulson, S., (2017). *Phytother. Res., 31*, 1316–1322. doi: 10.1002/ptr.5856.

Steiner, M., & Li, W., (2001). *J. Nutr., 131*, 980S–984S. doi: 10.1093/jn/131.3.980S.

Stillo, P., Icka, P., & Damo, R., (2018). *Amoracia rusticana* Gaertn, Mey. & Scherb. *A Neglected Multi Useful Species, 26*, 312–322. BSHN (UT).

Szulińska, M., Kręgielska-Narożna, M., Świątek, J., Styś, P., Kuźnar-Kamińska, B., Jakubowski, H., Walkowiak, J., & Bogdański, P., (2018). *Biomed. Pharmacother., 102*, 792–797. doi: 10.1016/j.biopha.2018.03.

Tasleem, F., Azhar, I., Ali, S. N., Perveen, S., & Mahmood, Z. A., (2014). *Asian Pac. J. Trop. Dis., 7*, S461–S468. doi: 10.1016/ S1995–7645(14)60275-3.

Thiagarajan, V. R. K., Shanmugam, P., Krishnan, U. M., Muthuraman, A., & Singh, N., (2013). Antinociceptive effect of *Butea monosperma* on vincristine-induced neuropathic pain model in rats. *Toxicol. Ind. Health, 29*, 3–13.

Thiagarajan, V. R., Shanmugam, P., Krishnan, U. M., Muthuraman, A., & Singh, N., (2012). Ameliorative potential of *Butea monosperma* on chronic constriction injury of sciatic nerve induced neuropathic pain in rats. *An Acad. Brascienc., 84*, 1091–1104.

Usharani, P., Mateen, A. A., Naidu, M. U., Raju, Y. S., & Chandra, N., (2008). *Drugs R. D., 9*, 243–250.

Varshney, R., & Budoff, M. J., (2016). *J. Nutr., 146*, 416S–421S. doi: 10.3945/jn.114.202333.

Velasco, P., Francisco, M., & Cartea, M. E., (2010). Glucosinolates in *Brassica* and cancer. In: Watson, R. R., & Preedy, V. R., (eds.), *Bioactive Foods and Extracts. Cancer Treatment and Prevention* (1st edn., pp. 3–29). CRC Press: Boca Raton, FL, USA.

Vijayakumar, R. S., & Nalini, N., (2006). *Cell. Biochem. Funct., 24*, 491–498. doi: 10.1002/ cbf.1331.

Walters, S. A., (2021). Horseradish: A neglected and underutilized plant species for improving human health. *Horticulturae, 7*, 167. https:// doi.org/10.3390/horticulturae7070167.

Wang, C., Zhang, Z., Wang, Y., & He, X., (2016). Cytotoxic constituents and mechanism from *Peganum harmala* L. *Chem. Biodivers., 13*(7), 199–212.

Wang, J., Zhang, X., Lan, H., & Wang, W., (2017). *Food Nutr. Res., 61*, 1377571. doi: 10.1080/16546628.2017.1377571.

Wang, S. L., Li, Y., Wen, Y., Chen, Y. F., Na, L. X., Li, S. T., & Sun, C. H., (2009). *Biomed. Environ. Sci., 22*, 32–39. doi: 10.1016/ S0895-3988(09)60019-2.

Wattanathorn, J., Chonpathompikunlert, P., Muchimapura, S., Priprem, A., & Tankamnerdthai, O., (2008). *Food Chem. Toxicol., 46*, 3106–3110. doi: 10.1016/j.fct.2008.06.014.

Wijeratne, S. S. K., & Cuppett, S. L., (2007). *J. Agric. Food Chem., 55*, 1193–1199. doi: 10.1021/jf063089m.

Wilborn, C., Taylor, L., Poole, C., Foster, C., Willoughby, D., & Kreider, R., (2010). *Int. J. Sport Nutr. Exerc. Metab., 20*, 457–465.

Williams, M. J., Sutherland, W. H., McCormick, M. P., Yeoman, D. J., & De Jong, S. A., (2005). *Phytother. Res., 19*, 314–319. doi: 10.1002/ptr.1663.

Wojcikowski, K., Myers, S., & Brooks, L., (2007). *Platelets, 18*, 29–34. doi: 10.1080/09537100600800636.

Womack, C. J., Lawton, D. J., Redmond, L., Todd, M. K., & Hargens, T. A., (2015). *J. Int. Soc. Sports Nutr., 12*, 23. doi: 10.1186/s12970-015-0084-9.

Yang, S., Liu, L., Meng, L., & Hu, X., (2019). *Chem.-Biol. Interact., 297*, 1–7.

Yu, J. J., Pei, L. B., Zhang, Y., Wen, Z. Y., & Yang, J. L., (2015). *J. Clin. Psychopharmacol., 35*, 406–410. doi: 10.1097/ JCP.0000000000000352.

Zanzer, Y. C., Plaza, M., Dougkas, A., Turner, C., & Östman, E., (2018). *Food Funct., 9*, 2774–2786. doi: 10.1039/C7FO01715D.

CHAPTER 2

NOVEL IMPORTANCE OF HERBS AND THEIR EFFECTS ON HUMAN HEALTH

SHASHI SONI,[1] UROOSA NOOR,[1] EHSAN AMIRI,[2] and ENA GUPTA[1]

[1]*Department of Family and Community Sciences, University of Allahabad, Uttar Pradesh, India*

[2]*Department of Phytopharmaceuticals (Traditional Pharmacy), Faculty of Pharmacy, Shiraz University of Medical Sciences, Shiraz, Iran*

ABSTRACT

In the era of COVID-19, herbal nutrition has gained much popularity because of its immunity-boosting properties ,which are helpful in maintaining primary health. Herbs have been used as an Ayurvedic medicine since ancient times and are economically important. Nowadays, people are moving toward natural treatment because of the negligible side effects and easy affordability. Some medicinal herbs such as *Stevia rebaudiana* (Stevia), *Tinospora cordifolia* (Giloy), *Asparagus racemosus* (Satavari), *Aloe barbadensis miller* (Aloe vera), and *Azadirachta indica* (Neem) provides protection against cancer, cardiovascular diseases (CVDs), diabetes, liver disorders, hypertension, and many other chronic diseases. Most of the herbs possess anti-inflammatory, anti-bacterial, antiviral, anti-fungal, antimalarial, antiprotozoal, antioxidant, antipyretic, immunomodulatory, diuretic, and immune boosting properties. This review chapter focuses on the pharmacological and therapeutic importance of commonly used Indian medicinal herbs on human health.

Immune-Boosting Nutraceuticals for Better Human Health: Novel Applications. Urmila Jarouliya, PhD, Raj K. Keservani, MPharm, Rajesh K. Kesharwani, PhD, Virendra K. Patel, PhD, and Adi D. Bharti, MPharm (Eds.)

2.1 INTRODUCTION

In India, herbal nutrition has taken a sharp turn during the era of COVID-19 due to its minimal ill effects and enormous beneficial effects on human health. Its prominent and potential role has been recorded in boosting up immunity. Data are given by WHO emphasize that approximately 80% of the population around the world are using herbal nutrition as an immune booster which also helps to maintain primary health. Since ancient times India is known as the home of economically important medicinal and aromatic herbs. Aromatic and medicinal plants occupy a major part of the flora, which is vastly used at commercial and domestic level in the pharmaceutical, cosmetic, and drug industries. Nowadays, most of the population is moving towards herbal (Sharma et al., 2020; Rane et al., 2020a–c) treatments because of its negligible side effects which are ultimately seen in the market with the expansion and increased demand of medicinal herbs. Positive health effects of medicinal and aromatic herbs have been proven worldwide. Data given by the World Health Organization (WHO) stated that in developing countries, around 90% of the population is using plant products in combination with western medicine as primary health care. These medicinal herbs exhibit enormous health benefits like antidiabetic, anticancer, antimicrobial, anti-inflammatory, antihypertensive, antioxidative, immune boosting and many more (D'souza, 2021; Namedo, 2021; Rathore & Mathur, 2019). This review chapter is focused on the imperative role of some commonly used Indian medicinal herbs on human health with major emphasis on *Tinospora cordifolia* (Giloy), *Stevia rebaudiana* (Stevia), *Aloe barbadensis miller* (Aloe vera), *Azadirachta indica* (Neem) and *Asparagus racemosus* (Shatavari).

2.2 *TINOSPORA CORDIFOLIA* (GILOY)

Tinospora cordifolia is a large, glabrous deciduous climbing shrub belonging to the family menispermaceae, which is well-known as Guduchi, Amrita, Gurach, and Tinospora. The stems are rather succulent with long filiform fleshy aerial roots from the branches (Singh, 2013). Giloy is considered to be one of the three Amrit plants meant for the nectar of God. In Sanskrit, it is named as "Amritavalli." This Ayurvedic herb is also known as the "Queen of all herbs" (Prajwala et al., 2019). This herb is widely spread at an altitude of 600 m, which are the tropical and subtropical zones of India, from Kumaon to Assam, in north ranging through Bihar, Kankan, Kerala, West Bengal, and

Karnataka. It is commonly distributed up to 1,200 m above sea level of the Indian tropical region and native in the areas of India, Sri Lanka, Myanmar, China, Philippines, Bangladesh, Indonesia, Borneo, Vietnam, North Africa, and South Africa. In deciduous and dry forests, it is naturally grown up to an elevation of 1,000 fits (Kumar et al., 2019). Genera *Tinospora* consists of about 15 species. Some of the medicinally important species are *T. cordifolia, T. crispa, T. malabarica, T. tomentosa, T. uliginosa*, etc. (Mesharm, 2013). Table 2.1 elucidates the systematic classification of *Tinospora cordifolia.*

TABLE 2.1 The Systematic Classification of Herb *Tinospora cordifolia*

Plant Profile	
English name	Giloy
Native name (Indian)	Guduchi
Classification	
Kingdom	*Plantae*
Division	*Magnoliophyta*
Class	*Magnoliophyta*
Order	*Ranunculales*
Family	*Menispermaceae*
Genus	*Tinospora*
Species	*T. cordifolia*

Many research studies have demonstrated the therapeutic potential of giloy and the occurrence of a number of bioactive compounds in its different parts. Polysaccharides (PS), glycosides, steroids, alkaloids, diterpenoid, lactone, squalene, eicosalone, stigmasterol, α-tocopherol, isopropylester, pentanoic acid, valproic acid and propyl ester, 9 linoeaidic acid, 3,5-dimethoxyaceto-phenone, neophytadiene, hexadecanoic acid, methyl ester, n-hexadecanoic acid, lupeol were found to be primarily responsible for its curing, renovation, and promotional effects on human health, detected to be present in the stems of *T. cordifolia* by GCMS technique (Gautam et al., 2020). In the current scenario of COVID-19, it is most commonly used herbal nutrition as primary health care in the form of decoction and additives in various food products. *T. cordifolia* is one of the most commonly used herbs in household as the first medication to treat many types of chronic and acute disorders like dengue fever as it increases the platelets count. Other benefits of *T. cordifolia* include anti-allergic, analgesic, digestive, controlling blood sugar, hepatoprotective, cardioprotective, antimicrobial, anti-inflammatory, anti-aging, etc. Scientific evaluations also confirm that it also heals the side-effects of chemotherapy by supporting intestinal system, immune system, skin, liver, and spleen (Gautam et al., 2021; Singh & Yadav, 2021; Verma et al., 2021; Pokharkar et al., 2021; Prajwala et al., 2019).

In a dose-dependent manner, aqueous extract of *T. cordifolia* extract was found to be motivating for apoptosis inducers in AW13516 cells and was even effective at a low concentration of 5 μg/ml. Induction of apoptosis was verified with caspase (cysteine-aspartic proteases) activity analysis. Treatment of cells with extract for 24 hours demonstrated a significant decrease in the appearance of EMT genes in a dose-dependent manner without an effect on metastatic genes. Thus, the aqueous extracts of *T. cordifolia* bring apoptosis-mediated cell death to the oral cancer cell line AW13516, while its potential for epithelial mesenchymal transition is attenuated (Chandra et al., 2020).

T. cordifolia, a tribal medicine actively fights with several diseases. The standard side of *T. cordifolia* (18-norcherone o-glucoside) exhibits antipyretic properties, which can fight against swine flu. One more chemical beberine plays an important role against arrhythmia and hyperlipidemia. It purifies the blood and fight against diabetes and many types of cancer. This herb exhibits negligible side effects along with antiperiodic and antispasmodic properties, which are again helpful in preventing infectious diseases like dengue, swine flu, malaria, etc. (Baghel et al., 2017). Common health benefits of *T. cordifolia* are summarized in Figure 2.1.

FIGURE 2.1 Health benefits of *Tinospora cordifolia* plant.

2.3 *STEVIA REBAUDIANA*

Stevia rebaudiana Bertoni commonly known as stevia belonging to the family Asteraceae is a native plant of Paraguay in the northeast region which is familiar for its imperative sweetening property. From its various parts, leaves are majorly valued for calorie-free sweetening potential which maintains body weight by reducing blood sugar level and energy intake. Recently, it has become more popular due to its disease preventing effect and beneficial effect on cellular dysfunction, including diabetes and beta cells. The systematic classification of *S. rebaudiana* is depicted in Table 2.2.

TABLE 2.2 The Systematic Classification of Herb *Stevia rebaudiana*

Plant Profile	
English name	Stevia
Native name (Indian)	Meethi tulsi
Classification	
Kingdom	*Plantae*
Division	*Tracheophyta*
Class	*Magnoliopsida*
Order	*Asterales*
Family	*Asteraceae*
Genus	*Stevia*
Species	*S. rebaudiana*

A unique natural complex substance present in the leaves of stevia are steviol glycosides, alkaloids, flavonoids, phenolic acid, xanthophylls, hydroxy-cinnamic acid, water-soluble chlorophylls, vitamins, and minerals. It exhibits a wide-ranging biological activity because of its phytochemical composition; each compound present in it shows multiple human health related positive effects. Steviol glycosides are considered as the molecule which accounts to its sweet flavor however presence of some of the non-sweetening molecules is considered for its insulinotropic potential. Leaves of the stevia consists more than 30 steviol glycosides such as steviolbioside, rebaudioside A-F, diterpene glycosides stevioside and dulcoside, among which stevioside, rebaudioside A and C, which gives it a superior place in food and beverage industry to be added as a natural herbal sweetener 100–300 times sweeter than table sugar and flavor enhancer without showing any adverse effect on health. Various *in vitro* and *in vivo* studies mentioned in the literature approves its anti-inflammatory, anti-hypertensive, anti-carcinogenic, and antidiabetic, antibacterial, antidiarrheal, anti-fertility, immunomodulatory, antiseptic, and diuretic actions (Iatridis et al., 2022; Jawad et al., 2022; Bugliani et al., 2021; Gupta et al., 2013, 2015) as shown in Figure 2.2.

Comparative study on *in vitro* antimicrobial activity of ethanolic and methanolic extracts of *S. rebaudiana* leaves through agar-well diffusion method with four different concentrations *viz* 25, 50, 75, and 100 mg/ml against eight bacterial strains was tested and result was compared with a standard drug tetracycline (10 μg/mL). From the two extracts, methanolic extract showed higher antibacterial activity at a concentration of 100 mg/ml on *Zymomonas* spp. with 25.93±0.16 mm inhibition zone.

Results of the phytochemical screening showed the presence of major bioactive compounds (viz. phenols, saponin, tannin, glycosides, phytosterols, and flavonoids) with excellent free radical scavenging activity (Keservani & Sharma, 2014; Gupta et al., 2017). Investigation of cytotoxicity, apoptosis induction and mechanism of action of steviol on MCF-7, a human breast cancer cell line, through Sulforhodamine-B and flow cytometer test confirmed the remarkable inhibition of growth rate on MCF-7 cells in a dose-dependent manner along with the abolishment of cell cycle progression with arresting of cells at G2/M phase and stimulating the cells for apoptosis and induction to produce reactive oxygen species (ROS) (Gupta et al., 2017).

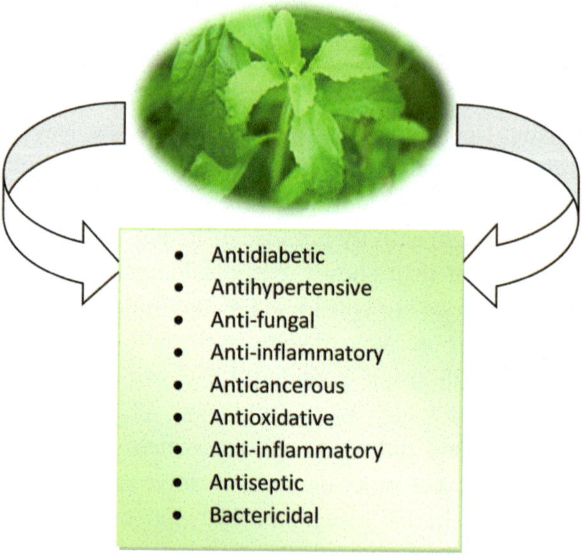

FIGURE 2.2 Common health benefits of *Stevia rebaudiana* plant.

2.4 *ALOE BARBADENSIS* MILLER

Aloe barbadensis Miller is commonly known as Aloe vera plant and is well known for its medicinal, healthcare, skin-care, and beauty properties since ancient times. It is a perennial, xerophytic, succulent, pea-green color plant belonging to the family of *Asphodelaceae* (*Liliaceae*), which includes around 360 species. Aloe vera is botanically known as *'Aloe barbadensismiller'* and their Sanskrit name is 'Ghee Kunwar.' The name Aloe vera is derived from the Arabic word 'alloeh' that means 'shinny bitter substance' while 'vera' is derived from the Latin word meaning 'true.' Aloe vera has lance-shaped, fleshy leaves along with sharp pointed edges, yellow cylindrical flowers and fruits that comprise several seeds. Every leaf of Aloe vera is made-up of three layers: first is an inner layer that contains clear gel like substance and composed of 99% water and remaining is made-up of glucomannans, amino acids, lipids, vitamins, and sterols; second or the middle latex layer is basically yellow bitter aloe juice or aloe sap and contain anthraquinones and glycosides; third and the outer layer contains 15–20 cells, called rind has protective function and synthesize protein and carbohydrates, this layer is also responsible for transportation of several substances like water (xylem) and starch (phloem) through vascular bundles located inside the rind. Around 20–30% weight of Aloe vera plant leaf is due to rind and approximately 65–80% weight represents pulp. Aloe vera is mainly grown in Asia, America, Africa, and Europe. In India, aloe vera is cultivated throughout but commonly found in Andhra Pradesh, Uttar Pradesh, Maharashtra, Rajasthan, Gujarat, and Tamil Nadu, etc. Around 2,000 years ago, *Aloe vera* was regarded as a universal panacea by the Greek scientists and Egyptians called it 'the plant of immortality.' By the early 1800s in United States Aloe vera is used as a laxative and later in the mid-1930s, it was successfully used to prevent chronic diseases and severe radiation dermatitis (Manvitha & Bidya, 2014; Surjushe et al., 2008; Boudreau & Beland, 2006). Table 2.3 elucidates the systematic classification of *Aloe barbadensis* Miller plant.

TABLE 2.3 The Systematic Classification of *Aloe barbadensis* Miller Plant

Plant Profile	
English name	Aloe vera
Native name (Indian)	Ghee Kunwar
Classification	
Kingdom	*Plantae*

TABLE 2.3 *(Continued)*

Plant Profile	
Division	*Magnoliophyta*
Class	*Liliopsida*
Order	*Liliales*
Family	*Liliaceae*
Genus	*Aloe*
Species	*A. Vera*

More than 200 compounds present in *Aloe barbadensis miller* from which around 75 have biological activity which include santhrones and their glycosides (e.g., 10-(1,5′-anhydroglucosyl)-aloe emodin-9-anthrone, also known as aloin A and B), chromones, anthraquinones (e.g., aloe-emodin), amino acids, glycoproteins, proteins, carbohydrates, lipids, sugars, organic acids, vitamins, minerals, hormones (auxins and gibberellins), enzymes (alkaline phosphatase, amylase, bradykinase, carboxypeptidase, catalase (CAT), cellulase, lipase, and peroxidase) and other compounds like salicylic acid, lignin, and saponins (Manvitha & Bidya, 2014; Surjushe et al., 2008; Hamman, 2008).

Aloe vera is used for numerous purposes but primarily used as a food preservative and medicine. It is the most common aloe species used in the processing industry around the world as a functional food (Keservani et al., 2010a, b, 2017, 2020) and as an ingredient in other food products like in the production of health drinks, beverages, and jellies, etc. It is also used in the cosmetic industry for the preparation of lotions, creams, moisturizers, soaps, shampoo, facial cleaner and hair styling gel, etc. (Manvitha & Bidya, 2014; Hamman, 2008). In pharmacological industry, it is used in the manufacturing of ointment and gel preparation along with tablets and capsules and mainly used as an immunity booster and detoxifier which are prescribed as a complementary therapy with antibiotics, chemotherapy, and NSAIDs (nonsteroidal anti-inflammatory drugs) to reduce the adverse effects induced by drugs like gastritis, etc. Previous studies revealed that different species of aloe found around the world have many pharmacological activities which acts as antioxidant, anticancer, antiulcer, antidiabetic, anti-inflammatory, antimicrobial, antiseptic, antibiotic, hepatoprotective, cardio-protective (Keservani et al., 2016a) and immunomodulatory (Keservani et al., 2015). It is also used to treat several skin conditions such as burns, cuts psoriasis, mycosis, fever, blisters, and eczema (Manvitha & Bidya, 2014; Sharma et al., 2014; Rajeswari et al., 2012; Hamman, 2008). Aloe vera is also used as a dietary supplement in case of osteoporosis, pre-, and post-operative patient and postmenopausal (Rajeswari et al., 2012). Common health benefits of *T. cordifolia* are summarized in Figure 2.3.

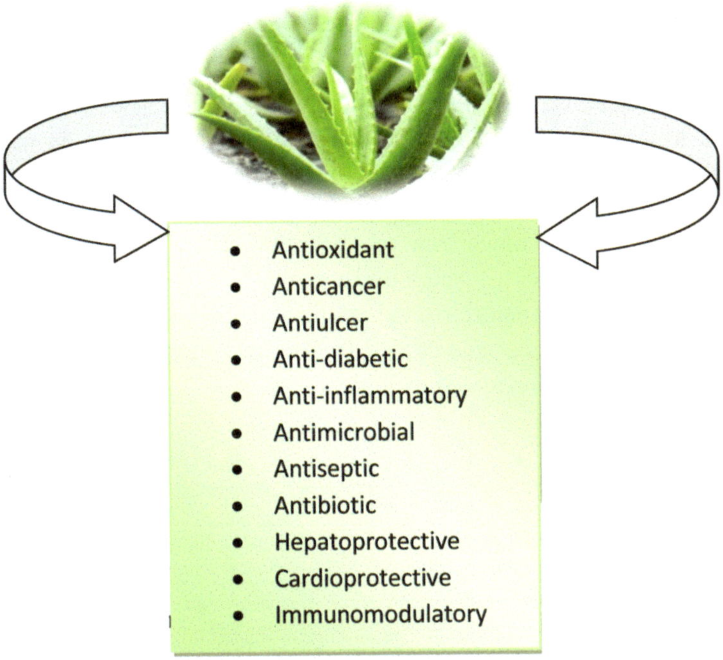

FIGURE 2.3 Common health benefits of *Aloe barbadensis* miller.

2.5 *AZADIRACHTA INDICA A. JUSS*

Azadirachta indica A. Juss is generally known as 'Neem' tree is native to Indian sub-continents and also found in tropical climate countries like Indonesia, Asia, Africa, Australia, and America. Neem is an evergreen tree which has been used as an Ayurvedic medicine for around more than 400 years due to its therapeutic potential. Neem plant is botanically known as *Azadirachta indica A. Juss.*, in Latin it is known as *Azadirachta indica* belonging to the family *Meliaceae* and sub-family *Melioideae*. In Sanskrit it is called 'arista' which means 'perfect, complete, and imperishable' and also known as 'nimba' the word arise from the term 'nimbati swasthyam dadati' meaning 'to give good health.' The benefits of neem are listed in the Indian documents as

'Charak-Samhita' and 'Susruta-Samhita' which is considered as a foundation of the natural treatment system, Ayurveda. The Persian name of neem tree is 'Azad-Darakht-E-Hind' which means 'free tree of India' which is deliberated as a part of India's genetic diversity. Neem is a tough, fast growing long-lasting tree which is about 25 meters in height having semi-straight to straight trunk, 3 meter width and long spreading branches forming an expansive crown around the tree. After 3–5 years the neem tree starts fruiting of yellowish ellipsoidal drupes which is the fruit of the tree, and around 10 years it becomes totally productive and survives for more than 200 years. It flourishes well in diverse climatic, topographical, and edaphic conditions and can grow in nearly all types of soils such as alkaline, saline, dry, clay, and rocky soils. Neem tree necessitate plenty of sun light with less water therefore it can grow in the areas where rainfall is in the range from to 1,200 mm, pH differ from 4 to 10 and temperature from 0 to 49°C. Neem tree contains a unique property of calcium mining which helps to neutralize the acidic soils (Girish & Shankara, 2008; Ogbuewu et al., 2011; Seriana et al., 2019). The systematic classification of *Azadirachta indica* is elucidated in Table 2.4.

TABLE 2.4 The Systematic Classification of *Azadirachta indica* Plant

Plant Profile	
English name	*Azadirachta indica*
Native name (Indian)	Neem
Classification	
Kingdom	*Plantae*
Order	Rutales
Family	Meliaceae
Genus	*Azadirachta*
Species	*A. indica*

More than 300 primary and secondary metabolites are present in the neem tree which includes carbohydrates, protein, and fat derivatives, whereas secondary metabolites are flavonoids, alkaloids, saponins, and steroids. Along with these the major bioactive compounds present in the tree are terpenes (isoazadirolide, 6 nimbocinolide, nibonolone, nimbonone, margosinone, and methylgrevillate) and limonoids (azadirachtin, I-tigloyl-3-acetyl-II-methoxyazadirachtin, 3-deacetyl-3-cinnamoylazadirachtin, 22,23-dihydro-23β-methoxyazadirachtin, 3-tigloylazadirachtol, 3-acetyl-1-tigloylazadirachtinin, nimbanal, margocilin, and magocinin, etc.).

Bioactive compounds present in neem increase the production of glutathione-s-transferase which helps to improve the ability of the liver to detoxify the chemical contamination in the body (Ogbuewu et al., 2011; Seriana et al., 2019). It is the most studied therapeutic tree and has many pharmacological applications.

Different parts of the neem tree are used to produce medicinal drugs for treating various acute and chronic diseases. For centuries humans used it to treat multiple diseases, due to its amazing properties. The United Nations named this tree as '21st century tree' and also denoted it as 'a tree to solve global problems.' Leaves of the neem is used as antifungal, antibacterial, antiviral, anxiolytic, analgesic, antimalarial, antiulcer, anti-cancer, anti-inflammatory, anti-hyperglycemic, antifertility, hepatoprotective, and immunomodulatory. It is also used as an herbal contraception through spermicidal and non-hormonal activities without causing any side-effects (Seriana et al., 2019; Giri et al., 2019). Neem is also used to treat various types of skin disorders like acute and chronic eczema, scabies, and ringworm, furunculosis, arsenical, herpes labialis infection, seborrheic dermatitis and ulcers. Stem bark and root bark of tree provide anti-hypertensive, spasmolytic, and diuretic activities (Brahmachari, 2004). Beside these activities, neem is also used in the cosmetic industry to prepare face cream, shampoo, conditioner, face wash, soap, nail polish and nail oil, etc. Neem cake is a by-product of neem-oil industry that is utilized as livestock-feed, natural pesticide and fertilizer (Girish & Shankara, 2008). Common health benefits of *Azadirachta indica* plant are summarized in Figure 2.4.

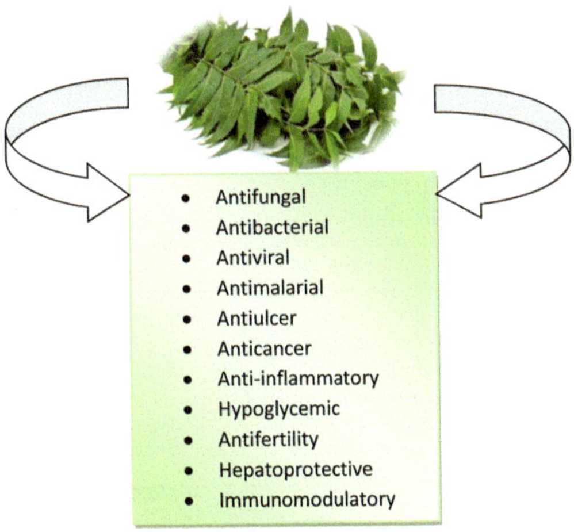

FIGURE 2.4 Common health benefits of *Azadirachta indica* plant.

2.6 *ASPARAGUS RACEMOSUS*

Asparagus racemosus is commonly known as 'shatawar' in Hindi, 'satmuli' (shata means 100 and muli means root) in the Himalayan region and 'shatavari' in Sanskrit which means 'able to having 1,00,000.' In Ayurveda, this wonderful herb is considered as 'queen of herbs' because of their potential to endorse love, special congeniality to women health and devotion to promoting the lovemaking ability. It is most extensively occurring plant having therapeutic potential and has been documented under six important Rasayana in Ayurvedic medicine that promotes good health by increasing the cellular vitality or resistance of an individual's especially important for women's overall health and strength. This plant belongs to the genus *Asparagus,* family *Asparagaceae*, subfamily *Asparagae.* The word *Asparagus* is derived from the Greek word meaning 'stalk' or 'shoot' (Kumar et al., 2008; Velavan et al., 2007; Alok et al., 2013; Selvaraj et al., 2019). *Asparagus racemosus* is a woody climber plant booming up to 1–2 meter in height, containing small, uniform, pine needle like leaves and white flowers having small spikes. The plant is mainly found in wild tropical and subtropical climates of India, Asia, Australia, and Africa. In India the crop is mainly cultivated in Andhra Pradesh, Kerala, Tamil Nadu and in the northern states. For the cultivation of *Asparagus racemosus* it requires heavy (clay), medium (loamy) and light (sandy) well drained and fertile soil at the climatic temperature ranging from 25–40°C. Around the world approximately 300 species of Asparagus has been identified, out of the 22 species is found in India from which various species such as *Asparagus racemosus, Asparagus gonaclades* and *Asparagus adsendens* are mainly cultivated, from which *Asparagus racemosus* is the most commonly used medicinal herb (Velavan et al., 2007). Table 2.5 depicts the systematic classification of *Asparagus racemosus* plant.

TABLE 2.5 The Systematic Classification of *Asparagus racemosus* Plant

Plant Profile	
English name	*Asparagus racemosus*
Native name (Indian)	Satawar or Satavari
Classification	
Kingdom	*Plantae*
Division	*Magnoliophyta*
Class	*Liliopsida*

TABLE 2.5 *(Continued)*

Plant Profile	
Order	*Asparagales*
Family	*Asparagaceae*
Genus	*Asparagus*
Species	*A. racemosus*

The main bioactive compounds present in the *Asparagus racemosus* is a group of steroidal saponins that contains Shatavarin I, II, III, IV, V, and VI–X phytoestrogen compounds, along with this several phytochemical constituents such as flavonoids (kaempferol, quercetin, rutin, and hyperoside), alkaloids, tannins, glycosides, phytosterols are also extracted from the root of the plant. It also contains several primary chemical compounds such as carbohydrates (PS, mucilage), proteins, crude fiber essential oils (EOs), vitamins, asparagine, arginine, tyrosine, resin, etc. (Selvaraj et al., 2019; Singh et al., 2018). Some trace elements such as calcium, potassium, zinc, manganese, copper, cobalt, and selenium are also detected in the fruit (Keservani et al., 2016b) and flower of the plant. Woody tuberous portion of the root contains kaempferol and sarapogenium in addition with this some essential fatty acids such as disogenin and gamma linoleinic acids are also obtained from the plant (Shaha & Bellankimath, 2017).

Asparagus racemosus has been used as an indigenous medicine system in India and British pharmacopeias since ancient times because of their phytoestrogen compounds. In Ayurvedic medicine it is recognized as a cooling operator and uterine tonic. It is mainly used to treat the problems related to gynecology, such as irregular menstrual cycle and sexual dysfunction (Shaha & Bellankimath, 2017). The root of the plant prevents hyper acidity, peptic ulcer and sooths the inner cavity of the stomach. Asparagine present in the plant acts as an anticancer agent and helps to treat leukemia. Steroidal saponins have the antispasmodic ability which relaxes the uterine muscle during pregnancy and helps to avoid premature birth and pre-term labor along with this it also helps to increase the secretion of milk in lactating mothers. Besides these activities it also possesses antibacterial, antiprotozoal, anti-inflammatory, antioxidant, antidepressant, antineoplastic, immunomodulatory, cardio-protective, hepato-protective, hypoglycemic, diuretic activities, etc. It is also used in the manufacturing of toilet sops and cleaners to avoid infections and to treat abnormalities of the uterine cavity (Alok et al., 2013; Selvaraj et al., 2019). Figure 2.5 summarizes the common health benefits of *Asparagus racemosus* plant.

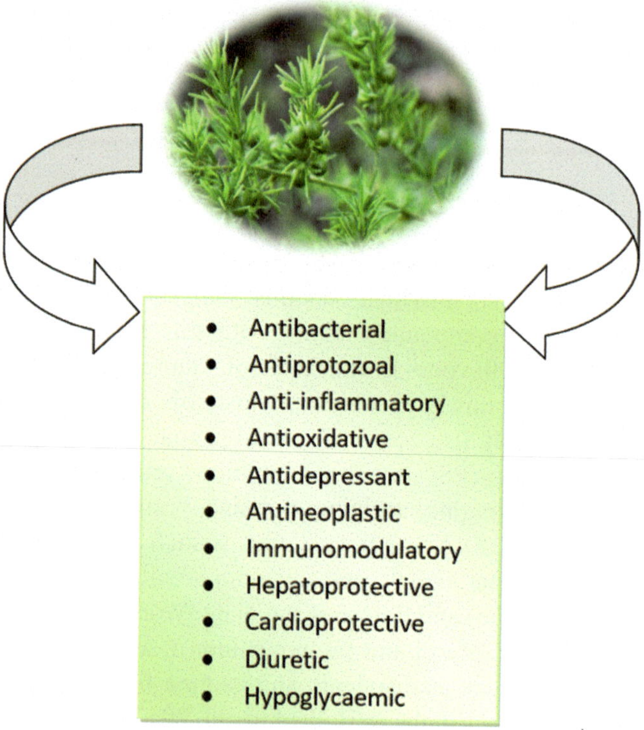

FIGURE 2.5 Common health benefits of *Asparagus racemosus* plant.

2.7 MARKET USE AND FUTURE PERSPECTIVE

Super high global acceptance rate of herbal nutrition in the current scenario of COVID-19 has been reported by various organizations and literature, which turns into high demand and supply of herbal plants in the market. Thus, increasing its cultivation and further processing at industrial level. Food and drug industries play a very crucial role in the marketing of these herbs for adding these highly nutritious additives in the diet of consumers. Value-added food products with important pharmacological and therapeutic potentials could be developed for easier consumption of herbs on a daily basis at adequate amounts. It is necessary to continue research on herbs to elucidate its disease preventing effect on some minor diseases other than the major ones which are already reported back in the literature survey.

2.8 CONCLUSION

In conclusion, it must be mentioned that herbal nutrition is a boon for today and tomorrow. India alone cultivates more than 100 species of herbs from which only five are described in this review chapter, namely *Tinospora cordifolia* (Giloy), *Stevia rebaudiana* (Stevia), *Aloe bardanesis miller* (Aloe vera), *Azadirachta indica* (Neem) and *Asparagus racemosus* (Shatavari). All of these five herbs are rich sources of secondary plant metabolites which gives an imperative health benefit. Each phytochemical present in the herbs shows specific health benefits in the management of many acute and chronic diseases. Overall, it is suggested to include herbal nutrition in day-to-day life for better health as it prevents the body from any fatal disease.

ACKNOWLEDGMENT

None declared.

CONFLICT OF INTEREST

None declared.

KEYWORDS

- ***Aloe barbadensis*** **miller**
- ***Asparagus racemosus***
- ***Azadirachta indica***
- ***Stevia rebaudiana***
- ***Tinospora cordifolia***

REFERENCES

Alok, S., Jain, S. K., Verma, A., Kumar, M., Mahor, A., & Sabharwal, M., (2013). Plant profile, phytochemistry and pharmacology of *Asparagus racemosus* (Shatavari): A review. *Asian Pacific Journal of Tropical Disease*, *3*(3), 242–251.

Baghel, A. S., Tiwari, A., Gupta, N., Gour, M., & Pinge, K., (2017). A review on swine flu treatment by tribal medicines using *Tinospora cordifolia. International Journal of Pharmacy & Life Sciences, 8*(11).

Boudreau, M. D., & Beland, F. A., (2006). An evaluation of the biological and toxicological properties of *Aloe barbadensis* (miller), *Aloe vera. Journal of Environmental Science and Health Part C, 24*(1), 103–154.

Brahmachari, G., (2004). Neem—An omnipotent plant: Retrospection. *Chembiochem, 5*(4), 408–421.

Bugliani, M., Tavarini, S., Grano, F., Tondi, S., Lacerenza, S., Giusti, L., & Angelini, L. G., (2022). Protective effects of *Stevia rebaudiana* extracts on beta cells in lipotoxic conditions. *Acta Diabetologica, 59*(1), 113–126.

Chandra, K., Jain, S., & Dwivedi, S., (2020). The role of herbal plants for the management of diabetes mellitus: A review. *Cardiology Today, 24*(2).

D'souza, M. R., (2021). Traditional Indian herbs for the management of diabetes mellitus and their herb–drug interaction potentials: An evidence-based review. In: Chen, H., & Zhang, M., (eds.), *Structure and Health Effects of Natural Products on Diabetes Mellitus*. Springer, Singapore. https://doi.org/10.1007/978-981-15-8791-7_16.

Gautam, A. S., Singh, A., & Kumar, K., (2020). Analysis of therapeutic value of *Tinospora cordifolia. Asia Pacific Journal of Multidisciplinary Research, 8*, 1–15.

Gautam, A., (2020). Therapeutic properties of *Syzygium cumini* (Jamun) and *Tinospora cordifolia* (Giloy) against various lethal diseases. In: *Advanced Pharmacological Uses of Medicinal Plants and Natural Products* (pp. 316–337). IGI Global.

Giri, R. P., Gangawane, A. K., & Giri, S. G., (2019). Neem the wonder herb: A short review. *International Journal of Trend in Scientific Research and Development (IJTSRD), 3*(3), 962–966.

Girish, K., & Shankara, B. S., (2008). Neem–a green treasure. *Electronic Journal of Biology, 4*(3), 102–111.

Gupta, E., Kaushik, S., Purwar, S., Sharma, R., Balapure, A. K., & Sundaram, S., (2017). Anticancer potential of steviol in MCF-7 human breast cancer cells. *Pharmacognosy Magazine, 13*(51), 345–350. https://doi.org/10.4103/pm.pm_29_17.

Gupta, E., Purwar, S., Singh, A., Sundaram, S., Rai, G. K., & Sundaram, S., (2015). Evaluation of nutritional, anti-nutritional and bioactive compounds in juice and powder of *Stevia rebaudiana. Indian Journal of Natural Sciences, 5*(28).

Gupta, E., Purwar, S., Sundaram, S., & Rai, G. K., (2013). Nutritional and therapeutic values of *Stevia rebaudiana*: A review. *Journal of Medicinal Plants Research, 7*(46), 3343–3353.

Gupta, E., Vajpayee, G., Purwar, S., Shakyawar, S., Alok, S., & Sundaram, S., (2017). Phytochemical screening and in-vitro studies of antioxidant and antimicrobial activity of extracts of dried *Stevia rebaudiana* leaves. *International Journal of Pharmaceutical Sciences and Research, 8*(8), 3354–3360.

Hamman, J. H., (2008). Composition and applications of *Aloe vera* leaf gel. *Molecules, 13*(8), 1599–1616.

Iatridis, N., Kougioumtzi, A., Vlataki, K., Papadaki, S., & Magklara, A., (2022). Anti-cancer properties of *Stevia rebaudiana*; More than a sweetener. *Molecules, 27*(4), 1362.

Jawad, M., Rizwan, B., Jawad, M., Khalid, F., & Ishaq, A., (2022). A nutraceutical and therapeutic potentials of *Stevia rebaudiana* Bertoni: Nutraceutical and therapeutic potentials of stevia. *Pakistan BioMedical Journal, 5*(1).

Keservani, R. K., & Sharma, A. K., (2014). Flavonoids: Emerging trends and potential health benefits. *Journal of Chinese Pharmaceutical Sciences, 23*(12), 815.

Keservani, R. K., Kesharwani, R. K., Sharma, A. K., & Jarouliya, U., (2015). Dietary supplements, nutraceutical and functional foods in immune response (immunomodulators). In: Debasis, B., Harry, G. P., & Anand, S., (eds.), *Nutraceutical and Functional Foods in Human Life and Disease Prevention* (pp. 343–358). Boca Raton, Fl. CRC Press, Taylor and Francis, Chapter 20. ISBN: 9781482237214.

Keservani, R. K., Kesharwani, R. K., Sharma, A. K., Vyas, N., & Chadoker, A., (2010b). Nutritional supplements: An overview. *International Journal of Current Pharmaceutical Review and Research, 1*(1), 59–75.

Keservani, R. K., Kesharwani, R. K., Vyas, N., Jain, S., Raghuvanshi, R., & Sharma, A. K., (2010a). Nutraceutical and functional food as future food: A review. *Der Pharmacia Lettre, 2*(1), 106–116.

Keservani, R. K., Sharma, A. K., & Kesharwani, R. K., (2016a). Nutraceutical and functional foods for cardiovascular health. In: *Food Process Engineering* (pp. 257–278, 291–312). Apple Academic Press, CRC Press. ISBN: 9781771884020.

Keservani, R. K., Sharma, A. K., & Kesharwani, R. K., (2016b). Medicinal effect of nutraceutical fruits for the cognition and brain health. *Scientifica*, 10. Article ID 3109254. doi: 10.1155/2016/3109254.

Keservani, R. K., Sharma, A. K., & Kesharwani, R. K., (2017). An overview and therapeutic applications of nutraceutical and functional foods. *Recent Advances in Drug Delivery Technology*, 160–201. ISBN:9781522507543.

Keservani, R. K., Sharma, A. K., & Kesharwani, R. K., (2020). *Nutraceuticals and Dietary Supplements: Applications in Health Improvement and Disease Management*. CRC Press. ISBN: 9781771888738.

Kumar, A., & Nagaraju, S., (2019). *Giloy [Tinospora cordifolia (Thunb.) Miers] 18*(1) 57–58.

Kumar, S., Mehla, R. K., & Dang, A. K., (2008). Use of shatavari (*Asparagus racemosus*) as a galactopoietic and therapeutic herb: A review. *Agricultural Reviews, 29*(2), 132–138.

Manvitha, K., & Bidya, B., (2014). *Aloe vera*: A wonder plant its history, cultivation and medicinal uses. *Journal of Pharmacognosy and Phytochemistry, 2*(5), 85–88.

Meshram, A., Bhagyawant, S. S., Gautam, S., & Shrivastava, N., (2013). Potential role of *Tinospora cordifolia* in pharmaceuticals. *World Journal of Pharmacy and Pharmaceutical Sciences, 2*(6), 4615–4625.

Namdeo, P., (2021). A review on herbal immunity booster and nutrition–to fight against COVID-19. *J Pharm. Adv. Res., 4*(5), 1226–1237.

Ogbuewu, I. P., Odoemenam, V. U., Obikaonu, H. O., Opara, M. N., Emnalom, O. O., Uchegbu, M. C., & Iloeje, M. U., (2011). The growing importance of neem (*Azdirachta indica* A. Juss) in agriculture, industry, medicine and environment: A review. *Research Journal of Medicinal Plant.*

Pokharkar, D., Choudhary, M., Choudhari, P., Bachate, V., & Chaudhari, L., (2021). *Immunity Booster Herbs Against Coronavirus, 10*(8), 1461–1467.

Prajwala, B., Raghu, N., Gopenath, T. S., & Basalingappa, K. M., (2019). Guduchi: Its medicinal properties. *J. Plant Physiol. Pathol., 7*(3), 2.

Rajeswari, R., Umadevi, M., Rahale, C. S., Pushpa, R., Selvavenkadesh, S., Kumar, K. S., & Bhowmik, D., (2012). *Aloe vera*: The miracle plant its medicinal and traditional uses in India. *Journal of Pharmacognosy and Phytochemistry, 1*(4), 118–124.

Rane, B. R., Bharath, M. S., Patil, R. R., Keservani, R. K., & Jain, A. S., (2020b). Novel approaches in nutraceuticals. In: Rajesh, K. K., Raj, K. K., & Anil, K. S., (eds.), *Enhancing the Therapeutic Efficacy of Herbal Formulations Through Novel Drug Delivery Systems* (pp. 241–266). IGI Global International Publisher, Pennsylvania, USA, Chapter 11. ISBN13: 9781799844532.

Rane, B. R., Patil, A. S., Keservani, R. K., & Jain, A. S., (2020a). Novel approaches in herbal formulation. In: Rajesh, K. K., Raj, K. K., & Anil, K. S., (eds.), *Enhancing the Therapeutic Efficacy of Herbal Formulations Through Novel Drug Delivery Systems* (pp. 43–68). IGI Global International Publisher, Pennsylvania, USA, Chapter 2. ISBN13: 9781799844532.

Rane, B. R., Tadavi, S. A., & Keservani, R. K., (2020c). Naturopathy. In: Anil, K. S., Raj, K. K., & Surya, P. G., (eds.), *Herbal Product Development* (pp. 321–347). Apple Academic Press, CRC Press, Taylor & Francis Group, Chapter 12. ISBN: 9781771888776.

Rathore, R., & Mathur, A., (2019). Scope of cultivation and value chain perspectives of medicinal herbs in India a case study on *Aloe vera* and isabgol. *J. Pharmacogn. Phytochem., 8*, 243–246.

Selvaraj, K., Sivakumar, G., Veeraraghavan, V. P., Dandannavar, V. S., Veeraraghavan, G. R., & Rengasamy, G., (2019). *Asparagus racemosus*: A review. *Systematic Reviews in Pharmacy, 10*(1), 87–89.

Seriana, I., Akmal, M., & Wahyuni, S., (2019). Neem leaves extract (*Azadirachta indica* A. Juss) on male reproductive system: A mini-review. In: *IOP Conference Series: Earth and Environmental Science* (Vol. 399, No. 1, p. 012106). IOP Publishing.

Shaha, P., & Bellankimath, A., (2017). Pharmacological profile of *Asparagus racemosus*: A review. *Int. J. Curr. Microbiol. App. Sci., 6*(11), 1215–1223.

Sharma, A. K., Keservani, R. K., & Gautam, S. P., (2020). *Herbal Product Development*. Apple Academic Press, CRC Press, Taylor & Francis Group. ISBN: 9781771888776.

Sharma, P., Kharkwal, A. C., Kharkwal, H., Abdin, M. Z., & Varma, A., (2014). A review on pharmacological properties of *Aloe vera*. *Int. J. Pharm. Sci. Rev. Res., 29*(2), 31–37.

Singh, A. K., Srivastava, A., Kumar, V., & Singh, K., (2018). Phytochemicals, medicinal and food applications of Shatavari (*Asparagus racemosus*): An updated review. *The Natural Products Journal, 8*(1), 32–44.

Singh, A., & Yadav, L., (2020). *Giloy: The Root of Immorlatity, 2*(1), (1–3).

Surjushe, A., Vasani, R., & Saple, D. G., (2008). *Aloe vera*: A short review. *Indian Journal of Dermatology, 53*(4), 163.

Tomar, H., Singhal, M., Juyal, P., Tyagi, M., Kumar, A., & Gaurav, N. (2021). *Potential Role of Giloy (an Important Traditional Aurvedic Medicinal Dioecious Creeper) as an Immunomodulator, 10*(6), 57–67.

Velavan, S., Nagulendran, K. R., Mahesh, R., & Begum, V. H., (2007). Phcog rev.: Plant review the chemistry, pharmacological and therapeutic applications of *Asparagus racemosus*: A review. *Pharmacognosy Reviews, 1*(2), 350–360.

Verma, D. K., Kumar, P., & El-Shazly, M., (2021). Unmasking the many faces of Giloy (*Tinospora cordifolia* L.): A fresh look on its phytochemical and medicinal properties. *Current Pharmaceutical Design*.

CHAPTER 3

CONSUMPTION OF CAROTENOID-RICH VEGETABLES AND FUNCTIONS OF HUMAN T-LYMPHOCYTES

SHANTI BHUSHAN MISHRA,[1] AMIT KUMAR SINGH,[2] SHRADHANJALI SINGH,[3] and DIVYA RANI SHARMA[4]

[1]*Department of Pharmacognosy, United Institute of Pharmacy, Prayagraj, Uttar Pradesh, India*

[2]*Department of Pharmaceutics, United Institute of Pharmacy, Prayagraj, Uttar Pradesh, India*

[3]*Department of Pharmaceutical Chemistry, United Institute of Pharmacy, Prayagraj, Uttar Pradesh, India*

[4]*Department of Pharmacology, United Institute of Pharmacy, Prayagraj, Uttar Pradesh, India*

ABSTRACT

Carotenoids and retinoids share a number of biological features, including antioxidant qualities, the ability to prevent malignant tumor development, and the ability to induce apoptosis. Carotenoids can affect cell development, gene expression, and immunological responses by augmenting with them. Epidemiological studies have linked high carotenoid consumption in the diet to a lower frequency of ovarian, breast, cervical, colorectal, and cardiovascular malformations, as well as cardiovascular and eye ailments. The effects of dietary carotenoids on growth factor signaling,

Immune-Boosting Nutraceuticals for Better Human Health: Novel Applications. Urmila Jarouliya, PhD, Raj K. Keservani, MPharm, Rajesh K. Kesharwani, PhD, Virendra K. Patel, PhD, and Adi D. Bharti, MPharm (Eds.)

intercellular communication, cell cycle progression, retinoid-like receptors, differentiation-related proteins, antioxidant response element, the AP-1 (activator protein-1) transcriptional complex, nuclear receptors, the Wnt/β-catenin pathway, and inflammatory cytokines are all involved in cancer chemoprevention. Furthermore, carotenoids can increase the production of B- and T-lymphocyte, cytotoxic T-cell activity, effector T-cell function, macrophage, and cytokine production. The antioxidant activities of the primary carotenoids, lycopene, crocin/crocetin, β-carotene, curcumin, lutein, and zeaxanthin, have recently been ascribed to the positive benefits of carotenoid-rich fruits and vegetables in health and in lowering the risk of specific illnesses. Carotenoids are hypothesized to have a time and dose-dependent action. The immunological and biological actions of the primary carotenoids used to treat various illnesses, as well as their probable mechanism of action, are briefly described in this chapter.

3.1 INTRODUCTION

Carotenoids are pigments present in the chloroplasts of all higher plants and algae, and they are presumed to be plant pigments. Carotenoids are liposoluble pigments that are brightly colored. They're most noticeable in non-photosynthetic tissues, where they're responsible for the yellow, orange, and red hues found in many fruits, flowers, and roots. Carotenoids are pigments present in microorganisms, including non-phototrophic bacteria and fungus, as well as photosynthetic membranes of photobacteria and cyanobacteria. They are also essential pigments in animals, particularly birds (yellow and red feathers and egg yolk), fish and crustaceans (orange-red skin and meat), and invertebrates, where they are responsible for a wide spectrum of colors in the free form or as complexes with proteins. These animal carotenoids, on the other hand, are derived from plants and are absorbed through the food (Britton, 1995).

There are around 600 carotenoids, grouped into carotenes, xanthophylls, and lycopene, with natural structural polymorphisms (Rutz et al., 2016; Jomova & Valko, 2013). Only approximately 40 carotenoids have been found in human blood and tissues, and only about 20 carotenoids are present in a typical human diet. α-carotene, β-carotene, lutein, lycopene, and cryptoxanthin are some of the carotenoids found in the human diet and body (Rao & Rao, 2007). Carotenoids are a kind of tetraterpene (a C40-based isoprenoid) that gives fruits, foliage, and flowers their yellow, orange, or red color (Tapiero et al., 2004; Kaulmann & Bohn, 2014). For example:

- Green vegetables encompass high proportions of the both hydrocarbon carotenes and xanthophylls;
- Lycopene is a lipophilic red colored pigment present in ripe tomatoes;
- β-carotene accounts for the orange color of carrots;
- Capsanthine is accountable for the bright red pigment of peppers; and
- Astaxanthin (ASTA) is responsible for the red/pink coloration of crustaceans (Astorg, 1997). The most of the carotenoids have alternating single and double bonds along with central carbon chain and cyclic or acyclic end groups (Stahl & Sies, 2005).

Recent contribution in carotenoids has been sparked by epidemiological studies that strongly recommend that eating foods high in carotenoids reduces the risk of cardiovascular diseases (CVDs), cancers, cataracts, diseases associated with low immune function, age-related macular degeneration (AMD) (Meydani et al., 1995; Benedich, 1994; Chew, 1995) infectious diseases like HIV infections, and other degenerative diseases. Because all carotenoids (e.g., lycopene, lutein, β-carotene, cryptoxanthin, and α-carotene) are considerably degraded in HIV patients, their plasma concentrations of carotenoids are generally low, and this is directly associated to an increased risk of mortality. The injection of 60 mg of β-carotene per day to HIV patients increases both CD4+ and CD8+ lymphocytes, and symptoms of the illness reduce over 2–3 years (Pechinskii & Kuregyan, 2014). Carotenoids have been shown to have a diversity of biological effects, including antioxidant activity, immunological boosting, the suppression of mutagenesis, and the regression of malignant lesions (Rao & Rao, 2007). The principal functions and methods of action of carotenoids are shown in Table 3.1; Figures 3.1 and 3.2.

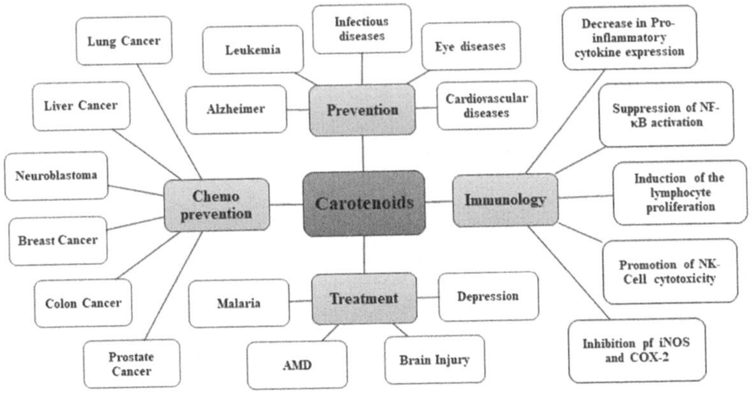

FIGURE 3.1 Biological activities of the main carotenoids used in the treatment of various diseases.

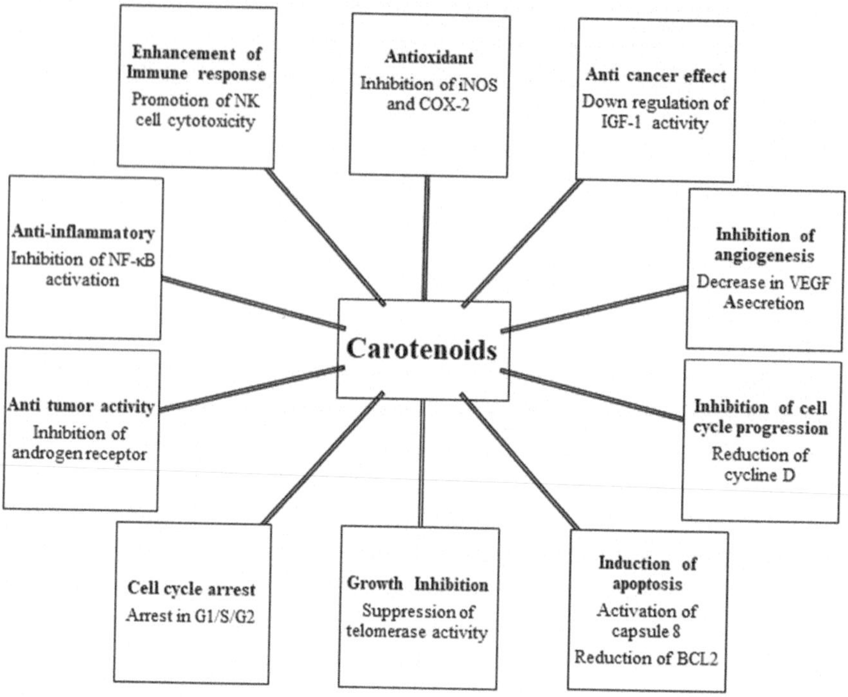

FIGURE 3.2 Molecular mechanisms of the main carotenoids used in the treatment of various diseases.

TABLE 3.1 Carotenoids Against Different Disease

Carotenoids	Disease	References
β-carotene	Leukemia	Upadhyaya et al. (2007)
	Colon cancer	Palozza et al. (2002)
	Gastric cancer	Jang et al. (2009)
	Adenocarcinoma	Palozza et al. (2001)
	Neuroblastoma	Kim et al. (2014)
	Hepatocarcinoma	Chen et al. (2012)
	Prostate cancer	Karppi et al. (2012)
	Cervical cancer	Guo et al. (2015)
	HCV infection	Kataria et al. (2016)
	Age-related macular degeneration (AMD)	Aoki et al. (2016)
	Breast cancer	Olmos et al. (2015)

TABLE 3.1 *(Continued)*

Carotenoids	Disease	References
β-carotene, canthaxanthin, phytoene	Skin tumors	Mathews-Roth (1982)
β-carotene and α-carotene	Hepatocarcinoma	Chen et al. (2013)
Crocin	Breast cancer	Chryssanthi et al. (2007)
	Depression	Noorbala et al. (2005)
Crocetin	Brain injury	Bie et al. (2011)
	Gastric cancer	Bathaie et al. (2013)
	Malignant cells	Abdullaev (1994)
	Lung cancer	Magesh et al. (2009)
	Colorectal cancer	Aung et al. (2007)
	Cervical cancer caused by HPV infections	Khavari et al. (2014)
Curcumin	Alzheimer	Lim et al. (2001)
	Head and neck cancer	LoTempio et al. (2005)
	Bladder cancer	Chadalapaka et al. (2008)
	Skin tumor	Limtrakul et al. (1997)
	Helicobacter pylori	De et al. (2009)
	Liver and small intestine cancer	Shen et al. (2006)
	Ovarian cancer	Pan et al. (2008)
	Pancreatic cancer	Parsons et al. (2016)
	Prostate cancer	Nakamura et al. (2002)
	Malaria	Reddy et al. (2005)
	Melanoma	Bush et al. (2001)
	Leukemia	Cui et al. (2006)
Curcuma biscuits	Cardiovascular disease	Madaric et al. (2013)
Thymoquinone, curcumin	Influenza	Umar et al. (2016)
Curcumin, tetrahydro curcumin (THC)	Colon cancer	Kim et al. (1998)
Hydrazinocurcumin	Endothelial liver cell	Sup Shim et al. (2002)
Hydrazinobenzoylcurcumin (HBC)	Lung cancer	Zhou et al. (2014)
Curcumin analogs	Tumor (human fibrosarcoma cells)	Hahm et al. (2004)

TABLE 3.1 *(Continued)*

Carotenoids	Disease	References
Zeaxanthin/lutein	Age-related macular degeneration (AMD)	Moeller et al. (2006),
	Age-related cataract	Chew et al. (2013)
	Diabetic retinopathy	Moshetova et al. (2015)
	Epidermal hyperproliferation and acute inflammation	Gonzalez et al. (2003)
	Skin cancer	Heinen et al. (2007)
	Atherosclerosis	Dwyer et al. (2001)
	Antioxidant activity	Firdous et al. (2010)
Astaxanthin	Colon cancer	Yasui et al. (2011)
Lycopene	Prostate cancer	Kucuk et al. (2001)
	Breast cancer	Preet et al. (2013)
	Oral cancer	Mayne et al. (2004)
	Digestive tract cancer	Franceschi et al. (1994)
	Lung cancer	Sup Shim et al. (2002)
	HPV infection	Sedjo et al. (2002)
	Atherosclerosis	Rissanen et al. (2003)
	Cardiovascular disease	Gajendragadkar et al. (2014)
	Teratogenicity	Christian et al. (2003); Hoppe et al. (2003)
Lycopene, α- and β-carotene	Breast cancer	Levy et al. (1995)
	Lung cancer	Gloria et al. (2014)

3.2 CAROTENOIDS IN VEGETABLES

The color of fruits and vegetables can signify the type and accessibility of carotenoids; for example, orange-yellow fruits and vegetables are often high in β-carotene and α-carotene. Orange fruits, such as mandarin, orange, and papaya, contain α-cryptoxanthin and zeinoxanthin. Similarly, lycopene pigment (liable for the bright red color) is a key constituent of tomatoes and tomato products. The predominant forms of carotenoids in green leafy vegetables are lutein (nearly 45%) and β-carotene (25–30%), followed by violaxanthin (10–15%) and neoxanthin (10–15%) (Lakshminarayana et al., 2005; Priyadarshani & Jansz, 2014), although the absolute concentration of each carotenoid Green leafy vegetables also include trace amounts of α-carotene, β-cryptoxanthin, zeaxanthin, antheraxanthin, and lutein 5,6-epoxide (luteoxanthin). When compared to its geometric isomer α-carotene, β-carotene predominates in the majority of fruits and vegetables. A small number of fruits and vegetables, such as sweet potatoes, carrots, pumpkin, and dark green vegetables such as green beans, spinach, and broccoli, contain significant amounts of α-carotene (Khoo et al., 2011). Nutritional experts would benefit from knowledge of carotenoid content in various edible sections and cultivars when selecting nutrient-rich plants for food fortification and suitable diet prescription. Table 3.2 show the compositions of key carotenoids in various vegetables. Numerous neglected green crops, such as *Lactuca indica* (Indian lettuce), *Moringa oleifera* (Drumstick tree), and *Oenanthe javanica* (Water dropwort), have been identified as extraordinarily rich resources of carotenoids in recent years (Kongkachuichai et al., 2015; Andarwulan et al., 2012; Saini et al., 2014). Among 25 popular and less common green vegetables, including fenugreek, amaranthus, and spinach, *M. oleifera* leaves had the greatest β-carotene concentration (19.7 mg/100 g) (Bhaskarachary et al., 1995, 1995). Similarly, the highest amount of β-carotene (14 mg/100 g) was found in the leaves of *Moringa pterygosperma* among underused vegetables from Indonesia (Andarwulan et al., 2012). Maximum levels of β-carotene and lutein were found in *L. indica* and *O. javanica*, respectively, in a review of native Thai vegetables (Kongkachuichai et al., 2015). International non-governmental organizations (NGOs) such as Educational Concerns for Hunger Organization (ECHO) and Trees for Life have advocated promoting Moringa leaves as "natural nourishment for the tropics." India is the world's greatest producer of Moringa, allowing it to give low-cost food to impoverished people. In the future, comprehensive carotenoids content in these neglected vegetables may provide important information for selecting nutrient-rich plants for food formulation. The flower stigmas of the saffron plant (*Crocus sativus*) and the waxy seed arils of the achiote shrub (*Bixa orellana*) are the largest sources of apocarotenoids in the plant kingdom (Rosati et al., 2009). Citrus fruits and paprika vegetables, on the other hand, contain trace amounts of apocarotenoids. Agócs et al. (2007) found 2–7% red apocarotenoids β-citraurin in citrus pulps. Fleshman et al. (2011) discovered 1.5% of β-apocarotenoids in cantaloupe (Cucumis melo Reticulatus Group) and Orange Dew (*Cucumis melo*, Inodorus Group). The seed pulp or aril of matured fruits of *Momordica cochinchinensis*, often known as gac fruit, is the greatest known source of carotenoids obtained from plants. Lycopene and β-carotene are the most abundant carotenoids in gac fruit, accounting for 408 g/g and 83.3 g/g fresh weight, respectively (Vuong et al., 2006). De Rosso & Mercadante (2007) assessed major and minor carotenoids in Amazonian native fruits such as mamey (*Mammea americana*), buriti (*Mauritia vinifera*), marimari (*Geoffrola striata*), physalis (*Physalis angulata*), peach palm (*Bactris gasipaes*), and tucuma (*Astrocaryum aculeatum*). According to the findings, a total of 60 different carotenoids were found, with all-E-β-carotene being the most abundant carotenoid in all fruits, and the overall carotenoid concentration ranging from 38 g/g in marimari to 514 g/g in buriti. Silva et al. (2014) investigated the carotenoids and other beneficial chemicals found in tropical fruit pulps and byproducts. On a dry weight basis, the pulp of Acerola (26.23 g/g) has the greatest level of β-carotene, followed by papaya (20.24 g/g) and Surinam cherry (*Eugenia uniflora*; 15.64 g/g). Traditional and non-traditional tropical fruits are abundant in carotenoids and other beneficial components, implying good prospects for the use of these fruit species and byproducts in prospective commodities.

TABLE 3.2 Carotenoid Contents of Common Leafy and Non-Leafy Vegetables

SL. No.	Vegetable	Botanical Name	Carotenoid	References
1.	Spring onion leaves	*Allium fistulosum*	β-Carotene, lutein	Tee et al. (1991)
2.	Garlic leaves	*Allium sativum*	β-Carotene, lutein	Rodriguez-Amaya (1997)
3.	Onion leaves	*Allium cepa*	β-Carotene, lutein	Bhaskarachary et al. (1995)
4.	Celery	*Apium graveolens*	β-Carotene, lutein, zeaxanthin	Gil et al. (2002)
5.	Coriander leaves	*Coriandrum sativum*	β-Carotene, lutein	Tee et al. (1991)
6.	Coriander	*Coriandrum sativum*	β-Carotene, lutein	Bhaskarachary et al. (1995)
7.	Spinach	*Spinacia oleracea*	β-Carotene, lutein, zeaxanthin	Tee et al. (1991)
8.	Lettuce	*Lactuca sativa*	β-Carotene, lutein	Tee et al. (1991); Bhaskarachary et al. (1995)
9.	Chinese mustard leaves	*Brassica juncea*	β-Carotene, lutein	Tee et al. (1991)
10.	Kale	*Brassica oleracea* var acephala	β-Carotene, lutein	Holden et al. (1999)

TABLE 3.2 *(Continued)*

SL. No.	Vegetable	Botanical Name	Carotenoid	References
11.	Chinese kale	*Brassica oleracea* var alboglabra	β-Carotene, lutein	Tee et al. (1991)
12.	Cabbage	*Brassica oleracea* var capitala	β-Carotene, lutein	Holden et al. (1999)
13.	Papaya leaves	*Brassica papaya*	α-Carotene, β-carotene, lutein	Hulshof et al. (1997)
14.	Swamp cabbage	*Brassica aquatica*	β-Carotene, lutein	Tee et al. (1991); Shi et al. (2002)
15.	Bitter melon leaves	*Momordica charantia*	β-Carotene, lutein	Speek et al. (1988)
16.	Fenugreek	*Trigonella foenum-graecum*	β-Carotene, lutein	Setiawan et al. (2001)
17.	Mint	*Mentha arvensis*	β-Carotene, lutein	Bhaskarachary et al. (1995)
18.	Neem tree leaves	*Azadirachta indica*	β-Carotene, lutein	Speek et al. (1988)
19.	Drumstick leaves	*Moringa oleifera*	β-Carotene, lutein, zeaxanthin	Tee et al. (1991)
20.	Black nightshade	*Solanum nigrum*	β-Carotene, lutein	Tee et al. (1991)
21.	Curry leaves	*Murraya sprengel*	β-Carotene, lutein	Tee et al. (1991); Bhaskarachary et al. (1995)
22.	Chives	*Allium schoenoprasum*	β-Carotene, lutein	Tee et al. (1991); Rodriguez-Amaya (1997)
23.	Carrot	*Daucus carota*	α-Carotene, β-carotene	Tee et al. (1991); Holden et al. (1999)
24.	Taro	*Colocasia esculenta*	Lutein, zeaxanthin	Aruna et al. (2009)
25.	Asparagus	*Asparagus officinalis*	α-Carotene, β-carotene	Holden et al. (1999)
26.	Broccoli	*Brassica oleracea* var calabrese	β-Carotene, lutein	Holden et al. (1999)
27.	Brussels sprout	*Brassica oleracea* var gemmiferae	β-Carotene, lutein	Rodriguez-Amaya (1997)
28.	Cauliflower	*Brassica oleracea* var botrytis	β-Carotene, lutein	Rodriguez-Amaya (1997); Holden et al. (1999); Bhaskarachary et al. (1995)

TABLE 3.2 *(Continued)*

SL. No.	Vegetable	Botanical Name	Carotenoid	References
29.	Sweet potato	*Ipomea batatas*	α-Carotene, β-carotene, lutein	Bhaskarachary et al. (1995)
30.	Ivy gourd	*Coccinea grandis*	β-Carotene, lutein	Rodriguez-Amaya (1997); Aruna et al. (2009)
31.	Cucumber	*Cucumis sativus*	α-Carotene, β-carotene, lutein, zeaxanthin	Holden et al. (1999); Aruna et al. (2009)
32.	Pumpkin	*Cucurbita maxima*	α-Carotene, β-carotene, lutein, zeaxanthin	Tee et al. (1991); Bhaskarachary et al. (1995),
33.	Bitter gourd	*Momordica charantia descourt*	β-Carotene	Lako et al. (2007)
34.	Tapioca shoot	*Manihot utilissima*	β-Carotene, lutein	Tee et al. (1991),
35.	French bean	*Phaseolus vulgaris*	β-Carotene, lutein, zeaxanthin	Tee et al. (1991); Aruna et al. (2009)
36.	Common bean	*Phaseolus vulgaris* var red	α-Carotene, β-carotene	Lako et al. (2007)
37.	Cowpea	*Vigma unguiculata subsp. Unguiculata*	Lutein, zeaxanthin	Aruna et al. (2009)
38.	Long bean	*Vigma unguiculata* subsp. *Sesquipedalis*	β-Carotene, lutein	Tee et al. (1991)
39.	Okra	*Abelmoschus esculentus*	α-Carotene, β-carotene, lutein, zeaxanthin	Holden et al. (1999)
40.	Arrowroot	*Marana arundinacea*	β-Carotene	Holden et al. (1999)
41.	Maize	*Zea mays*	β-Carotene, β-cryptoxanthin, lutein, zeaxanthin	Holden et al. (1999)
42.	Chilies	*Capsicum annum* var cayenne	α-Carotene, β-carotene, β-cryptoxanthin, lutein, zeaxanthin	Tee et al. (1991); Aruna et al. (2009)
43.	Capsicum	*Capsicum annum* var grossa	β-Carotene, lutein, zeaxanthin	Bhaskarachary et al. (1995); Aruna et al. (2009)

TABLE 3.2 *(Continued)*

SL. No.	Vegetable	Botanical Name	Carotenoid	References
44.	Pepper	*Capsicum annuum sp*	α-Carotene, β-carotene, β-cryptoxanthin, lutein	Tee et al. (1991); Holden et al. (1999)
45.	Tree tomato	*Solanum betaceum* Cav	β-Carotene, β-cryptoxanthin	Tee et al. (1991)
46.	Tomatoes	*Solanum lycopersicum*	α-Carotene, β-carotene, lutein, zeaxanthin	Tee et al. (1991); Aruna et al. (2009)
47.	Red eggplant	*Solanum melongena*	Lutein, zeaxanthin	Aruna et al. (2009)
48.	Potato	*Solanum tuberosum*	β-Carotene	Holden et al. (1999)

3.3 CAROTENOIDS – CHEMISTRY AND OCCURRENCE

Carotenoids are tetra-terpenoid C40 compounds made up of eight isoprenoid (IP) units. The IP units are connected head-to-tail, but the order is flipped at the molecular core. Thus, the two center methyl side chains have a 1,6-positional association, while the other non-terminal methyl groups have a 1,5-positional relationship. All carotenoids (carotenes and xanthophylls) may be synthesized from acyclic ($C_{40}H_{56}$) lycopene by:

- Hydrogenation;
- Dehydrogenation;
- Cyclization;
- Oxygen insertion;
- Double bond migration;
- Methyl migration;
- Chain elongation; and
- Chain shortening.

Carotenoids are generally classified into two types:

- Carotenes; and
- Xanthophylls.

Carotenes (for example, lycopene from tomatoesandβ-carotene from carrots) are hydrocarbons. Oxygenated carotenoids with alcohol, carbonyl, or other functional groups are known as xanthophylls. Colorants and nutrients such as xanthophylls, lutein, and capsanthin are widely employed in the food and animal feed sectors. These carotenoids are generally obtained from the blooms of paprika (*Capsicum annuum* L.) and marigold (*Tagetes erecta* L.). Epoxy carotenoids (such as violaxanthin and antheraxanthin) are a prominent class of xanthophylls, although their physiological relevance is unknown. Apocarotenoids are carotenoids that lack an end group and have their carbon skeleton reduced.

Carotenoids can be acyclic (like lycopene) or have a ring of six carbons at one or both ends of the molecule (likeβ-carotene). Carotenoids have traditionally been given trivial names following discovery, which in most cases correspond to the natural sources from which they were separated, such as zeaxanthin from maize (*Zea mays*) and lycopene from tomato (*Lycopersicon esculentum* L.). Because this system lacks structural information, a semi-systematic nomenclature was developed that includes structural information as well as a reference to the parent carotene. The International Union of Pure and Applied Chemists (IUPAC) endorsed this method, which was published in a compendium of guidelines for carotenoids nomenclature (Britton, 1995). The end groups that may be present in the carotenoid molecule are denoted by Greek letters.

The presence of a significant number of conjugated double bonds in the carotenoid molecule allows for the formation of a large number of geometric isomers. However, in actuality, most carotenoids are naturally present in the transform (E). Some are naturally present in cis form (Z), such as bixin in annatto seeds (*Bixa orellana*) or pro-lycopene (a carotenoid with multiple double bonds and Z-configuration) in some tomato types.

Carotenoids are present in a wide range of living organisms, including plants and animals, although they are more abundant and diverse in plants, as well as specific bacteria, which are the only organisms capable of synthesizing them.

Carotenogenesis is strongly involved with the ripening process of tomatoes, peppers, and citrus fruits. There are both qualitative and quantitative changes in pigment production during the ripening process. It should be emphasized that as plants age, many changes occur: (a) the degradation of chlorophyll; (b) the unmasking of other pigments; and (c) the denovo synthesis of carotenoids and anthocyanins.

Carotenoids are prevalent in green plant tissues, but their color is mainly obscured by the abundance of green chlorophyll. β-carotene, lutein, violaxanthin, and neoxanthin are the most common carotenoids. Small quantities of zeaxanthin, γ-carotene, β-cryptoxanthin, and antheraxanthin are also present.

Xanthophylls are typically present in larger quantities than carotenoids in fruits. For example, lutein and zeaxanthin are the most abundant pigments in maize, whereas β-cryptoxanthin and zeaxanthin predominate in mango (*Mangifera indica*) and persimmon (*Diospyros kaki*). One exception is tomato, where the primary color is a carotene (lycopene). Certain carotenoids are found only in one plant species. Capsanthin and capsorubin are present almost entirely in mature Capsicum fruits and are responsible for their appealing red color (Weedon & Moss, 1995). The orange (*Citrus sinensis*) includes various levels of apo-carotenoids β-citraurin and β-citranaxanthin, as well as β-cryptoxanthin, lutein, antheraxanthin, violaxanthin, and traces of their carotene precursors (Farin et al., 1983).

Xanthophylls can be found in free form in the green plant tissues of the vegetables we eat. However, in non-dietary green plant tissues, carotenoids undergo esterification with various fatty acids as a result of leaf senescence and ripening, which coincides with the change of chloroplasts into chromoplasts. The chromatic properties of carotenoids are not altered by esterification. Furthermore, esterification is linked to the plant's ability to overproduce and accumulate carotenoids (particularly in fruits and flowers). The shift in esterification profile of xanthophylls in red pepper fruits has recently been recommended as a ripening indicator (Hornero-Méndez et al., 2000).

Carotenoids are lipophilic molecules and so insoluble in aqueous media, with the exception of those with very polar functional groups, such as norbixin, a carotenoid with a dicarbonyl acid structure.

The difference in structure between bicyclic β-carotene and acyclic lycopene is that β-carotene can be digested into a succession of shorter molecules, ending in the synthesis of retinoic acid via an excentric cleavage pathway or retinal via a central cleavage pathway.

One of the most distinguishing aspects of carotenoids is light absorption, which is caused by the existence of a long, broad system of conjugated double bonds (or polyene chain). A chromophore with seven or more double bonds has the ability to absorb light in the visible range, allowing for the observation of colors ranging from yellow to red via a wide spectrum of orange tones. Furthermore, because of the polyene chain, the carotenoid molecule is very vulnerable to isomerizing and oxidizing circumstances such as light, heat, or acids.

Because of the interaction with the physiochemical environment (mostly lipids and proteins) around the pigments, the characteristics of carotenoid pigments *in vitro* may differ from those *in vivo*. This is especially important when it comes to the functioning and effect of carotenoids *in vivo*.

3.4 IMMUNOPHARMACOLOGICAL ROLE OF CAROTENOIDS

The effects of distinct carotenoids on immune responses, such as cytokine release, lymphocyte proliferation, natural killer (NK) cell cytotoxic activity, phagocytic and microbicidal capacities, and inflammation, have been studied *in vitro* and *in vivo*. In this section, we describe the immunopharmacological roles of individual carotenoids.

3.4.1 ASTAXANTHIN (ASTA)

ASTA (3,3-dihydroxy-β, β-carotene-4, 4-dione) improves neutrophil phagocytic and microbicidal ability while increasing intracellular calcium concentration (Ca^{2+}) and NO generation. These actions were linked to lower levels of hydrogen peroxide (H_2O_2), superoxide anion, IL-6, and TNF-cytokines. Indeed, ASTA therapy dramatically reduces oxidative damage in proteins and lipids. Furthermore, therapy with 5 μM ASTA has been demonstrated to boost neutrophil phagocytic (30%) and fungicide (28%) (Macedo et al., 2010). Another study discovered that ASTA therapy reduced the release of pro-inflammatory cytokines in activated U937 cells. These findings suggest that ASTA inhibits H_2O_2-mediated NF-κB activation and cytokine release through regulating SHP-1 expression. SHP-1 is a PTP that functions as a cytokine signaling pathway negative regulator (Speranza et al., 2012). The researchers also demonstrated that at a concentration of 20 μmol/L, all three stereoisomers of ASTA, (3S, 3'S)-trans-ASTA, (3R, 3'R)-trans-ASTA, and meso-trans-ASTA, substantially increased phagocytic capacity of peritoneal exudates cells, lymphocyte proliferation, and cytotoxic activity of natural killer (NK) cells. Furthermore, *in vitro*, the 3S, 3 'S enantiomer displayed greater immunoregulatory action than the other two enantiomers (Weihong et al., 2016). ASTA, a carotenoid that lacks vitamin A activity, increased the synthesis of human immunoglobulins (IgM, IgA, and IgG) by peripheral blood mononuclear cells (PBMCs) in response to T-cell-dependent stimuli (polyclonal stimulants) (Jyonouchi et al., 1995). In young healthy women, dietary ASTA was shown to diminish a biomarker of DNA damage and acute phase protein (plasma C-reactive protein (CRP)) and boost immune responses (e.g., NK cell cytotoxic activity and total T- and beta-cell subpopulations)

(Park et al., 2010). Furthermore, dietary ASTA increased IgG, IgM, and beta cell population levels while decreasing plasma amounts of C reactive protein and DNA damage in female Beagle dogs (Chew et al., 2011). In cats, ASTA was demonstrated to have comparable effects, serving as a powerful antioxidant and regulating the immunological response (Park et al., 2011). ASTA supplementation also alleviated the immunological dysfunction, or reduction of NK cell activity, caused by stress in mice (Kurihara et al., 2002). Furthermore, this carotenoid has been shown to improve salivary IgA response and reduce muscle damage in young soccer players, avoiding inflammation caused by strenuous physical exercise or situations of elevated oxidative stress (OS) (Baralic et al., 2015).

3.4.2 CROCIN AND CROCETIN

Crocin and crocetin carotenoids, which are present in saffron, have a number of pharmacological effects. Crocin, produced from the stigma of *Crocus sativus*, has been shown to be a potent antioxidant, even more potent than α-tocopherol (Bolhassani et al., 2014). Indeed, the antioxidant capabilities of crocin and crocetin are primarily responsible for their neuroprotective benefits (Ochiai et al., 2007). Crocin raises intracellular glutathione (GSH) levels, preventing cell death in hypoxic PC12 cells, a cell culture model for brain ischemia (Ochiai et al., 2007). Crocetin also reduces TNFα, IL-1 β, and iNOS mRNA expression in the liver and increases overall survival in a hemorrhagic shock model (Yang et al., 2006). Crocin and crocetin substantially reduce LPS-induced NO release from cultured BV2 mouse brain microglial cells while also lowering NFκB activation, TNFα and IL-1 β cytokine levels, and intracellular ROS. Crocin and crocetin have been found to mediate neuroprotection by reducing the synthesis of several neurotoxic chemicals by activated microglia caused by IFN-γ and A β (Nam et al., 2010).

3.4.3 CURCUMIN

Curcumin is a safe and effective immune system immunomodulator. Under identical settings (e.g., incubation with 25 μM curcumin for 24 h), curcumin can trigger apoptosis in tumor cells but not in initial cultures or non-transformed cells. Curcumin also influences adaptive immunity by increasing T-cell proliferation (Varalakshmi et al., 2008). These findings were supported by a considerable increase in macrophage phagocytic activity as well as immunostimulatory activity in curcumin-treated Balb/c rats (Antony et al., 1999). Curcumin (diferuloylmethane) can reduce the cellular mitogenic response generated by *H. pylori* infection in epithelial cells. Curcumin inhibited IκBα degradation, IkB kinase a and b (IKKa and b) activity, and NF-κB DNA binding in an AGS culture infected with *H. pylori*. Curcumin (~40 M) was observed to inhibit *H. pylori*-induced NF-κB activation and IL-8 production (Foryst-Ludwig et al., 2004).

In another study, curcumin directly reduced T-cell-dependent inflammatory stress *in vitro* by modulating CD4+ T-cell activation at various levels, including:

- Inhibition of CD2/CD3/CD28-initiated CD4+ T-cell proliferation;
- Enhancement of CD69, CCR7, L-selectin, and TGFβ1 expression; and
- Dose-dependent suppression of IL-12 production (Kim et al., 2013).

In fact, curcumin-induced reduction of IL-12 production in LPS-stimulated mice macrophages might explain numerous of curcumin's biological actions, including its antiinflammatory benefits in chronic inflammatory disorders. Curcumin has the potential to directly alter the NF-κB-DNA interaction by creating a complex with NF-κB that is incapable of binding kB sites (Kang et al., 1999). A comparable study found that pretreatment with curcumin dramatically reduces IL-12 production by mice macrophages activated with either LPS or head-killed Listeria monocytogenes; this suggests that curcumin may have a therapeutic impact on Th1-mediated immunological disorders. Curcumin most likely reduces IL-12 synthesis in macrophages by decreasing the action of NF-κB on the IL-12 p40 gene, which is recognized as the highly inducible component of IL-12 (Kang et al., 1999). Curcumin inhibited the production of Th1 cytokines such as IL-2 and IFN-γ in macrophages or splenic T-cells (Gao et al., 2004). In DO11.10 transgenic mice, dietary curcumin and limonin were reported to decrease CD4+ T-cell proliferation, IL-2 production, and NF-κB p65 nuclear translocation in activated CD4+ T-cells (Kim et al., 2009). Furthermore, curcumin-treated dendritic cells (DCs) are very efficient at antigen (Ag) acquisition via mannose receptor-mediated endocytosis, and this effect is dose-dependent. Curcumin suppresses LPS-induced MAPK activation and NF-κBp65 translocation. Furthermore, it strongly decreased CD80, CD86, and MHC class II expression in DCs while not inhibiting MHC class I expression. At a dosage of µ25 M, curcumin inhibited the maturation of bone marrow derived murine DC (Kim et al., 2005), and other investigations have indicated that curcumin inhibits DC movement and chemokine release. In a chemotaxis experiment, curcumin inhibited migration toward CCL19 and CCL21 while also lowering levels of the chemokines fractalkine (CX3CL1) and interferon (IFN) generating factor (IP-10). Both fractalkine and IP-10 have the ability to recruit inflammatory cells to areas of inflammation (Shirley et al., 2008).

In CNS inflammatory illnesses such as multiple sclerosis, activated astrocytes have been demonstrated to have a dual functional role. Curcumin inhibited the production of IL-6 and the activity of the enzyme MMP-9 in an astrocyte cell line (U373-MG) in an *in vitro* LPS-induced inflammatory condition. Curcumin's anti-inflammatory effect on neuroinflammation resulted in CNS healing (Seyedzadeh et al., 2014). Curcumin also reduced the dimerization of TLR 4, the degradation of IRAK-1, and the activation of NF-κB by LPS in a dose-dependent manner (~20 μM). Curcumin suppressed both MyD88- and TRIF-dependent pathways in TLR4 signaling generated by LPS (Youn et al., 2006). Curcumin's immunomodulatory activities have been studied in a number of investigations. Curcumin decreased phytohemagglutinin (PHA)-induced T-cell proliferation, IL-2 production, NO generation, LPS-induced NF-κB, and increased NK-cell cytotoxicity in general. These findings imply that curcumin, which has been proven to have powerful immunosuppressive characteristics *in vitro*, might have a therapeutic impact in Th1-mediated autoimmune diseases. Curcumin is a powerful scavenger of reactive oxygen species (ROS) such as hydroxyl radicals and superoxide anions (Yadav et al., 2005), and it has the ability to influence both endoplasmic reticulum (ER) stress and mitochondrial functional pathways. Indeed, curcumin-induced apoptosis in activated T-cells has been linked to severe ER stress (Zheng et al., 2013). Curcumin may boost the therapeutic efficacy of adoptive T-cell therapy (the *ex vivo* growth and subsequent transfusion of tumor-specific T-cells) in the treatment of cancer. Curcumin, when coupled with adoptive therapy in male tumor-bearing C57BL/6 mice, was demonstrated to increase the cytotoxicity of Ag-specific CD8+ T-cells against tumors via altering the tumor microenvironment during treatment. T-cell activity was increased by the combination therapy due to the inhibition of different immunosuppressors (e.g., regulatory T-cells, indoleamine 2,3-dioxygenase, TGF-β) (Chang et al., 2012). Several studies have shown that curcumin regulates T-cell proliferation and activation in a dose-dependent way. Indeed, in a mouse model, low-dose curcumin increases splenic lymphocyte proliferation while high-dose curcumin decreases this effect. Curcumin can successfully restore CD4+ and CD8+ cell numbers in the tumor microenvironment while also inhibiting the loss of central and/or effector memory T-cells. Curcumin dramatically reduces the levels of IL10, TGF-β, and Treg cell population; curcumin pretreatment of CD4+ CD25+ Treg cells reduced their immunosuppressive activity (Bose et al., 2015). Tumor-derived exosomal proteins have been shown to decrease IL-2-induced NK-cell activity in breast cancer, and Zhang et al. demonstrated that curcumin promoted proteasomal degradation of these proteins, partially restoring NK-cell activity against the tumor. As a result, curcumin can target immune escape methods that are critical for immunological responses (Zhang et al., 2007). Furthermore, oral curcumin (50 $mg.kg^{1}$) dosing reduced the mast cell-dependent IgE response in Ag-induced local passive cutaneous anaphylaxis (Srivastava et al., 2011).

3.4.4 LUTEIN (LU) AND β-CRYPTOXANTHIN (βCR)

Lutein (30 M) significantly reduces several skin inflammatory responses, including increased expression of IL-6 from LPS-treated macrophages, up-regulation of COX-2 from IFN/TNF treated aneuploid immortal keratinocyte cells (HaCaT cells), and enhancement of MMP-9 levels (an acute inflammation marker) in UV-irradiated keratinocytes (Oh et al., 2013). Lutein is a powerful antioxidant that lowers OS caused by benzo(a)pyrene (Vijayapadma et al., 2014), a high-cholesterol diet (Kim et al., 2012), H_2O_2 (Gao et al., 2011), and D-galactose (Mai et al., 2010). For three months, lutein supplements (~20 mg/day^{1}) resulted in a substantial reduction in blood IL-6 and monocyte chemoattractant protein-1 (MCP-1), triglyceride (TG), and LDL in patients with early atherosclerosis (Xu et al., 2013).

Yellow silkworm cocoons have been shown to be a rich source of dietary lutein, and this silk lutein extract has been shown to improve both innate and adaptive immunological capabilities. In female BALB/c mice, the silk extract increased IL-2 and IFN-γ production, the populations of CD3+ and CD4+ CD3+ cells, antibody production, and NK-cell activity (Promphet et al., 2014). On weeks 8 and 12, domestic cats given 10 mg lutein had a high plasma IgG level and higher percentages of CD4 and CD21 lymphocytes, indicating that lutein has an immunomodulatory effect (Kim et al., 2000). In domestic dogs, dietary lutein triggered both cell mediated and humoral immune responses (IgG) and increased the percentages of cells expressing CD5, CD4, CD8, and MHC II markers (Kim et al., 2000). Lutein-supplemented birds had higher pigmentation efficiency than curcumin-supplemented birds; nevertheless, both groups of birds had lower lipid oxidation in the liver (Rajput et al., 2013). Lutein's anti-oxidative, anti-inflammatory, and anti-apoptotic properties protect the retina against ischemic/hypoxic damage (Li et al., 2012). Dietary lutein reduced breast tumor development and anti-apoptotic Bcl2 gene expression while boosting mRNA expression of pro-apoptotic genes (e.g., p53 and Bax) and the Bax: Bcl-2 ratio in tumors (Chew & Park, 2004). Lutein and β-cryptoxanthin may inhibit factors implicated in rheumatoid arthritis and osteoarthritis (OA) inflammation. β Cr inhibited the production of IFN-γ, IL-1α, and IL-2 cytokines, whereas Lu enhanced their levels. Lu enhanced the amounts of IL-4 and IL-10 whereas βCr lowered them. Both Lu and Cr inhibited the synthesis of NF-κB p50 (Di Filippo et al., 2012). On the murine macrophage cell line, several of the immunity-related characteristics of β-cryptoxanthin and lutein were investigated (RAW264). β-cryptoxanthin (10 M) dramatically reduced IL-1 mRNA levels but had no effect on IL-6 or IL-12 p40 mRNA levels. Lutein had no effect on the transcription of these three cytokines (Katsuura et al., 2009). In animals, β-cryptoxanthin has been demonstrated to increase the number of CD4+ cells in the blood as well as serum IgG, IgM, and IgA levels (i.e., humoral immunity) (Ghodratizadeh et al., 2014). According to Nishi et al. (2002), βCr treatment increases IL-6 levels but has no effect on IL-4 or IFN-levels, indicating a large rise in immunoglobulin levels and CD4+ cell count (Nishi et al., 2002., 2012). Oxygenated carotenoids, such as βCr, may be esterified with different fatty acid chains, and the esterified form of this xanthophyll is more bioavailable than the free form (Fu et al., 2010). The esterified βCr with a fatty acid chain may be effective in enhancing βCr's immune stimulatory function (Ghodratizadeh et al., 2014).

3.4.5 LYCOPENE

Lycopene inhibits LPS-induced pro-inflammatory responses through a variety of mechanisms, including:

- Enhancement of vascular barrier integrity;
- Inhibition of barrier permeability and expression of cell adhesion molecules (CAM); and
- Prevention of leukocyte adhesion and trans endothelial migration.

Indeed, lycopene's anti-inflammatory properties are mediated by the suppression of NF-κB expression and TNF-α production (Bae & Bae, 2011). Lycopene inhibited the expression of TNF-α induced intercellular adhesion molecule-1 (ICAM-1) in HUVEC, but not the expression of COX-2 or platelet-endothelial cell (EC) adhesion molecule. Lycopene inhibited

TNF-induced IκB phosphorylation, NF-κB expression, and NF-κB p65 translocation from the cytosol to the nucleus, implying that its anti-inflammatory effect could be used to prevent CVDs. Furthermore, lycopene had no effect on IFN-γ induced ICAM-1 expression, indicating that lycopene first influences the TNF-α induced signaling pathway (Hung et al., 2008). Lycopene pretreatment of human THP-1 macrophages resulted in a significant inhibition of cigarette smoke extract-induced IL-8 expression at both the RNA and protein levels; the molecular mechanism underlying this effect is mediated by NF-κB inactivation. Inactivation of NF-κB has been linked to inhibition of redox signaling and activation of PPAR signaling (Simone et al., 2011). At a concentration of 5 μM, lycopene improves NK-cell viability and cytotoxicity. Its anti-apoptosis effect on NK cells is linked to a decrease in the expression of caspase 3 and 9 genes. Furthermore, lycopene had no effect on the expression of functional receptors on NK cells such as NKG2A, NKG2D, NKp30, and NKp44. After 7 days, lycopene treatment increased IFN-γ expression at both the gene and protein levels (Li et al., 2014). Lycopene has been shown to increase IL-1β and TNF-α production in a dose dependent manner while decreasing IL-2, IL-10, and IFN-γ secretion in human PBMCs, while IL-6 and the IL-1 receptor antagonist were unaffected. Lycopene can enhance inflammatory responses by increasing the production of pro-inflammatory cytokines (IL-1β and TNF-α) while decreasing the production of anti-inflammatory cytokines (IL-10) (Bessler et al., 2008). Lycopene, on the other hand, has been shown in a variety of adipose tissues and adipocyte models to reduce the expression of pro-inflammatory cytokines and chemokines (e.g., IL-6, IL-1β, and MCP-1) at both the mRNA and protein levels (Gouranton et al., 2011). A high concentration of lycopene (10 μmol/L^1) had a strong anti-angiogenic effect, which may be related to the upregulation of IL-12 (163%) and IFN-γ (531%) in HUVEC (Huang et al., 2013). Lycopene may be useful in the prevention or treatment of acute pancreatitis by suppressing NF-κB activation and the expression of inflammatory cytokines (IL-6), which is mediated by a decrease in intracellular ROS levels in pancreatic acinar cells (Kang et al., 2011). Lycopene's anti-inflammatory effects on RAW 264.7 macrophages were linked to a reduction in LPS-stimulated migration (Marcotorchino et al., 2012). The studies revealed that pre-incubating mouse peritoneal macrophages with lycopene (1 μM), lutein (1 μM), and β-carotene (2 μM) before the addition of LPS caused a synergistic inhibition of NO, PGE2, and superoxide production, resulting from the down-regulation of iNOS, COX-2, and NADPH oxidase (NOX) at both the mRNA and protein expression levels, and a synergistic inhibition of TNF-α secretion (Hadad

& Levy, 2012). Indeed, lycopene (10 μM) treatment reduced LPS-induced iNOS protein and NO production (40%) in RAW 264.7 in a dose dependent manner (Rafi et al., 2007).

Lycopene supplementation, on the other hand, has been shown to reduce allergic inflammation systemically, particularly in the lungs, by suppressing Th2 cytokine responses. Lycopene inhibited the infiltration of inflammatory leukocytes (neutrophils, eosinophils, lymphocytes, and macrophages) into bronchoalveolar lavage fluid, which was linked to the suppression of the Th2 transcription factor GATA-3, the cytokine IL-4, and the activity of eosinophil peroxidase and MMP-9 (Lee et al., 2008; Hazlewood et al., 2011). In a mouse model, lycopene has been shown to reduce inflammation and apoptosis (e.g., caspase-3, -8, and -9 expression) in post myocardial infarction remodeling by inhibiting the NF-κB signaling pathway (e.g., NF-κB p65 phosphorylation) (He et al., 2015). Lycopene has been shown to inhibit LPS-mediated release of high mobility group 1 (HMGB1) and HMGB1-mediated pro-inflammatory signaling responses in both primary HUVEC and animals by down-regulating cell surface expression of CAMs, HMGB1 receptors, TLR-2 and -4, and receptors for advanced glycation end (AGEs) products, as well as to initiate pro-inflammatory responses in ECs (Lee et al., 2012). Xanthophylls (which contain 60% zeaxanthin and 40% lutein) have been shown to regulate anti-inflammatory and pro-inflammatory cytokine expression in chick and hen tissues. Dietary xanthophyll reduces the expression of pro-inflammatory cytokines (e.g., IFN-γ, IL-6, IL-1β, and LITAF) in the liver, duodenum, and jejunum of hens while increasing the expression of anti-inflammatory cytokines (e.g., IL-4 and IL-10) in the liver, jejunum, and ileum (Gao et al., 2012). Furthermore, in Chinese hamster lung fibroblasts, lycopene reduces peroxynitrite (ONOO) or OS-induced DNA damage (Muzandu et al., 2006; Kim, 2011).

3.4.6 VIOLAXANTHIN

Violaxanthin of Chlorella ellipsoidea, a non-synthetic natural product, has been proposed as a safe and effective anti-inflammatory agent that can also function as an adjuvant. The anti-inflammatory activity of violaxanthin may be based on the inhibition of LPS-mediated NF-κB p65 subunit translocation into the nucleus and, as a result, the inactivation of the NF-κB pathways. Violaxanthin (60 μM) inhibited NO production as well as the expression of iNOS and COX-2 in LPS-treated RAW 264.7 macrophages in a concentration-dependent manner (Soontornchaiboon et al., 2012).

3.4.7 β-CAROTENE

Human NK-cells treated with β-carotene at doses ranging from 2 to 200 ng/mL[1] have a decreased tumorolytic effect *in vitro*; however, treatment with β-carotene at doses ranging from 0.1 to 10 μgm/L[1] has a significant increase in tumorolytic function. However, other studies have shown that α-tocopherol increases the tumorolytic effect of mouse splenocytes, whereas β-carotene inhibits this effect. Indeed, β-carotene inhibits the beneficial effects of α-tocopherol in mice, but the mechanism by which this occurs is unknown (Ashfaq et al., 2000). The accumulation of β-carotene in murine macrophage cells (RAW264) was found to be dose and time dependently related to cellular lipid peroxidation, indicating pro-oxidative activity of β-carotene and the synthesis of GSH as an intracellular antioxidant. Furthermore, β-carotene supplementation inhibited the transcription of IL-1β, IL-6, and IL-12 p40 cytokines (Katsuura et al., 2009). The number of lymphoid cells with surface markers for NK-cells, IL-2, and transferrin receptors (or the number of Th and T-inducer lymphocytes) has been shown to be increased in PBMCs from people who take β-carotene orally (Watson et al., 1991). Seifter et al. also discovered that β-carotene stimulates thymus gland growth and increases the number of thymic small lymphocytes (Seifter et al., 1981).

In cultured Peyer's patch (PP) cells stimulated with the mitogen concanavalin A, oral administration of three carotenoids, β-carotene, β-cryptoxanthin, and lycopene, resulted in modulated Th cytokine production. In fact, oral administration of β-carotene at 1 or 5 mg/kg[1]/day[1] significantly increased IL-2 production in PP cells after 72 hours, while IL-4 levels remained unchanged. In addition, combining β-carotene with capsaicin (both at 5 mgkg[1]) significantly increased IFN- and IL-5 levels, which were three times higher than those induced by capsaicin alone (Yamaguchi et al., 2010). *In vitro* cell cultures were used to study the immunomodulatory activities of β-carotene and its associated carotenoids, such as canthaxanthin and ASTA, on the proliferation and functions of murine immunocompetent cells. The findings show that canthaxanthin, β-carotene, and ASTA all have potent but distinct stimulatory effects on the proliferation of BALB/c thymocytes and spleen cells. At certain concentrations, all three carotenoids significantly increased the release of IL-1 and TNF-α from murine peritoneal adherent cells.

However, the cytokine-inducing activities of ASTA were higher than those of canthaxanthin and β-carotene (Okai & Higashi-Okai, 1996). Yang et al. demonstrated for the first time that a *Dunaliella salina* carotenoid extract containing lutein, zeaxanthin, β-carotene, and α-carotene inhibits the production of NO, PGE2, and pro-inflammatory cytokines (TNF-α, IL-1, and IL-6) as well as the expression of iNOS and COX-2 in LPS-activated RAW264.7 cells. Indeed, the extract inhibited NF-κB activation and JNK (c-Jun N-terminal kinase) phosphorylation, resulting in anti-inflammatory effects (Yang et al., 2013). Meso-Zeaxanthin is a xanthophyll carotenoid with strong antioxidant activity that, like zeaxanthin, has an anti-inflammatory effect in the BALB/c mouse model (Firdous et al., 2015). Other research has found that lutein and β-carotene increase *in vivo* antibody production against T-dependent Ags but not T-independent Ags in old B6 mice, though at a lower level than ASTA. The generation of antibody against TD Ags *in vivo* or *in vitro* was significantly lower in old mice than in young mice (Jyonouchi et al., 1994).

Furthermore, β-carotene supplementation effectively increased the level of mucosal IgA (both the number and concentration of IgA antibody-secreting cells) in weanling mice's jejunum and ileum. These effects were primarily due to retinoid X receptor mRNA expression, retinoic acid receptors [(RAR) α and (RAR)γ], and retinoic acid (RA)-mediated immune response (Nishida et al., 2014). Selenium improves immune function (NK-cell cytotoxicity) as well as the phenotypic expression of T-cell subsets, whereas β-carotene only affects immune function. As a result, supplemental selenium and β-carotene may improve immune function in elderly people by increasing the total percentage of T-cells and decreasing NK-cell activity (Wood et al., 2000). Nonetheless, because of its ability to inhibit cytokine expression in Suid herpesvirus-induced inflammation via NF-B inactivation, β-carotene could be considered a potential anti-inflammatory agent for DNA-virus infections, particularly for the human herpes simplex virus (HSV) (Lin et al., 2012; Guedes et al., 2014).

3.5 CONCLUSION

Carotenoids have long been thought to have important health-promoting properties. Their biological role in the prevention and possibly treatment of human chronic diseases is currently being researched and acknowledged. Although some carotenoids have been explored for their antioxidant properties, other mechanisms such as their pro-vitamin A activity; immune, endocrine, and metabolic activities; and their role in cell cycle regulation, apoptosis, and cell differentiation are also being investigated. In vertebrates, carotenoids can play important roles in immunoregulation and immuno-stimulation. However, there are numerous key difficulties to consider when using dietary carotenoids to prevent or cure various illnesses, including (a) the efficiency of individual carotenoids may be dependent on quantities of other carotenoids. Indeed, carotenoids may interact synergistically, and supplementation with a single carotenoid may be ineffective; (b) these compounds are highly sensitive to oxidation, chemical, or enzymatic pathways, and as a result, they can be converted to other products, the effects of which are unknown; (c) variations in response to treatment with carotenoids among individuals with similar dietary intakes; and (d) the efficiency of carotenoid. Finally, further research into the safety, metabolism, and molecular biological features of carotenoids should be conducted before they are utilized to prevent cancer.

KEYWORDS

- **antioxidant**
- **carotenoid**
- **curcumin**
- **T-lymphocytes**
- **vegetables**

REFERENCES

Abdullaev, F. I., (1994). Inhibitory effect of crocetin on intracellular nucleic acid and protein synthesis in malignant cells. *Toxicol Lett., 70*, 243–251.

Agócs, A., Nagy, V., Szabó, Z., Márk, L., Ohmacht, R., & Deli, J., (2007). Comparative study on the carotenoid composition of the peel and the pulp of different citrus species. *Innova. Food Sci. Emerge. Technol., 8*(3), 390–394.

Andarwulan, N., Kurniasih, D., Apriady, R. A., Rahmat, H., Roto, A. V., & Bolling, B. W., (2012). Polyphenols, carotenoids, and ascorbic acid in underutilized medicinal vegetables. *J. Funct. Foods, 4*(1), 339–347.

Antony, S., Kuttan, R., & Kuttan, G., (1999). Immunomodulatory activity of curcumin. *Immunol. Invest., 28*, 291–303.

Aoki, A., Inoue, M., Nguyen, E., Obata, R., Kadonosono, K., Shinkai, S., et al., (2016). Dietary n-3 fatty acid, α-tocopherol, zinc, vitamin D, vitamin C, and β-carotene are associated with age-related macular degeneration in Japan. *Sci. Rep., 6*, 1–7.

Aruna, G., Mamatha, B. S., & Baskaran, V., (2009). Lutein content of selected Indian vegetables and vegetable oils determined by HPLC. *J. Food Compos. Anal., 22*, 632–636.

Ashfaq, M. K., Zuberi, H. S., & Waqar, M. A., (2000). Vitamin E and β-carotene affect natural killer cell function. *Int. J. Food Sci. Nutr., 51*, S13–S20.

Astorg, P., (1997). Food carotenoids and cancer prevention: An overview of current research. *Trends Food Sci. Technol., 8*, 406–413.

Aung, H. H., Wang, C. Z., Ni, M., Fishbein, A., Mehendale, S. R., Xie, J. T., et al., (2007). Crocin from *Crocus sativus* possesses significant anti-proliferation effects on human colorectal cancer cells. *Exp. Oncol., 29*, 175–180.

Bae, J. W., & Bae, J. S., (2011). Barrier protective effects of lycopene in human endothelial cells. *Inflamm. Res., 60*, 751–758.

Baralic, I., Andjelkovic, M., Djordjevic, B., Dikic, N., Radivojevic, N., Suzin, Z. V., et al., (2015). *Effect of Astaxanthin Supplementation on Salivary IgA, Oxidative Stress, and Inflammation in Young Soccer Players* (pp. 1–9). Hindawi Publishing Corporation.

Bathaie, S. Z., Hoshyar, R., Miri, H., & Sadeghizadeh, M., (2013). Anticancer effects of crocetin in both human adenocarcinoma gastric cancer cells and rat model of gastric cancer. *Biochem. Cell Biol., 91*, 397–403.

Benedich, A., (1994). Recent advances in clinical research involving carotenoids. *Pure Appl. Chem., 66*, 1017–1024.

Bessler, H., Salman, H., Bergman, M., Alcalay, Y., & Djaldetti, M., (2008). *In vitro* effect of lycopene on cytokine production by human peripheral blood mononuclear cells. *Immunol. Invest., 37*, 183–190.

Bhaskarachary, K., Rao, D. S. S., Deosthale, Y. G., & Reddy, V., (1995). Carotene content of some common and less familiar foods of plant origin. *Food Chem., 54*(2), 189–193.

Bie, X., Chen, Y., Zheng, X., & Dai, H., (2011). The role of crocetin in protection following cerebral contusion and in the enhancement of angiogenesis in rats. *Fitoterapia, 82*, 997–1002.

Bolhassani, A., Khavari, A., & Bathaie, S. Z., (2014). Saffron and natural carotenoids: Biochemical activities and anti-tumor effects. *Biochim et Biophys Acta, 1845*, 20–30.

Bose, S., Panda, A. K., Mukherjee, S., & Sa, G., (2015). Curcumin and tumor immune-editing: Resurrecting the immune system (RW). *Cell Div., 10*, 1–13.

Britton, G., (1995). Structure and properties of carotenoids in relation to function. *Faseb J., 9*, 1551–1558.

Bush, J. A., Cheung, K. J., & Li, G., (2001). Curcumin induces apoptosis in human melanoma cells through a Fas receptor/caspase-8 pathway independent of p53. *Exp Cell Res., 271*, 305–314.

Chadalapaka, G., Jutooru, I., Chintharlapalli, S., Papineni, S., Smith, R., Li, X., et al., (2008). Curcumin decreases specificity protein expression in bladder cancer cells. *Cancer Res., 68*, 5345–5354.

Chang, Y. F., Chuang, H. Y., Hsu, C. H., Liu, R. S., Gambhir, S. S., & Hwang, J. J., (2012). Immunomodulation of curcumin on adoptive therapy with T cell functional imaging in mice. *Cancer Prev. Res., 5*, 444–452.

Chen, H. Y., Huang, S. M., Yang, C. M., & Hu, M. L., (2012). Diverse effects of β-carotene on secretion and expression of VEGF in human hepatocarcinoma and prostate tumor cells. *Molecules, 17*, 3981–3988.

Chen, H. Y., Yueh, T. C., Chen, Y. C., Huang, C. H., Yang, C. M., & Hu, M. L., (2013). Antimetastatic effects of α-carotene and possible mechanisms of action in human hepatocarcinoma SK-Hep-1 cells. *J. Agric Food Chem., 61*, 10368–10376.

Chew, B. P., & Park, J. S., (2004). Carotenoid action on the immune response. *J. Nutr., 134*, 257S–261S.

Chew, B. P., (1995). Antioxidant vitamins affect food animal immunity and health. *J. Nutr., 125*, 1804S–1808S.

Chew, B. P., Mathison, B. D., Hayek, M. G., Massimino, S., Reinhart, G. A., & Parka, J. S., (2011). Dietary astaxanthin enhances immune response in dogs. *Vet. Immunol. Immunopathol., 140*, 199–206.

Chew, E. Y., SanGiovanni, J. P., Ferris, F. L., Wong, W. T., Agron, E., Clemons, T. E., et al., (2013). Lutein/zeaxanthin for the treatment of age-related cataract: AREDS2 randomized trial report. *JAMA Ophthalmol., 131*, 843–850.

Chryssanthi, D. G., Lamari, F. N., Iatrou, G., Pylara, A., Karamanos, N. K., & Cordopatis, P., (2007). Inhibition of breast cancer cell proliferation by style constituents of different *Crocus* species. *Anticancer Res., 27*, 357–362.

Cui, S. X., Qu, X. J., Xie, Y. Y., Zhou, L., Nakata, M., Makuuchi, M., et al., (2006). Curcumin inhibits telomerase activity in human cancer cell lines. *Int. J. Mol. Med., 18*, 227–231.

De Rosso, V. V., & Mercadante, A. Z., (2007). Identification and quantification of carotenoids, By HPLC–PDA–MS/MS, from Amazonian fruits. *J. Agri. Food Chem., 55*(13), 5062–5072.

De, R., Kundu, P., Swarnakar, S., Ramamurthy, T., Chowdhury, A., Nair, G. B., et al., (2009). Antimicrobial activity of curcumin against *Helicobacter pylori* isolates from India and during infections in mice. *Antimicrob Agents Chemother, 53*, 1592–1597.

Di Filippo, M. M., Mathison, B. D., Park, J. S., & Chew, B. P., (2012). Lutein and β-cryptoxanthin inhibit inflammatory mediators in human chondrosarcoma cells induced with IL-1β. *Open Nutr. J, 6*, 41–47.

Dwyer, J. H., Navab, M., Dwyer, K. M., Hassan, K., Sun, P., Shircore, A., et al., (2001). Oxygenated carotenoid lutein and progression of early atherosclerosis: The Los Angeles atherosclerosis study. *Circulation, 103*, 2922–2927.

Farin, D., Ikan, R., & Gross, J., (1983). The carotenoid pigments in the juice and flavedo of mandarin hybrid (*Citrus reticulate*) cv Michal during ripening. *Phytochemistry, 22*, 403–408.

Firdous, A. P., Kuttan, G., & Kuttan, R., (2015). Anti-inflammatory potential of carotenoid meso-zeaxanthin and its mode of action. *Pharm Biol., 53*, 961–967.

Firdous, A. P., Preethi, K. C., & Kuttan, R., (2010). Antioxidant potential of meso-zeaxanthin a semi-synthetic carotenoid. *Food Chem., 119*, 1096–1101.

Fleshman, M. K., Lester, G. E., Riedl, K. M., Kopec, R. E., Narayanasamy, S., Curley, R. W., et al., (2011). Carotene and novel apocarotenoid concentrations in orange-fleshed Cucumis melo melons: Determinations of β-carotene bio-accessibility and bioavailability. *J. Agri. Food Chem., 59*(9), 4448–4454.

Foryst-Ludwig, A., Neumann, M., Schneider-Brachert, W., & Naumanna, M., (2004). Curcumin blocks NF-κB and the motogenic response in *Helicobacter pylori*-infected epithelial cells. *Biochem. Biophys. Res. Commun., 316*, 1065–1072.

Franceschi, S., Bidoli, E., La, Vecchia. C., Talamini, R., D'Avanzo, B., & Negri, E., (1994). Tomatoes and risk of digestive-tract cancers. *Int. J. Cancer, 59*, 181–184.

Fu, H., Xie, B., Fan, G., Ma, S., Zhu, X., & Pan, S., (2010). Effect of esterification with fatty acid of β-cryptoxanthin on its thermal stability and antioxidant activity by chemiluminescence method. *Food Chem., 122*, 602–609.

Gajendragadkar, P. R., Hubsch, A., Maki-Petaja, K. M., Serg, M., Wilkinson, I. B., & Cheriyan, J., (2014). Effects of oral lycopene supplementation on vascular function in patients with cardiovascular disease and healthy volunteers: A randomized controlled trial. *PLoS One, 9*, e99070.

Gao, Y. Y., Xie, Q. M., Jin, L., Sun, B. L., Ji, J., Chen, F., et al., (2012). Supplementation of xanthophylls decreased pro-inflammatory and increased anti-inflammatory cytokines in hens and chicks. *Br. J. Nutr., 108*, 1746–1755.

Ghodratizadeh, S., Kanbak, G., Beyramzadeh, M., Dikmen, Z. G., Memarzadeh, S., & Habibian, R., (2014). Effect of carotenoid βcryptoxanthin on cellular and humoral immune response in rabbit. *Vet. Res. Commun, 38*, 59–62.

Gil, M. I., Tomás-Barberán, F. A., Hess-Pierce, B., & Kader, A. A., (2002). Antioxidant capacities, phenolic compounds, carotenoids, and vitamin C contents of nectarine, peach, and plum cultivars from California. *J. Agric. Food Chem., 50*, 4976–4982.

Gloria, N. F., Soares, N., Brand, C., Oliveira, F. L., Borojevic, R., & Teodoro, A. J., (2014). Lycopene and beta-carotene induce cell-cycle arrest and apoptosis in human breast cancer cell lines. *Anticancer Res., 34*, 1377–1386.

Gonzalez, S., Astner, S., A, W., Goukassian, D., & Pathak, M. A., (2003). Dietary lutein/zeaxanthin decreases ultraviolet B-induced epidermal hyperproliferation and acute inflammation in hairless mice. *J. Invest. Dermatol., 121*, 399–405.

Gouranton, E., Thabuis, C., Riollet, C., Malezet-Desmoulins, C., Yazidi, C. E., Amiot, M. J., et al., Lycopene inhibits proinflammatory cytokine and chemokine expression in adipose tissue. *J. Nutr. Biochem., 22*, 642–648.

Guedes, A. C., Amaro, H. M., Sousa-Pinto, I., & Malcata, F. X., (2014). Bioactive carotenoids from microalgae. *Bioactive Compounds from Marine Foods: Plant and Animal* Sources, 130–151.

Guo, L., Zhu, H., Lin, C., Che, J., Tian, X., Han, S., et al., (2015). Associations between antioxidant vitamins and the risk of invasive cervical cancer in Chinese women: A case–control study. *Sci. Rep., 5*, 13607.

Hadad, N., & Levy, R., (2012). The synergistic anti-inflammatory effects of lycopene, lutein, β-carotene, and carnosic acid combinations via redox-based inhibition of NF-κB signaling. *Free Radic. Biol. Med., 53*, 1381–1391.

Hahm, E. R., Gho, Y. S., Park, S., Park, C., Kim, K. W., & Yang, C. H., (2004). Synthetic curcumin analogs inhibit activator protein-1 transcription and tumor-induced angiogenesis. *Biochem. Biophys. Res. Commun., 321*, 337–344.

Hazlewood, L. C., Wood, L. G., Hansbro, P. M., & Foster, P. S., (2011). Dietary lycopene supplementation suppresses Th2 responses and lung eosinophilia in a mouse model of allergic asthma. *J. Nutr. Biochem., 22*, 95–100.

Heinen, M. M., Hughes, M. C., Ibiebele, T. I., Marks, G. C., Green, A. C., & Van Der, P. J. C., (2007). Intake of antioxidant nutrients and the risk of skin cancer. *Eur. J. Cancer, 43*, 2707–2716.

Holden, J. M., Eldridge, A. L., Beecher, G. R., Buzzard, I. M., Bhagwat, A. S., Davis, C. S., Douglass, L. W., et al., (1999). Carotenoid content of U.S. foods: An update of the database. *J. Food Compos. Anal., 12*, 169–196.

Hornero-Méndez, D., & Mínguez-Mosquera, M. I., (2000). Xanthophyll esterification accompanying carotenoid overaccumulation in chromoplast of *Capsicum annuum* ripening fruits is a constitutive process and useful for ripening index. *J. Agric. Food Chem., 48*, 1617–1622.

Huang, C. S., Chuang, C. H., Loa, T. F., & Huc, M. L., (2013). Anti-angiogenic effects of lycopene through immunomodulation of cytokine secretion in human peripheral blood mononuclear cells. *J. Nutr. Biochem., 24*, 428–434.

Hulshof, P. J. M., Chao, X., Van De Bovenkamp, P., Muhilal., & West, C. E., (1997). Application of a validated method for the determination of provitamin A carotenoids in Indonesia foods of different maturity and origin. *J. Agric. Food Chem., 45*, 1174–1179.

Hung, C. F., Huang, T. F., Chen, B. H., Shieh, J. M., Wu, P. H., & Wu, W. B., (2008). IFNγ-induced ICAM-1 expression at 8 and 16 h were not affected by lycopene, suggesting that lycopene primarily affects TNF-α-induced signaling pathway. *Eur. J. Pharmacol., 586*, 275–282.

Jang, S. H., Lim, J. W., & Kim, H., (2009). Mechanism of β-carotene-induced apoptosis of gastric cancer cells: Involvement of ataxia-telangiectasia mutated. *Ann. N Y Acad. Sci., 1171*, 156–162.

Jomova, K., & Valko, M., (2013). Health protective effects of carotenoids and their interactions with other biological antioxidants. *Eur. J. Med. Chem., 70*, 102–110.

Jyonouchi, H., Sun, S., & Gross, M., (1995). Effect of carotenoids on *in vitro* immunoglobulin production by human peripheral blood mononuclear cells: Astaxanthin, a carotenoid without vitamin A activity, enhances *in vitro* immunoglobulin production in response to a T-dependent stimulant and antigen. *Nutr. Cancer, 23*, 171–183.

Jyonouchi, H., Zhang, L., Gross, M., & Tomita, Y., (1994). Immunomodulating actions of carotenoids: Enhancement of *in vivo* and *in vitro* antibody production to T-dependent antigens. *Nutr. Cancer, 21*, 47–58.

Kang, B. Y., Chung, S. W., Chung, W. J., Im, S. Y., Hwang, S. Y., & Kim, T. S., (1999). Inhibition of interleukin-12 production in lipopolysaccharide activated macrophages by curcumin. *Eur. J. Pharmacol., 384*, 191–195.

Kang, M., Park, K. S., Seo, J. Y., & Kim, H., (2011). Lycopene inhibits IL-6 expression in cerulein-stimulated pancreatic acinar cells. *Genes Nutr., 6*, 117–123.

Karppi, J., Kurl, S., Laukkanen, J. A., & Kauhanen, J., (2012). Serum betacarotene in relation to risk of prostate cancer: The kuopio ischemic heart disease risk factor study. *Nutr. Cancer, 64*, 361–367.

Kataria, Y., Deaton, R. J., Enk, E., Jin, M., Petrauskaite, M., Dong, L., et al., (2016). Retinoid and carotenoid status in serum and liver among patients at high-risk for liver cancer. *BMC Gastroenterol., 16*, 1–12.

Katsuura, S., Imamura, T., Bando, N., & Yamanishi, R., (2009). β-carotene and β-cryptoxanthin but not lutein evokes redox and immune changes in RAW264 murine macrophages. *Mol. Nutr. Food Res., 53*, 1396–1405.

Kaulmann, A., & Bohn, T., (2014). Carotenoids, inflammation, and oxidative stress-implications of cellular signaling pathways and relation to chronic disease prevention. *Nutr. Res., 34*, 907–929.

Khavari, A., Bolhassani, A., Alizadeh, F., Bathaie, S. Z., Balaram, P., Agi, E., et al., (2014). Chemo-immunotherapy using saffron and its ingredients followed by E7-NT (gp96) DNA vaccine generates different anti-tumor effects against tumors expressing the E7 protein of human papillomavirus. *Arch Virol., 160*, 499–508.

Khoo, H. E., Prasad, K. N., Kong, K. W., Jiang, Y., & Ismail, A., (2011). Carotenoids and their isomers: Color pigments in fruits and vegetables. *Molecules, 16*(2), 1710–1738.

Kim, G. Y., Kim, K. H., Lee, S. H., Yoon, M. S., Lee, H. J., Moon, D. O., et al., (2005). Curcumin inhibits immunostimulatory function of dendritic cells: MAPKs and translocation of NF-κB as potential targets. *J. Immunol., 174*, 8116–8124.

Kim, G., Jang, M. S., Son, Y. M., Seo, M. J., Ji, S. Y., Han, S. H., et al., (2013). Curcumin inhibits $CD4^+$ T cell activation, but augments CD69 expression and TGF-β1-mediated generation of regulatory T cells at late phase. *PLoS One, 8*, e62300.

Kim, H. W., Chew, B. P., Wong, T. S., Park, J. S., Weng, B. B. C., Byrne, K. M., et al., (2000). Modulation of humoral and cell-mediated immune responses by dietary lutein in cats. *Vet. Immunol. Immunopathol., 73*, 331–341.

Kim, J. M., Araki, S., Kim, D. J., Park, C. B., Takasuka, N., Baba-Toriyama, H., et al., (1998). Chemopreventive effects of carotenoids and curcumins on mouse colon carcinogenesis after 1, *2*-dimethylhydrazine initiation. *Carcinogenesis, 19*, 81–85.

Kim, W., Fan, Y. Y., Smith, R., Patil, B., Jayaprakasha, G. K., McMurray, D. N., et al., (2009). Dietary curcumin and limonin suppress $CD4^+$ T-cell proliferation and interleukin-2 production in mice. *J. Nutr., 139*, 1042–1048.

Kim, Y. S., Lee, H. A., Lim, J. Y., Kim, Y., Jung, C. H., Yoo, S. H., et al., (2014). β-Carotene inhibits neuroblastoma cell invasion and metastasis *in vitro* and *in vivo* by decreasing level of hypoxia-inducible factor-1α. *J. Nutr. Biochem., 25*, 655–664.

Kongkachuichai, R., Charoensiri, R., Yakoh, K., Kringkasemsee, A., & Insung, P., (2015). Nutrients value and antioxidant content of indigenous vegetables from Southern Thailand. *Food Chem., 173*, 838–846.

Kucuk, O., Sarkar, F. H., Sakr, W., Djuric, Z., Pollak, M. N., Khachik, F., et al., (2001). Phase II randomized clinical trial of lycopene supplementation before radical prostatectomy. *Cancer Epidemiol. Biomarkers Prev., 10*, 861–868.

Kurihara, H., Koda, H., Asami, S., Kiso, Y., & Tanaka, T., (2002). Contribution of the antioxidative property of astaxanthin to its protective effect on the promotion of cancer metastasis in mice treated with restraint stress. *Life Sci., 70*, 2509–2520.

Lako, J., Trenerry, V. C., Wahlqvist, M., Wattanapenpaiboon, N., Sotheeswaran, S., & Premier, R., (2007). Phytochemical flavonols, carotenoids and the antioxidant properties of a wide selection of Fijian fruit, vegetables and other readily available foods. *Food Chem., 101*, 1727–1741.

Lakshminarayana, R., Raju, M., Krishnakantha, T. P., & Baskaran, V., (2005). Determination of major carotenoids in a few Indian leafy vegetables by high-performance liquid chromatography. *J. Agri. Food Chem., 53*(8), 2838–2842.

Lee, C. M., Chang, J. H., Moon, D. O., Hyun, C. Y., Choi, I. W., Park, Y. M., et al., (2008). Lycopene suppresses ovalbumin-induced airway inflammation in a murine model of asthma. *Biochem. Biophys. Res. Commun., 374*, 248–252.

Lee, W., Ku, S. K., Bae, J. W., & Bae, J. S., (2012). Inhibitory effects of lycopene on HMGB1-mediated pro-inflammatory responses in both cellular and animal models. *Food Chem Toxicol., 50*, 1826–1833.

Levy, J., Bosin, E., Feldman, B., Giat, Y., Miinster, A., Danilenko, M., et al., (1995). Lycopene is a more potent inhibitor of human cancer cell proliferation than either alpha-carotene or beta-carotene. *Nutr. Cancer, 24*, 257–266.

Li, Q., Huyan, T., Ye, L. J., Ren, H. L., Shi, J. L., & Huang, Q. S., (2014). Lycopene prolongs the lifespan and enhances the cytotoxicity of NK cells after *ex vivo* expansion. *Bioeng. Biosci., 2*, 23–30.

Lim, G. P., Chu, T., Yang, F., Beech, W., Frautschy, S. A., & Cole, G. M., (2001). The curry spice curcumin reduces oxidative damage and amyloid pathology in an Alzheimer transgenic mouse. *J. Neurosci., 21*, 8370–8377.

Limtrakul, P., Lipigorngoson, S., Namwong, O., Apisariyakul, A., & Dunn, F. W., (1997). Inhibitory effect of dietary curcumin on skin carcinogenesis in mice. *Cancer Lett., 116*, 197–203.

Lin, H. W., Chang, T. J., Yang, D. J., Chen, Y. C., Wang, M., & Chang, Y. Y., (2012). Regulation of virus induced inflammatory response by β-carotene in RAW264.7 cells. *Food Chem., 134*, 2169–2175.

LoTempio, M. M., Veena, M. S., Steele, H. L., Ramamurthy, B., Ramalingam, T. S., Cohen, A. N., et al., (2005). Curcumin suppresses growth of head and neck squamous cell carcinoma. *Clin Cancer Res., 11*, 6994–7002.

Macedo, R. C., Bolin, A. P., Marin, D. P., & Otton, R., (2010). Astaxanthin addition improves human neutrophil's function: *In vitro* study. *Eur. J. Nutr., 49*, 447–457.

Madaric, A., Kadrabova, J., Krajcovicova-Kudlackova, M., Valachovicova, M., Spustova, V., Mislanova, C., et al., (2013). The effect of bioactive complex of quercetin, selenium, catechins and curcumin on cardiovascular risk markers in healthy population after a two-month consumption. *Bratisl. Lek. Listy., 114*, 84–87.

Magesh, V., DurgaBhavani, K., Senthilnathan, P., Rajendran, P., & Sakthisekaran, D., (2009). *In vivo* protective effect of crocetin on benzo (α) pyrene-induced lung cancer in Swiss albino mice. *Phytother. Res., 23*, 533–539.

Mai, J., Shen, X., Shi, D., Wei, Y., Shen, H., & Wu, M., (2010). Effect of lutein on relieving oxidative stress in mice induced by D-galatose. *Wei Sheng Yan Jiu, 39*, 430–432.

Marcotorchino, J., Romier, B., Gouranton, E., Riollet, C., Gleize, B., Malezet, D. C., et al., (2012). Lycopene attenuates LPS-induced TNF secretion in macrophages and inflammatory markers in adipocytes exposed to macrophage-conditioned media. *Mol. Nutr. Food Res., 56*, 725–732.

Mathews-Roth, M. M., (1982). Antitumor activity of beta-carotene, canthaxanthin and phytoene. *Oncology, 39*, 33–37.

Mayne, S. T., Cartmel, B., Lin, H., Zheng, T., & Goodwin, W. J., (2004). Low plasma lycopene concentration is associated with increased mortality in a cohort of patients with prior oral, pharynx or larynx cancers. *J. Am. Coll. Nutr., 23*, 34–42.

Meydani, S. N., Wu, D., Santos, M. S., & Hayek, M. G., (1995). Antioxidants and immune response in aged persons: Overview of present evidence. *Am. J. Clin. Nutr., 62*, 1462S–1476S.

Moeller, S. M., Parekh, N., Tinker, L., Ritenbaugh, C., Blodi, B., Wallace, R. B., et al., (2006). Associations between intermediate age-related macular degeneration and lutein and zeaxanthin in the carotenoids in age-related eye disease study (CAREDS): Ancillary study of the Women's health initiative. *Arch Ophthalmol., 124*, 1151–1162.

Moshetova, L. K., Vorob'eva, I. V., Alekseev, I. B., & Mikhaleva, L. G., (2015). Results of the use of antioxidant and angioprotective agents in type 2 diabetes patients with diabetic retinopathy and age-related macular degeneration. *Vestn. Oftalmol., 131*, 34–40.

Muzandu, K., Ishizuka, M., Sakamoto, K. Q., Shaban, Z., El Bohi, K., Kazusaka, A., et al., (2006). Effect of lycopene and beta-carotene on peroxynitrite mediated cellular modifications. *Toxicol. Appl. Pharmacol., 215*, 330–340.

Nakamura, K., Yasunaga, Y., Segawa, T., Ko, D., Moul, J. W., Srivastava, S., et al., (2002). Curcumin down-regulates AR gene expression and activation in prostate cancer cell lines. *Int. J. Oncol., 21*, 825–830.

Nam, K. N., Park, Y. M., Jung, H. J., Lee, J. Y., Min, B. D., Park, S. U., et al., (2010). Anti-inflammatory effects of crocin and crocetin in rat brain microglial cells. *Eur. J. Pharmacol., 648*, 110–116.

Nishi, K., Muranaka, A., Nishimoto, S., Kadotac, A., & Sugahara, T., (2012). Immunostimulatory effect of β-cryptoxanthin *in vitro* and *in vivo*. *J. Funct. Foods, 4*, 618–625.

Nishida, K., Sugimoto, M., Ikeda, S., & Kume, S., (2014). Effects of supplemental β-carotene on mucosal IgA induction in the Jejunum and Ileum of mice after weaning. *Br. J. Nutr., 111*, 247–253.

Noorbala, A. A., Akhondzadeh, S., Tahmacebi-Pour, N., & Jamshidi, A. H., (2005). Hydro-alcoholic extract of *Crocus sativus* L. versus fluoxetine in the treatment of mild to moderate depression: A double-blind, randomized pilot trial. *J. Ethnopharmacol., 97*, 281–284.

Ochiai, T., Shimeno, H., Mishima, K., Iwasaki, K., Fujiwara, M., Tanaka, H., et al., (2007). Protective effects of carotenoids from saffron on neuronal injury *in vitro* and *in vivo*. *Biochim. Biophys. Acta, 1770*, 578–584.

Oh, J., Kim, J. H., Park, J. G., Yi, Y. S., Park, K. W., Rho, H. S., et al., (2013). *Radical Scavenging Activity-Based and AP-1-Targeted Anti-Inflammatory Effects of Lutein in Macrophage-Like and Skin Keratinocytic Cells* (pp. 1–8). Hindawi Publishing Corporation, Mediators of Inflammation.

Okai, Y., & Higashi-Okai, K., (1996). Possible immunomodulating activities of carotenoids in *in vitro* cell culture experiments. *Int. J. Immunogenet., 18*, 753–758.

Olmos, J., Gómez, R., & Rubio, V., (2015). Apoptosis comparison effects between synthetic and natural β-Carotene from *Dunaliella salina* on MDA-MB-231 breast cancer cells. *J. Microb. Biochem. Technol., 7*, 051–056.

Palozza, P., Calviello, G., Serini, S., Maggiano, N., Lanza, P., Ranelletti, F. O., et al., (2001). β-carotene at high concentrations induces apoptosis by enhancing oxy-radical production in human adenocarcinoma cells. *Free Radic. Biol. Med., 30*, 1000–1007.

Palozza, P., Serini, S., Maggiano, N., Angelini, M., Boninsegna, A., Di Nicuolo, F., et al., (2002). Induction of cell cycle arrest and apoptosis in human colon adenocarcinoma cell lines by β-Carotene through down-regulation of cyclin A and Bcl-2 family proteins. *Carcinogenesis, 23*, 11–18.

Pan, W., Yang, H., Cao, C., Song, X., Wallin, B., Kivlin, R., et al., (2008). AMPK Mediates Curcumin-Induced Cell Death in CaOV3 Ovarian Cancer cell. *Oncology Reports, 20*(6), 1553–1559.

Park, J. S., Chyun, J. H., Kim, Y. K., Line, L. L., & Chew, B. P., (2010). Astaxanthin decreased oxidative stress and inflammation and enhanced immune response in humans. *Nutr. Metab., 7*, 1–10.

Park, J. S., Mathison, B. D., Hayek, M. G., Massimino, S., Reinhart, G. A., Chew, B. P., et al., (2011). Astaxanthin stimulates cell-mediated and humoral immune responses in cats. *Vet Immunol Immunopathol., 144*, 455–461.

Parsons, H. A., Baracos, V. E., Hong, D. S., Abbruzzese, J., Bruera, E., & Kurzrock, R., (2016). The effects of curcumin (diferuloylmethane) on body composition of patients with advanced pancreatic cancer. *Oncol. Rep., 20*, 1553–1559.

Pechinskii, S. V., & Kuregyan, A. G., (2014). Molecular-biological problems of drug design and mechanism of drug action. *Pharmaceut. Chem. J., 47*, 509–513.

Preet, R., Mohapatra, P., Das, D., Satapathy, S. R., Choudhuri, T., Wyatt, M. D., et al., (2013). Lycopene synergistically enhances quinacrine action to inhibit Wnt-TCF signaling in breast cancer cells through APC. *Carcinogenesis, 34*, 277–286.

Priyadarshani, A. M. B., & Jansz, E. R., (2014). A critical review on carotenoid research in Sri Lankan context and its outcomes. *Crit. Rev. Food Sci. Nutr, 54*(5), 561–571.

Promphet, P., Bunarsa, S., Sutheerawattananonda, M., & Kunthalert, D., (2014). Immune enhancement activities of silk lutein extract from *Bombyx mori* cocoons. *Biol. Res., 47*, 1–10.

Rafi, M. M., Yadav, P. N., & Reyes, M., (2007). Lycopene inhibits LPS-induced proinflammatory mediator inducible nitric oxide synthase in mouse macrophage cells. *J. Food Sci., 72*, S69–S74.

Rajput, N., Naeem, M., Ali, S., Zhang, J. F., Zhang, L., & Wang, T., (2013). The effect of dietary supplementation with the natural carotenoid's curcumin and lutein on broiler pigmentation and immunity. *Poult. Sci., 92*, 1177–1185.

Rao, A. V., & Rao, L. G., (2007). Carotenoids and human health. *Pharmacol. Res., 55*, 207–216.

Reddy, R. C., Vatsala, P. G., Keshamouni, V. G., Padmanaban, G., & Rangarajan, P. N., (2005). Curcumin for malaria therapy. *Biochem. Biophys. Res. Commun., 326*, 472–474.

Rissanen, T. H., Voutilainen, S., Nyyssonen, K., Salonen, R., Kaplan, G. A., & Salonen, J. T., (2003). Serum lycopene concentrations and carotid atherosclerosis: The Kuopio ischemic heart disease risk factor study. *Am. J. Clin. Nutr., 77*, 133–138.

Rodriguez-Amaya, D. B., (1997). Carotenoids and food preparation. In: *The Retention of Provitamin a Carotenoids in Prepared, Processed, and Stored Foods*. John Snow Inc: Rio de Janeiro, Brazil.

Rosati, C., Diretto, G., & Giuliano, G., (2009). Biosynthesis and engineering of carotenoids and apocarotenoids in plants: State of the art and future prospects. *Biotechnol. Genet. Eng. Rev., 26*(1), 139–162.

Rutz, J. K., Borges, C. D., Zambiazi, R. C., Da Rosa, C. G., & Da Silva, M. M., (2016). Elaboration of microparticles of carotenoids from natural and synthetic sources for applications in food. *Food Chem., 202*, 324–333.

Saini, R. K., Shetty, N. P., & Giridhar, P., (2014). Carotenoid content in vegetative and reproductive parts of commercially grown *Moringa oleifera* Lam. cultivars from India by LC–APCI–MS. *Eur. Food Res. Technol., 238*(6), 971–978.

Sedjo, R. L., Roe, D. J., Abrahamsen, M., Harris, R. B., Craft, N., Baldwin, S., et al., (2002). Vitamin A, carotenoids, and risk of persistent oncogenic human papillomavirus infection. *Cancer Epidemiol. Biomarkers Prev., 11*, 876–884.

Seifter, E., Rettura, G., & Levenson, S. M., (1981). Carotenoids and cell mediated immune responses. In: *The Quality of Foods and Beverages, Chemistry and Technology* (Vol. 2, pp. 335–347). Academic Press: New York.

Setiawan, B., Sulaeman, A., Giraud, D. W., & Driskell, J. A., (2001). Carotenoid content of selected Indonesian fruits. *J. Food Compos. Anal., 14*, 169–176.

Seyedzadeh, M. H., Safari, Z., Zare, A., Gholizadeh, Navashenaq, J., Razavi, S. A., Kardar, G. A., et al., (2014). Study of curcumin immunomodulatory effects on reactive astrocyte cell function. *Int. Immunopharmacol., 22*, 230–235.

Shen, G., Xu, C., Hu, R., Jain, M. R., Gopalkrishnan, A., Nair, S., et al., (2006). Modulation of nuclear factor E2-related factor 2-mediated gene expression in mice liver and small intestine by cancer chemopreventive agent curcumin. *Mol. Cancer Ther., 5*, 39–51.

Shi, J., Maguer, M. L., & Bryan, M., (2002). Lycopene from tomatoes. In: Shi, J., Mazza, G., & Maguer, M. L., (eds.), *Functional Food: Biochemical and Processing Aspects* (Vol. 2, pp. 135–167). CRC Press LCC: Danvers, MA, USA.

Shirley, S. A., Montpetit, A. J., Lockey, R. F., & Mohapatra, S. S., (2008). Curcumin prevents human dendritic cell response to immune stimulants. *Biochem. Biophys. Res. Commun., 374*, 431–436.

Silva, L. M. R. Da., Figueiredo, E. A. T. De., Ricardo, N. M. P. S., Vieira, I. G. P., Figueiredo, R. W. De., Brasil, I. M., et al., (2014). Quantification of bioactive compounds in pulps and byproducts of tropical fruits from Brazil. *Food Chemistry, 143*, 398–404.

Simone, R. E., Russo, M., Catalano, A., Monego, G., Froehlich, K., Boehm, V., et al., (2011). Lycopene inhibits NF-kB-mediated IL-8 expression and changes redox and PPARc signaling in cigarette smoke-stimulated macrophages. *PLoS One, 6*, e19652.

Soontornchaiboon, W., Joo, S. S., & Kim, S. M., (2012). Anti-inflammatory effects of violaxanthin isolated from microalga *Chlorella ellipsoidea* in RAW 264.7 macrophages. *Biol Pharm Bull, 35*, 1137–1144.

Speek, A. J., Speek-Saichua, S., & Schreurs, W. H. P., (1988). Total carotenoids and β-carotene contents of Thai vegetables and effects of processing. *Food Chem., 27*, 245–251.

Speranza, L., Pesce, M., Patruno, A., Franceschelli, S., De Lutiis, M. A., Grilli, A., et al., (2012). Astaxanthin treatment reduced oxidative induced pro-inflammatory cytokines secretion in U937: SHP-1 as a novel biological target. *Mar. Drugs, 10*, 890–899.

Srivastava, R. M., Singh, S., Dubey, S. K., Misra, K., & Khar, A., (2011). Immunomodulatory and therapeutic activity of curcumin. *Int. Immunopharmacol., 11*, 331–341.

Stahl, W., & Sies, H., (2005). Bioactivity and protective effects of natural carotenoids. *Biochim. Biophys. Acta, 1740*, 101–107.

Sup, S. J., Hoon, K. D., Jung, H. J., Hee, Kim, J., Lim, D., Lee, S. K., et al., (2002). Hydrazinocurcumin, a novel synthetic curcumin derivative, is a potent inhibitor of endothelial cell proliferation. *Bioorg. Med. Chem., 10*, 2439–2444.

Tapiero, H., Townsend, D. M., & Tew, K. D., (2004). The role of carotenoids in the prevention of human pathologies. *Biomed. Pharmacother., 58*, 100–110.

Tee, E. S., & Lim, C. L., (1991). Carotenoid composition and content of Malaysian vegetables and fruits by the AOAC and HPLC methods. *Food Chem., 41*, 309–339.

Umar, S., Shah, M. A., Munir, M. T., Yaqoob, M., Fiaz, M., Anjum, S., et al., (2016). Synergistic effects of thymoquinone and curcumin on immune response and anti-viral activity against avian influenza virus (H9N2) in turkeys. *Poult. Sci.*

Upadhyaya, K. R., Radha, K. S., & Madhyastha, H. K., (2007). Cell cycle regulation and induction of apoptosis by β-carotene in U937 and HL60 leukemia cells. *J. Biochem. Mol. Biol., 40*, 1009–1015.

Varalakshmi, C., Mubarak, A. A., Pardhasaradhi, B. V. V., Srivastava, R. M., Singh, S., & Khar, A., (2008). Immunomodulatory effects of curcumin: *In vivo*. *Int. Immunopharmacol., 8*, 688–700.

Vijayapadma, V., Ramyaa, P., Pavithra, D., & Krishnaswamy, R., (2014). Protective effect of lutein against benzo(a)pyrene-induced oxidative stress in human erythrocytes. *Toxicol. Ind. Health., 30*, 284–293.

Vuong, L. T., Franke, A. A., Custer, L. J., & Murphy, S. P., (2006). *Momordica cochinchinensis* Spreng. (gac) fruit carotenoids reevaluated. *J. Food Composition Anal, 19*(6, 7), 664–668.

Watson, R. R., Prabhala, R. H., Plzia, P. M., & Alberts, D. S., (1991). Effects of βcarotene on lymphocyte subpopulations in elderly humans: Evidence for a dose–response relationship. *Am. J. Clin. Nutr., 53*, 90–94.

Weedon, B. C. L., & Moss, G. P., (1995). Structure and nomenclature. In: Britton, G., Liaaen-Jensen, S., & Pfander, H., (eds.), *Carotenoids: Isolation and Analysis* (Vol. 1A, pp. 27–70). Birkhäuser Verlag: Basel, Switzerland.

Weihong, S., Lihong, X., Hong, L., Kailiang, L., Yuxiu, Z., & Xiaofang, L., (2016). Assessment and comparison of *in vitro* immunoregulatory activity of three astaxanthin stereoisomers. *J. Ocean Univ. China, 15*, 283–287.

Wood, S. M., Beckham, C., Yosioka, A., Darban, H., & Watson, R. R., (2000). Carotene and selenium supplementation enhances immune responses in aged humans. *Intern. Med., 2*, 85–92.

Xu, X. R., Zou, Z. Y., Xiao, X., Huang, Y. M., Wang, X., & Lin, X. M., (2013). Effects of lutein supplement on serum inflammatory cytokines, apoE and lipid profiles in early atherosclerosis population. *J. Atheroscler. Thromb., 20*, 170–177.

Yadav, V. S., Mishra, K. P., Singh, D. P., Mehrotra, S., & Singh, V. K., (2005). Immunomodulatory effects of curcumin. *Immunopharmacol. Immunotoxicol., 27*, 485–497.

Yamaguchi, M., Hasegawa, I., Yahagi, N., Ishigaki, Y., Akano, F., & Ohta, T., (2010). Carotenoids modulate cytokine production in Peyer's patch cells *ex vivo. J. Agric. Food Chem., 58*, 8566–8572.

Yang, D. J., Lin, J. T., Chen, Y. C., Liu, S. C., Lu, F. J., Chang, T. J., et al., (2013). Suppressive effect of carotenoid extract of *Dunaliella salina* alga on production of LPS-stimulated pro-inflammatory mediators in RAW264.7 cells via NF-κB and JNK inactivation. *J. Funct. Foods, 5*, 607–615.

Yang, R., Tan, X., Thomas, A. M., Shen, J., Qureshi, N., Morrison, D. C., et al., (2006). Crocetin inhibits mRNA expression for tumor necrosis factor-alpha, interleukin-1beta, and inducible nitric oxide synthase in hemorrhagic shock. *J. Parenter. Enteral. Nutr., 30*, 297–301.

Yasui, Y., Hosokawa, M., Mikami, N., Miyashita, K., & Tanaka, T., (2011). Dietary astaxanthin inhibits colitis and colitis-associated colon carcinogenesis in mice via modulation of the inflammatory cytokines. *Chem. Biol. Interact., 193*, 79–87.

Youn, H. S., Saitoh, S. I., Miyake, K., & Hwang, D. H., (2006). Inhibition of homodimerization of toll-like receptor 4 by curcumin. *Biochem. Pharmacol., 72*, 62–69.

Zhang, H. G., Kim, H., Liu, C., Yu, S., Wang, J., Grizzle, W. E., et al., (2007). Curcumin reverses breast tumor exosomes mediated immune suppression of NK cell tumor cytotoxicity. *Biochim. Biophys. Acta, 1773*, 1116–1123.

Zheng, M., Zhang, Q., Joe, Y., Lee, B. H., Ryu, D. G., Kwon, K. B., et al., (2013). Curcumin induces apoptotic cell death of activated human $CD4^+$ T cells via increasing endoplasmic reticulum stress and mitochondrial dysfunction. *Int. Immunopharmacol., 15*, 517–523.

Zhou, G. Z., Zhang, S. N., Zhang, L., Sun, G. C., & Chen, X. B., (2014). A synthetic curcumin derivative hydrazinobenzoylcurcumin induces autophagy in A549 lung cancer cells. *Pharm. Biol., 52*, 111–116.

CHAPTER 4

ROLE OF VEGETABLES CONTAINING CAROTENOIDS ON HUMAN IMMUNE FUNCTIONS

DEEPTI DWIVEDI, AARTI TIWARI, AJAY KUMAR SHUKLA, SATYA PRAKASH SINGH, and ANKUR SRIVASTAVA

Institute of Pharmacy, Dr. Ram Manohar Lohia Avadh University, Ayodhya, Uttar Pradesh, India

ABSTRACT

Chronic illness risks associated with oxidative stress (OS), obesity, cancer, cardiovascular disease (CVD), osteoporosis, and diabetes are just a few of the illnesses that may be prevented by eating more fruits and vegetables, according to dietary recommendations. Antioxidant phytochemicals found in fruits and vegetables protect the body from the detrimental effects of OS. Phytochemicals are carotenoids that are responsible for a variety of food's hues. Carotenoids are playing a role in the prevention, and treatment of human illnesses. Carotenoids are powerful antioxidants that also contribute to the body's supply of vitamin A. Carotenoids have been shown scientifically that evidence suggests that phytochemicals may help prevent several different illnesses. Despite extensive research on the chemistry of carotenoids, it's been a long time since their bioavailability, metabolism, and biological activities have been studied thoroughly. Carotenoids are a recent topic of study. Lycopene's impact on human health has been the subject of many studies. Carotenoids such as lycopene do not promote vitamin A production, unlike certain other carotenoids. A result of unsaturated lycopene is a

Immune-Boosting Nutraceuticals for Better Human Health: Novel Applications. Urmila Jarouliya, PhD, Raj K. Keservani, MPharm, Rajesh K. Kesharwani, PhD, Virendra K. Patel, PhD, and Adi D. Bharti, MPharm (Eds.)

powerful antioxidant and singlet oxygen quencher because of its structure. Carotenoids are the subject of this essay for their function in human health in general and lycopene in particular.

4.1 INTRODUCTION OF CAROTENOIDS

The link between nutrition and chronic illnesses is associated with each other. It has been found in clinical data. Dietary guidelines offered by clinical agencies for the prevention and management of diseases are cancer, cardiac disease, diabetes, and osteoporosis. It depends on such evidence. Plant foods are like fruit, vegetables, etc. These are rich sources of carotenoids and other physiologically active phytochemicals; it is one of the keys recommendations of these dietary guidelines. The beneficial benefits of fruits and vegetables are related to several pathways, including metabolism, immunological regulation, and hormone induction. The oxidative stress (OS) conditions are caused by reactive oxygen species (ROS), produced by normal aerobic metabolism as well as lifestyle variables like smoking, exercise, and food, which have recently been linked to the onset and development of some chronic illnesses. Recent research has focused on antioxidants that can reduce the harmful effects of ROS. Researchers, health professionals, and regulatory organizations are all interested in carotenoids because of their antioxidant capabilities. This review will concentrate on the role of carotenoids in animal health, and lycopene in particular, which has received much attention in recent years due to its potential role in chronic disease prevention.

Plants and microbes, but not mammals, generate carotenoids are colored chemicals. They help the photosynthetic apparatus of plants and protect them from light harm. The chief sources of carotenes in the human diet are fruits and vegetables (Mangels et al., 1993). They are present in fruits and vegetables as micro-components and are liable for their yellow, orange, and red colors. Carotenoids are assumed to be responsible for fruits and vegetables' therapeutic characteristics in avoiding human illnesses such as coronary heart disease, cancer, and other chronic disorders (Paiva & Russell, 2000). They're good sources of vitamin A in the diet. The antioxidant capabilities of carotenoids have been a prominent focus of research in recent years (Paiva & Russell, 2000). More than 600 carotenes have been discovered in nature thus far. In a usual human diet, however, only around 40 are present. Around 20 of these 40 βcarotenoids are in the adult bloodstream. It has been discovered in mature blood and tissues. β-carotene, lycopene, lutein, and cryptoxanthin make up over 90% of the carotenoids in the human diet and body (Astrog et al., 1999; Dias et al., 2022; Martini et al., 2022; Zhou et al., 2020).

4.1.1 CHEMISTRY AND DIETARY SOURCES

All carotene has a polyisoprenoid structure. It is long linked chain with double bonds and is near bilaterally symmetrical around the core double bond in terms of chemical properties (Britton, 1995). Different carotenoids are formed by cyclizing the end groups. Attaching oxygen plays a role in structures, giving them their different colors, and antioxidant properties. Due to the conjugated double bonds, carotenoids may alter conformation and become cis-trans isomers, and their trans isoforms are more common and stable in foods. It is known biological ramifications of carotenoid esterification reaction in human health. Carotenoids are available in several foods. Darkly colored fruits, juices, and vegetables are the richest sources of these nutrients. The majority of the protein and β-carotene is found in yellow-orange fruits and vegetables, while lutein and orange cryptoxanthin are found in fruits. The other sources of cryptoxanthin are green foods, and sources of lycopene are tomatoes and tomato foods (Ong & Tee, 1992). Table 4.1 lists common dietary sources and the number of carotenoids they contain. Lutein and zeaxanthin, for example, are found in significant amounts in egg yolks. Because carotenoids are unsaturated, they are susceptible to oxidation-related alterations. Other elements such as temperature, light, and pH, on the other hand, can cause changes in the color of meals as well as their nutrient benefits (Melendez-Martinez et al., 2004; Sarkar et al., 2020; Saini et al., 2018; Rearte et al., 2020). In general, conventional domestic cooking methods such as heating, steaming, and boiling have no effect on the carotenoid content of meals, while excessive heat can cause oxidative degradation of carotenoids (Thane & Reddy, 1997). Although multiple nutritional databases give estimates of human carotenoids consumption, these figures vary widely due to the sensitive nature and specificity of the various analytical techniques employed to detect these phytochemicals. Seasonal fluctuations and the methods of preparation of foods are altering the amount of carotenoids. Even though the fact that carotenoids have a positive function in human health, they are not considered essential nutrients and hence do not have a recommended dietary intake (DRI) value.

4.2 T-LYMPHOCYTES

Lymphocytes should be able to react quickly to infections, switching from a dormant state to an extremely active state in hours of activation. The majority of research papers have focused on the signal transduction that leads to cell cycle control genes and innate immune molecules. The macromolecular synthesis must also be rise for the cell to carry out the transcriptional programs of growth, division, and effector function, it has recently been discovered. The regulation of lymphocyte metabolism to sustain the bio-energetically demanding processes of growth, on the other hand, has received very little study. These higher synthetic activities, activated cells must drastically modify their metabolism. The cellular metabolism is homeostatically run through growing demands for energy or synthesis intermediates. It is causing catabolic pathways previously called in response to rising ATP to ADP conversion. It is affiliated with signaling pathways and transcription, evidence is overwhelming that extracellular signals are required to support lymphocyte nutrient uptake and metabolism to support cellular proliferation, proliferation, and secretion of cytokines. As a consequence, cell metabolic signaling pathways may become potential therapeutic targets. Lymphocytic metabolism during rest T-lymphocytes exits the thymus as little dormant cells and enters the bloodstream. These resting T-cells absorb just a little amount of glucose and other important nutrients (Thane & Reddy, 1997), giving enough energy to keep the home in order (Buttgereit, 2000). Glucose is used in the synthesis of oligosaccharides (glycosylation processes), lactate formation (glycolysis alone), and CO_2 oxidation in almost equal amounts. T-cells require external cues, such as cytokines and low-level activation through the TCR, to sustain even this baseline metabolic rate. T-cells lower their capacity to import sugar to levels below those needed to maintain cellular homeostasis in the absence of such signals (Rathmell et al., 2000). The availability of trophic cues, rather than the availability of nutrients, limits the metabolism of resting lymphocytes.

4.3 THE ANTIOXIDANT AND BIOLOGICAL PROPERTIES OF THE CAROTENOIDS

4.3.1 PROPERTIES OF CAROTENOIDS

Several previous reviews have looked at the biochemical features of carotenoids (Krinsky et al., 1994). The biological activities of this intriguing class of chemicals appear to be derived from their physiochemical features, which makes sense. The bulk of the 600 carotenes found in nature has a length of 40 carbons and are either pure hydrocarbons (carotenes) or have oxygenated functional groups (xanthophylls). Many of the carotenoid and xanthophylls identified in human serum have their structures illustrated. These chemicals' capacity to absorb visible light is due to their long chain conjugated polyene structure, which also renders them prone to oxidation. This last characteristic is linked to their potential to serve as antioxidants.

4.3.2 ANTIOXIDANT ACTIONS OF CAROTENOIDS

Several studies have been done on the role and importance of carotenoids. Carotenoids are capable of protecting from radical-initiated processes, notably those that result in lipid peroxidation, in addition to quenching $1O_2$. These antioxidant effects have been studied before (Krinsky et al., 1989), and current findings imply that this protection differs from that provided by traditional dietary antioxidants such as tocopherol. Burton and Ingold have studied the antioxidant properties of carotene in 1984 (Burton & Ingold, 1984). Currently, investigations of the antioxidant properties of β-carotene through the methods used *in-vitro* method and represented outstanding positive results like certain carotenoids are a stronger antioxidant effect than β-tocopherol (Miller et al., 1996). The chemical interaction between radical species and carotenoids produces protective compounds (Handelman et al., 1991). Both substitution products, as well as carotene-radical adduct (Handelman et al., 1991) both substitution products, as well as carotene-radical adduct (Liebler & Mcclure, 1996).

4.3.3 PROOXIDANT ACTIONS OF CAROTENOIDS

Burton and Ingold were the first ones to show that -carotene may operate as a prooxidant during radical-induced lipid peroxidation, but only at 100% oxygen (Burton et al., 1989). Since then, several studies were published carotene works as a prooxidant when oxygen tension is at 100%. Palazzo and colleagues were found reported β-carotene suitable for preserving liver microsomes (Palozza et al., 2008), and mouse thymocytes (Palozza et al., 2009) from radical-induced lipid peroxidation in the presence of oxygen. Truscott has established a theoretical basis for β-carotene's antioxidant and prooxidant activities under low and high oxygen pressures (Truscott et al., 1996).

4.3.4 CAROTENOIDS AND ANTI-CANCER PROPERTIES

Carotenoids are beneficial in the treatment of numerous cancers. Carotenoids with pro-vitamin activity include β-carotene, β-carotene, β-carotene, and cryptoxanthin. Plant cell walls and carotene-protein complexes are difficult to digest and degradation, resulting in lower bioavailability. Thermal processing has been shown to improve β-carotene availability and absorption when compared to industrial processing (Van het Hof et al., 2000). In women, higher blood levels of these carotenoids have been related to a decreased risk of breast cancer. According to the findings, lycopene and carotene treatment of several human breast cell lines for 48 and 96 hours inhibits cell growth, stops the cell cycle at various phases, and increases apoptosis (Eliassen et al., 2012).

4.4 MODULATION OF HUMAN T-LYMPHOCYTE FUNCTIONS BY THE CONSUMPTION OF CAROTENOID-RICH VEGETABLES

The effect of consumption of carotenoid-rich foods on the immunological system has been investigated and found nonsmoking subjects (23 men) not limited in their daily diet except that they had to avoid high-carotene fruits and vegetables for the duration of the study. The research divided into four two-week periods: Weeks 1–2: reduced period; weeks 3–8: daily ingestion of 330 ml tomato soup (40 mg lycopene/d, 15 mg b-carotene/d, 0 mg lutein/d) (weeks 3–4), 330 ml carrot juice (216 mg b-carotene/d, 15 mg a-carotene/d, 0 mg lutein/d After a 12-hour fast, blood was taken from individuals once a week. Proliferation and production of immunoreactive cytokines were used to evaluate T-lymphocyte function. The proliferation of mononuclear cells (PBMC) grown with concanavalin A was considerably inhibited when a low-carotene diet was consumed. When compared to lymphocyte proliferation after the conclusion of the depletion period, lymphocyte proliferation did not change significantly after two weeks of tomato juice intake and until the completion of the intervention period. The dietary intervention impacted the secretion of cytokines by T-helper-1-like lymphocytes (interleukin (IL)-2) and T-helper-2-like lymphocytes (IL-4). The low-carotene diet significantly reduced IL-2 and IL-4 secretion (P,0001 and P,005, respectively, as compared to baseline). The cytokine secretion capability of PBMC after carrot juice and spinach powders consumption was not substantially different from that after the conclusion of the depletion phase. To summarize, the findings of this study show that a low β-carotenoids diet lowers T-lymphocyte functions, but adding tomato juice restores these capabilities. Changes in plasma carotenoid concentrations could not explain this modulation. The active ingredients of tomato juice, and the biological relevance of the immunomodulation, are still unknown (Mangels et al., 1993). Numerous studies on the effects of vitamin A and carotene on various animal species as well as several human trials have been conducted (Bendich et al., 1989). Pure β-carotene has been shown to have an immunostimulating effect in animal studies (Bendich, 1991), but human trials have had conflicting results. According to studies (Watson et al., 2010), supplemental β-carotene stimulates a range of immunological activities.

The majority of human research has employed pure b-carotene at doses varying from 15 to 300 mg/d for short and exterior periods (Bendich, 1991). Results from depletion trials with reduced diets have been mixed. Daudu et al. (1994) found that a low-b-carotene diet ingested for 68 days did not affect the effect on T-lymphocytes, but Kramer & Burri in 1997 found that lymphocyte proliferative responsiveness decreased in human inhuman human patients who followed a low-b-carotene diet (0.5 mg/d) for 60 days. Only lycopene has been investigated in human beings, aside from b-carotene. There were no significant changes in monocyte surface marker expression after 28 days of treatment with 15 mg lycopene/d (Hughes et al., 1997).

Other carotenoids' immunomodulatory action in human individuals is unknown. One study looked into the impact of lycopene vegetable extracts on lymphocyte function in human participants (Clevidence et al., 1997).

4.5 MECHANISM OF CAROTENOIDS ACTION

Carotenoids control immunity through mainly unknown and mainly unknown processes. The antioxidant effect of carotenoids is the most well-known mode of action (Krinsky et al., 1989). In biological systems, singlet oxygen radicals and peroxyl radicals are constantly produced. These reactive substances are capable of causing DNA single-strand breaks and altering cell membrane function (Weinberg et al., 1835). 3-carotene has significant scavenging and quenching abilities for singlet oxygen and peroxyl radicals, especially when oxygen tension is low (Burton et al., 1984). Many naturally occurring carotenoids and xanthophylls outperformed J3-carotene in terms of singlet oxygen quenching ability (Di Mascio, 2009).

4.6 DISTRIBUTION OF CAROTENOIDS IN VEGETABLES AND FRUITS

The classification of an item as a fruit or a vegetable can cause some confusion. Anatomical precision and culinary application are sometimes at odds. Tomatoes and peppers, for example, are technically fruits, but they are commonly utilized in the form of vegetables. For such cases, the phrase 'fruit veggies' has been employed in several evaluations (Rodriguez-Amaya, 1999). Some vegetables are blossoms (broccoli, cauliflower), whereas others are seeds.

4.6.1 GREEN VEGETABLES AND FRUITS

Chlorophyll is present in the photosynthesis structures, the chloroplasts, all green plant tissues, including leaves, stems, green fruit, pods, and seeds of legumes such as beans and lentils, are green. Carotenoids are found in the chloroplast colorant complexes of photosystems 1 and 2. The carotenoid makeup of plant chloroplasts is extremely consistent, with -carotene (25–30%), lutein (40–50%), xanthophylls (15%), and neoxanthin as the primary components (15%). Other carotenoids, such as β-carotene, zeaxanthin, utilizing appropriate and lutein 5,6-epoxide, may be detected in trace levels. Lactucaxanthin may replace a few of the lutein in rare cases, with lettuce being the most common example.

Even though the carotenoid makeup is nearly consistent, the carotenoid quantitative content varies greatly. However, there is a definite link: darker green suggests a large number of chloroplasts and thus a higher proportion of carotenoids. The deepest green and consequently the highest carotenoid content may be found on the outer leaves of plants with quite some closely packed leaves, including lettuce and mustard family members. Inner leaves that are not exposed to light may be pale green or almost white with very little carotenoid, or yellow (etiolated) with a separate carotenoid composition, generally with little or no carotenoids and a different xanthophyll structure. This variance must be taken into consideration when predicting the carotenoid composition and content of a given vegetable to be ingested (Khachik et al., 1986).

4.6.2 CAROTENOIDS IN FOOD

A review reported the qualitative dispersion of the major carotenoid in some of the most regularly eaten green fruits and vegetables. These carotenoids are always characterized by major levels of their geometric isomers. The geometrical isomers 9Z and 13Z, as well as the 15Z isomer, are the most common geometric isomers in foods for symmetrical carotenoids. 9Z, 9′Z, 13Z, 13′Z, and 15Z isomers of unsymmetrical carotenoids may all be available in different amounts. Carotenoids that are di-Z isomers are uncommon in foods.

4.6.3 YELLOW, ORANGE, AND RED FRUITS AND VEGETABLES

Many of the greatest carotene sources aren't green. High concentrations of cyanobacteria can be found in yellow, orange, and red vegetative tissues, such as fruits, flowers, roots, and seeds. It's crucial to remember that these colors aren't always caused by carotenoids; anthocyanins, betalains, and quinones are just a few examples:

1. **Fruits:** The fruit has been extensively investigated as a key dietary source of carotenoids. While some fruits have minimal levels of carotenoids or just modest amounts of carotenoids detected in chloroplasts, others have greater concentrations of various carotenoids. According to Goodwin (1980), several distinct patterns have been found to be present in a variety of fruits. These patterns include (i) significant amounts of cryptoxanthin and zeaxanthin, which are hydroxy derivatives of beta-carotene; (ii) but also lutein, which is a yellow-orange color; and (iii) but also β-carotene and/or its hydroxy speculative products. Because of the identification and knowledge of the genes associated with carotenoid synthesis, these observations may now be explained. Chloroplasts are found in green, unripe fruits, and they contain the characteristic array of carotenoids seen in chloroplasts. As the fruit grows, the chloroplast hues may remain or be eliminated. In many situations, however, color changes occur when the fruits age and acquire chromoplasts, intracellular organelles that supplement and may be generated from chloroplasts. Carotenoids are generated in the chromoplasts and accumulate there. Activation of the group of genesis in control of synthesis is a crucial part of the ripening stage. The carotenoid concentration of mature fruit is determined by the presence and activity of ripening-specific genes. Lycopene is generated if the phytoene synthesis and desaturase genes are active; if the cyclase and hydroxylase genes are also involved, dicyclic β-carotene as well as its hydroxy derivatives, or monocyclic β-carotene and its hydroxy counterpart rubixanthin accumulate (category ii). When

β-cyclase and β-hydroxylase are present, β-carotene and lutein are produced (category iii). Similarly, having an active performance in a specific gene puts you in group iv. Other genes might be present, resulting in a variety of end products (category v).

Because of this knowledge, it is now feasible to genetically change carotenoid profiles or content in a variety of agricultural species.

Carotenes are abundant in a variety of yellow/red fruits and vegetables, whereas xanthophylls (Khachik et al., 1987) are present in very trace amounts. Some species, such as apricot, have a very distinct carotenoids profile. Lycopene and the metabolic intermediates substance that AIDS (acquired immunodeficiency syndrome), phytofluene, and β-carotene, as well as neurosporene, β-carotene are abundant in red foods such as tomatoes. Lycopene content in tomatoes is 5–10 mg/100 g. Tomato-based foods, including ketchup, soup, and sauces, sustain a high level of (Khachik et al., 1992).

2. **Roots:** The pigments are generated and concentrated in chromoplasts, including carrots and sweet potatoes. The amount of β-carotene generated and the ratio of -carotene to β-carotene in carrots differs significantly between strains. -Carotene content can vary from 5% to 50% of total carotene content. A larger amount of -carotene is found in varieties with a richer orange-red tint. In general, the concentration in exterior tissues (phloem) is higher than in the inner core (xylem). Lycopene, which might be the predominant pigment in some red cultivars, is also present. Instead of β-carotene, a yellow variant has a high concentration of lutein. Sweet potatoes also acquire -carotene, though usually in small amounts. Low quantities of common chloroplast xanthophylls are found in common potatoes, especially yellow types. Potatoes that have been genetically modified to collect carotene have been developed. Other yellow root vegetables, such as swede (Brassica rutabaga), may include low concentrations of carotenoids, including some lycopene (Joyce et al., 1954).
3. **Seeds:** Lutein, -carotene, zeaxanthin, and cryptoxanthins are the primary pigments that give sweetcorn (*Zea mays*) its yellow-orange color. The invention of a genetically modified 'Golden' rice (Potrykus, 2001) that accumulates -carotene in the butter-colored endosperm is the result of a lengthy research program (Humphries, 2003). Grain and pasta derivatives have been studied. Lutein and zeaxanthin were the only carotenoids of note, with lutein dominating. In Australian grass wheat, the contents were usually very low (ng/g), but slightly

higher (g/g). Because lutein and zeaxanthin from eggs are employed in the processing, wheat pasta has a greater carotenoid concentration. Durum wheat (*Triticum durum*) pasta has higher lutein because of the durum flour. Other grains and flours have relatively little carotenoid. β-carotene and chloroplast xanthophylls are found in the green seeds of legumes, such as peas. Bixin, an apocarotenoid, accumulates in abundance in the seeds of the plant *Bixa Orellana*. This product (annatto) is commonly used as a food colorant; however, it is unknown whether it is harmful to human health.

4. **Flowers:** These are not often consumed, with cauliflower and broccoli being the most well-known examples. White cauliflowers have little or no carotenoid, but an orange variety that accumulates β-carotene and was first discovered in a field of white cauliflowers is now commercially accessible (Ong & Goh, 2002). Broccoli and calabrese flower heads are picked before the petals unfold. The yellow florets, which are high in lutein esters, may appear in older or preserved samples.
5. **Oils:** Some oil palm fruits generate and accumulate significant levels of -carotene and -carotene, which are preserved in the pressed fruit oil. Red palm oil from *Elaeis guineensis* is refined in many forms and used as a cooking oil and component in manufactured foods on a wide scale (Godoy & Rodriguez-Amaya, 1995). Other palm fruits, such as the 'Buriti' (*Mauritia vinifera*) from South America, have a high carotenoid concentration of Shewmaker et al. (2008). Canola (rapeseed, Brassica napus) oil generally contains little or no carotenoid, but GM techniques can be used to introduce the ability to synthesize carotenoids. The oil of the 'Gac' fruit (*Momordica cochinchinnensis*) from Southeast Asia includes a high concentration of β-carotene, but it has only been used locally so far. The common sources of carotenoid are shown in Table 4.1.

TABLE 4.1 Common Sources of Carotenoid

Common Name	Content
Loquat	Low-moderate
Papaya	Moderate-high
Pepper (red, orange)	Moderate
Persimmon	High
Pitanga	High
Squash/pumpkin	Moderate-high
Tree tomato	Moderate-high

4.7 EFFECTS OF STORAGE, PROCESSING, AND COOKING

The stability of carotenoids in fruit and vegetables is more stable than pure crystal form. It is related to the preventive role of special conditions inside the tissues due to molecular conversations with proteins, molecular accumulation, and crystallization, and the use of organic antioxidants, such as antioxidative enzymes like superoxide dismutase (SOD). Any disruption of the tissues, like during preparations or cooking, or natural aging, might compromise this barrier, exposing the carotenoids to toxic substances and change. Cut, diced, shredded, or pulped fruits and vegetables are exposed to more oxygen, and the barriers that keep carotenoids and oxidizing enzymes like lipoxygenase apart are broken down. When foods are kept, processed, or cooked, the properties of carotenoids imply that extended exposure to air, bright light, elevated heat, or acid causes the largest losses and changes. Carotenoids become unstable from warmth and exposure to intense light and air must be avoided. Food transportation, storage, and processing must all be managed to avoid or reduce quality loss while preserving nutritional advantages. Some losses are inevitable, such as the removal of a lycopene peel or skin that is unattractive. Carotenoids can also be lost or changed during processing and storage as a result of enzymatic or non-enzymatic oxidation, geometrical isomerization, rearrangement, or other processes. During industrial processing, these issues may be handled and controlled, not during home preparations, where losses can be significant and difficult to control (Rodriguez-Amaya, 1997; Ram et al., 2020; Elik et al., 2020).

4.8 CONCLUSIONS

Carotenoids, including beta-carotene and other carotenoids, regulate the immune system. Furthermore, provitamin A has little effect on the bioactivities of carotenoids likely. Carotenoids in the diet help the body's defenses, resulting in greater health. As a consequence, further study is needed to validate this connection. Carotenoids' effects on resistance and health have mostly concentrated on beta-carotene, with the effects of other carotenoids generally ignored. Although lutein, zeaxanthin, cryptoxanthin, carotene, and 'Y-carotene are present in bovine blood and milk. Several other carotenoids, such as lutein, zeaxanthin, cryptoxanthin, carotene, and 'Y-carotene, are also present in high concentrations. Future research should look at these possibly necessary carotenoids.

KEYWORDS

- **carotenoids rich vegetables**
- **phytochemicals**
- **T-lymphocytes**

REFERENCES

Agarwal, S., & Rao, A. V., (2000). Carotenoids and chronic diseases. *Drug Metab Drug Interact., 17*(1–4), 189–210.

Bendich, A., (1989). Carotenoids, and the immune response. *Journal of Nutrition, 119*, 112–115.

Bendich, A., (1991). Beta-carotene, and the immune response. *Proceedings of the Nutrition Society, 50*, 263–274.

Britton, G., (1995). Structure and properties of carotenoids about function. *FASEB J., 9*, 1551–1558.

Burton, G. W., & Ingold, K. U., (1984). Beta-carotene: An unusual type of lipid antioxidant. *Science (Washington, DC), 224*, 569.

Burton, G. W., & Ingold, K. U., (1984). β-Carotene: An unusual type of lipid antioxidant. *Science, 224*, 569–573.

Burton, G. W., (1989). Antioxidant action of carotenoids. *J. Nutr., 119*, 109–111.

Buttgereit, F., Burmester, G. R., & Brand M. D., (2000). Bioenergetics of immune functions: Fundamental and therapeutic aspects. *Immunol. Today, 21*, 192.

Clevidence, B. A., Khachik, F., Brown, E. D., Nair, P. P., Wiley, E. R., Prior, R. L., Cao, G., et al., (1997). Human consumption of carotenoid-rich vegetables. In: Aruoma, O. I., & Cuppett, S. L., (eds.), *AOCS Antioxidant Methodology in Vivo and in Vitro Concepts* (pp. 53–63). Press.

Daudu, P. A., Kelly, D. S., Taylor, P. C., Burri, B. J., & Wu, M. M., (1994). Effect of a low beta-carotene diet on the immune functions of adult women. *American Journal of Clinical Nutrition, 60*, 969–972.

Dias, M. G., Borge, G. I. A., Kljak, K., Mandić, A. I., Mapelli-Brahm, P., Olmedilla-Alonso, B. O. B., Pintea, A. M., Francisco, R., et al., (2022). European database of carotenoid levels in foods. *Factors Affecting Carotenoid Content. Foods*, 1–31.

Donohoe, E. G., & Kong, F., (2014). Beta-carotene: Digestion, microencapsulation, and *vitro* bioavailability. *Food Bioprocess Tech., 7*, 338–354.

Eliassen, A. H., Hendrickson, S. J., Brinton, L. A., Buring, J. E., Campos, H., Dai, G., et al., (2012). Circulating carotenoids and risk of breast cancer: A pooled analysis of eight prospective studies. *J. Natl. Cancer Inst., 104*, 1905–1916.

Elik, A., Yanık, D. K., & Göğüş, F., (2020). Microwave-assisted extraction of carotenoids from carrot juice processing waste using flaxseed oil as a solvent. *LWT – Food Science and Technology, 123*, 109100.

Frauwirth, K. A., Riley, J. L., Harris, M. H., Parry, R. V., Rathmell, J. C., Plas, D. R., Elstrom, R. L., et al., (2002). The CD28 signaling pathway regulates glucose metabolism. *Immunity, 16*, 769.

Gerster, H., (1997). The potential role of lycopene for human health. *J. Am. Coll. Nutr., 16*, 109–126.

Goad, L. J., & Goodwin, T. W., (1970). In: Hulm¢, A. C., (ed.), *Biochemistry of J' Units and Their Products* (Vol. 1, p. 305). London, Academic Press.

Godoy, H. T., & Rodriguez-Amaya, D. B., (1995). *Arq. Biol. Tecnol., 38*, 109.

Goodwin, T. W., (1980). Fungi. In: Goodwin, T. W., (ed.), *The Biochemistry of the Carotenoids, Plants* (2nd edn, Vol. 1, pp. 257–290). Chapman and Hall, London.

Handelman, G. J., Van, K. F. J. G. M., Chatterjee, A., & Krinsky, N. I., (1991). Characterization of products formed during the autoxidation of β-carotene. *Free Radical Biol. Med., 10*, 427–437.

Hughes, D. A., Wright, A. J. A., Finglas, P. M., Peerless, A. C. J., Bailey, A. L., Astley, S. B., Pinder, A. C., & Southon, S., (1997). Comparison of effects of beta-carotene and lycopene supplementation on the expression of functionally associated molecules on human monocytes. *Biochemical Society Transactions, 25*, 206.

Hume, D. A., Radik, J. L., Ferber, E., & Weidemann, M. J., (1978). Aerobic glycolysis and lymphocyte transformation. *Biochem. J., 174*, 703.

Humphries, J. H., & Khachik, F. J., (2003). *Agric. Food Chem., 51*, 1322.

Johnson, E. J., (2002). The role of carotenoids in human health. *Nutr. Clin. Care, 5*(2), 47–49.

Joyce, A. E., (1954). *Nature, 173*, 311.

Kennedy, T. A., & Liebler, D. C., (1991). Peroxyl radical oxidation of β-carotene: Formation of β-carotene epoxides. *Chem. Res. Toxicol., 4*, 290–295.

Khachik, F., & Beecher, G. R., (1987). *J. Agric. Food Chem., 35*, 732.

Khachik, F., Beecher, G. R., & Lusby, W. R., (1989). *J. Agric. Food Chem., 37*, 1465.

Khachik, F., Beecher, G. R., & Whittaker, N. F., (1986). *J. Agric. Food Chem., 34*, 603.

Khachik, F., Beecher, G. R., Goli, M. B., & Lusby, W. R., (1991). *Pure Appl. Chem., 63*, 71.

Khachik, F., Beecher, G. R., Goli, M. B., & Lusby, W. R., (1992). *Meth. Enzymol., 213*, 347.

Khachik, F., Goli, M. B., Beecher, G. R., Holden, J., Lusby, W. R., Tenorio, M. D., & Barrera, M. R., (1992). *J. Agric. Food Chem., 40*, 390.

Krinsky, N. I., (1989). Antioxidant functions of carotenoids. *Free Radical Biol. Med., 7*, 617–635.

Krinsky, N. I., (1994). The biological properties of carotenoids. *Pure Appl. Chem., 66*, 1003–1010.

Krinsky, N. I., (1998). Carotenoid properties define primary biological actions and metabolism defines secondary biological actions. In: Özben, T., (ed.), *Free Radicals, Oxidative Stress, and Antioxidants: Pathological and Physiological Significance* (pp. 323–332).

Laferriere, L., & Gabelman, W. H., (1968). *Proc. Am. Soc. Hort. Sci., 93*, 408.

Liebler, D. C., & Mcclure, T. D., (1996). Antioxidant reactions of β-carotene: Identification of carotenoid-radical adducts. *Chem. Res. Toxicol., 9*, 8–11.

Mangels, A. R., Holden, J. M., Beecher, G. R., Forman, M. R., & Lanza, E., (1993). Carotenoid contents of fruits and vegetables: An evaluation of analytical data. *J. Am. Diet. Assoc., 93*, 284–296.

Martini, D., Negrini, L., Marino, M., Riso, P., Del Bo, C., & Porrini, M., (2022). What is the current – direction of the research on carotenoids and human health: An overview of registered clinical trials. *Nutrients, 14*, 1191.

Mascio, D. I., Kaiser, P. S., & Sies, H., (1989). Lycopene is the most efficient biological carotenoid singlet oxygen quencher. *Arch. Biochem. Biophys., 274*, 532.

Melendez-Martinez, A. J., Vicario, I. M., & Heredia, F. J., (2004). Stability of carotenoid pigments in foods. *Archivos Latinoamericanos de Nutricion, 54*(2), 209–215.

Meydani, S. N., Santos, M. S., Ribaya-Mercado, J. D., Leka, L., Han, S. N., & Russell, R. M., (1995). Effect of b-carotene on the immune response of elderly women. *FASEB Journal, 9*, 170.

Miller, N. J., Sampson, J., Candeias, L. P., Bramley, P. M., & Rice-Evans, C. A., (1996). Antioxidant activities of carotenes and xanthophylls. *FEBS Lett., 384*, 240–242.

Ong, A. S. H., & Goh, S. H., (2002). *Food Nutr. Bull., 23*, 11.

Ong, A. S. H., & Tee, E. S., (1992). Natural sources of carotenoids from plants and oils. *Methods Enzymol., 213*, 142–167.

Paiva, S., & Russell, R., (1999). Beta carotene and other carotenoids as antioxidants. *J. Am. Coll. Nutr., 18*, 426–433.

Palazzo, P., Luberto, C., Calvillo, G., Ricci, P., & Bartoli, G. M., (1997). Antioxidant and prooxidant role of β-carotene in murine normal and tumor thymocytes: Effects of oxygen partial pressure. *Free Radical Biol. Med., 22*, 1065–1073.

Palozza, P., & Krinsky, N. I., (1992). Antioxidant effects of carotenoids *in vitro* and *in vivo*: An overview. *Methods Enzymol., 213*, 403–420.

Palozza, P., & Krinsky, N. T., (1991). The inhibition of radical-initiated peroxidation of microsomal lipids by both a-tocopherol and /3-carotene. *Free Radical Bio. Med., 11*, 407.

Palozza, P., Calviello, G., & Bartoli, G. M., (1995). Prooxidant activity of β-carotene under100% oxygen pressure in rat liver microsomes. *Free Radical Biol. Med., 19*, 887–892.

Potrykus, (2001). *Plant Physiol., 125*, 157.

Quackenbush, F. W., Frisch, J. G., Brunson, A. M., & House, L. R., (1963). *Cereal Chem., 40*, 250.

Ram, S., Mitra, M., Shah, F., Tirkey, S. R., Mishra, S., Rani, S., & Mishra, S., (2020). Bacteria as an alternate biofactory for carotenoid production: A review of its applications, opportunities, and challenges. *Journal of Functional Foods, 67*(103867), 103867.

Rathmell, J. C., Farkash, E. A., Gao, W., & Thompson, C. B., (2001). IL-7 enhances the survival and maintains the size of naive T cells. *J. Immunol., 167*, 6869.

Rathmell, J. C., Vander, H. M. G., Harris, M. H., Frauwirth, K. A., & Thompson, C. B., (2000). In the absence of extrinsic signals, nutrient utilization by lymphocytes is insufficient to maintain either cell size or viability. *Mol. Cell, 6*, 683.

Rearte, T. A., Figueroa, F. L., Gómez-Serrano, C., Vélez, C. G., Marsili, S., Iorio, A. D. F., González-López, C. V., et al., (2020). Optimization of the production of lipids and carotenoids in the microalga *Golenkinia* aff. *brevispicula*. *Algal Research, 51*, 102004.

Rodriguez-Amaya, D. B., (1993). In: Charalambous, G., (ed.), *Shelf-life Studies of Foods and Beverages: Chemical, Physical and Nutritional Aspects*. Elsevier, Amsterdam.

Rodriguez-Amaya, D. B., (1997). *Carotenoids, and Food Preparation: The Retention of Provitamin a Carotenoids in Prepared Processed and Stored Foods*. OMNI, and Arlington.

Rodriguez-Amaya, D. B., (1999). *A Guide to Carotenoid Analysis in Foods*. ILSI, Washington DC.

Saini, R. K., & Keum, Y. S., (2018). Carotenoid extraction methods: A review of recent developments. *Food Chemistry, 240*, 90–103.

Sarkar, S., Mannan, M. S., Bhowmick, T. K., & Gayen, K., (2020). Extraction of chlorophylls and carotenoids from dry and wet biomass of isolated *Chlorella thermophila*: Optimization of process parameters and modeling by an artificial neural network. *Process Biochemistry, 96*, 58–72.

Shewmaker, C. K., Sheehy, J. A., Daley, M., Colburn, S., & Ke, D. Y., (1999). *Plant J., 20*, 401.

Strong, P., Gradelet, S., Berges, R., & Suschetet, M., (1997). Dietary lycopene decreases initiation of liver preneoplastic foci by diethylnitrosamine in the rat. *Nutr. Cancer, 29*, 60–68.

Thane, C., & Reddy, S., (1997). Processing of fruits and vegetables: Effect on carotenoids. *Nutr. Food Sci., 2,* 58–65.

Tollefsbol, T. O., & Cohen, H. J., (1985). Culture kinetics of glycolytic enzyme induction, glucose utilization, and thymidine incorporation of extended-exposure phytohemagglutinin-stimulated human lymphocytes. *J. Cell Physiol., 122*, 98.

Tonucci, L. H., Holden, J. M., Beecher, G. R., Khachik, F., Davis, C. S., & Mulokozi, G., (1995). *J. Agric. Food Chem., 43*, 579.

Truscott, T. G., (1996). β-Carotene and disease: A suggested pro-oxidant and anti-oxidant mechanism and speculations concerning its role in cigarette smoking. *J. Photochem. Photobiol. B: Biol., 35*, 233–235.

Van, H. H. K. H., De Boer, B. C., Tijburg, L. B., Lucius, B. R., Zijp, I., West, C., et al., (2000). Carotenoid bioavailability in humans from tomatoes processed in different ways determined from the carotenoid response in the triglyceride-rich lipoprotein fraction of plasma after a single consumption and in plasma after four days of consumption. *J. Nutr., 130*, 1189–1196.

Vuong, L. T., (2000). *Food Nutr. Bull., 21*, 173.

Watson, R. R., Prabhala, R. H., Plezia, P. M., & Alberts, D. S., (1991). Effects of beta-carotene on lymphocyte subpopulations in elderly humans: Evidence for a dose-response relationship. *American Journal of Clinical Nutrition, 53*, 90–94.

Weitberg, A. B., Weitzarnn, S. A., Clark, E. P., & Stossel, T. P., (1985). Effects of antioxidants on oxidant induced sister chromatic exchange formation. *J. Clin. Invest., 75*, 1835.

Zhang, R., Lebovka, N., Marchal, L., Vorobiev, E., & Grimm, N., (2020). Multistage aqueous and non-aqueous extraction of bio-molecules from microalga *Phaeodactylum tricornutum. Innovative Food Science & Emerging Technologies, 62*, 102367.

Zhou, W., Niu, Y., Ding, X., Zhao, S., Li, Y., Fan, G., Zhang, S., & Liao, K., (2020). Analysis of carotenoid content and diversity in apricots (*Prunus armeniaca* L.) grown in China. *Food Chemistry, 330,* 127223.

PART II

Nutritional Components to Boost Immunity

CHAPTER 5

PROBIOTICS AND IMMUNE RESPONSE IN DISEASE PREVENTION AND TREATMENT

MEENU JAIN,[1] ANSHU AGARWAL,[2] KAVITA SINGH,[3] and URMILA JAROULIYA[4]

[1]*Gajra Raja Medical College, Gwalior, Madhya Pradesh, India*

[2]*Department of Zoology, Agra College, Dr. B. R. Ambedkar University, Agra, Uttar Pradesh, India*

[3]*National Institute of Immunology, New Delhi, India*

[4]*SOS in Biochemistry, Jiwaji University, Gwalior, Madhya Pradesh, India*

ABSTRACT

Probiotics are beneficial live microorganisms that strengthen the foodstuffs to encourage the health of AIDS by stimulating immune response and acting as therapeutic potential in various diseases. It constructively influences T-lymphocyte count, induces antibody production, affects cytokine storm, encourages function of intestinal epithelial cell (IEC), reduces the physiological stress and maintain metabolic homeostasis. Probiotics not only stimulate nonspecific cellular immune response (innate immunity) through NK cells and macrophages, but also generate specific immune response (adaptive immunity) via lymphocytes (B- and T-cells). Probiotics induces the dendritic cells (DCs) that establish the bridge between innate

Immune-Boosting Nutraceuticals for Better Human Health: Novel Applications. Urmila Jarouliya, PhD, Raj K. Keservani, MPharm, Rajesh K. Kesharwani, PhD, Virendra K. Patel, PhD, and Adi D. Bharti, MPharm (Eds.)

and adaptive immune response. They also regulate the inflammatory processes by balancing between pro-inflammatory and anti-inflammatory cytokine. Immunomodulation has been executed by probiotic organisms in species-specific, strain-specific, single/mixed type (Gram-positive and Gram-negative) as well as in a dose-dependent manner. Probiotics have been accepted as a better preventive and therapeutic option against the variety of diseases that decline the duration/symptoms of the diseases, and also serve as an alternative choice to antibiotics.

5.1 INTRODUCTION

Gastrointestinal tract (GIT) provides a keynote niche for commensal microorganisms enriched by probiotics, that beneficially imitate the nutrition and food digestion. Gut microbiota communicate with the mucosa and maintain the intestinal immunity and metabolic equilibrium (Garrett et al., 2010). Probiotics are defined as "Live microorganisms, which, when consumed in adequate amounts, confer a health benefit on the host" by the Food and Agricultural Organization (FAO) of the World Health Organization (WHO) (FAO/WHO, 2001). Probiotics not only modulate the gut environment through immune regulation but also ameliorate diseases (Figure 5.1). Probiotics persuade the different signaling networks and instigate immune system by bacterial components like cell wall structure (peptidoglycan and lipoteichoic acid) and certain proteins (Bron et al., 2011; Maldonado et al., 2019). Probiotics also secrete some metabolites as short chain fatty acids (SCFA) are recognized to develop immunomodulatory effects (Venegas et al., 2019). These bacterial structure and metabolites bind with Toll-Like Receptors (TLRs) present on epithelial and immune cells and stimulate immune response (Venegas et al., 2019). Intestinal epithelial cells (IECs) layer is influenced by probiotics via various means, interfering pathogenic adherence and viability, regulating cell survival and different signaling pathways, increasing barrier function and mucin secretion, affecting cytokine production, and promoting antimicrobial production (Wehkamp et al., 2004; Schlee et al., 2008; Chen et al., 2006; Mack et al., 1999, 2003; Zyrek et al., 2007; Seth et al., 2008). Clinical health beneficial properties and immunomodulatory effects of the probiotics have attracted the focus of the researchers towards the treatment against the various pathogens, degenerative diseases, diabetes, obesity, allergies, cancers, inflammatory diseases, autoimmune disorders, and neurological diseases. Many studies suggested that numerous molecular mechanisms involving in the prevention and control of the diseases and disorders, including cytokine secretion, IgA production, generation of antibacterial substances, strengthen the intestinal barrier against invasion of the bacteria, denotes competition with pathogens, maintain colonic pH, stimulation of innate and adaptive immune responses (immunomodulation) (Azad et al., 2018). Probiotics provided by fermented milk product, aggravate the IgA and cytokine production in the intestine and convalesce the gut health (Ashraf & Shah, 2014). *Lactobacillus* and *Bifidobacterium* species are the most commonly studied probiotics that participated in the stimulation of immune responses (Azad et al., 2018). *L. casei* Shirota is the most commonly used probiotics because it denotes immunopotential and has several health benefits. In this chapter, we focused on the probiotic's health benefits, prebiotics, immune responses (innate and adaptive), pro- and anti-inflammatory cytokine's effects, mucosal immunity, mucin production, probiotic microorganisms that generate immune responses (non-specific, species specific and strain-specific manner), and preventive effect of probiotics in different diseases (viral, bacterial, cancer, neurological diseases and other metabolic disorder).

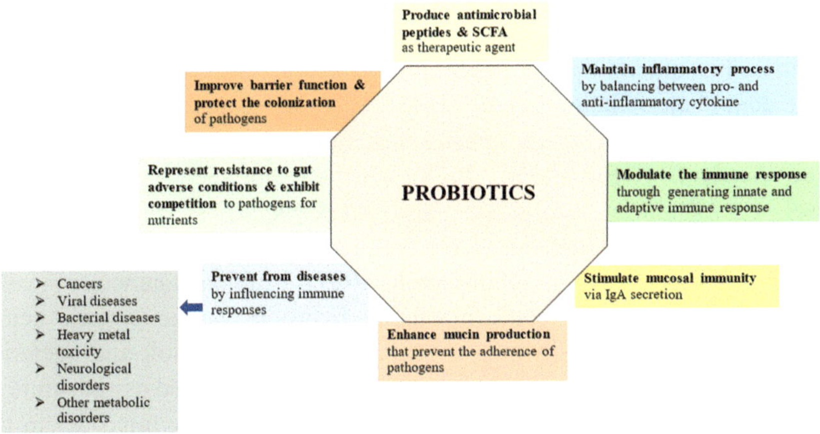

FIGURE 5.1 Probiotics beneficial and immunological properties.

5.2 PROBIOTICS AND THEIR HEALTH BENEFITS

GI tract is the initial site of action for the probiotics (Pot et al., 2013). It is mainly provided by the food supplements (e.g., fermented foods, cheese, yogurt, etc.). The "probiotics" term first proposed by Lilly & Stillwell (1965). It is originated from the Latin word "pro" and the Greek word "bios" means "for life" (Zivkovic, 1999), and it denotes the beneficial live microorganisms supplied as food additives that deliberate the health advantages to the host by maintaining the gut microbiome. Several definitions have been proposed by the other scientists for probiotics (Parker, 1974; Fuller, 1989). The modified definition in context of immunomodulatory effect of probiotics is "microbial dietary adjuvants that beneficially affect the host physiology by modulating mucosal and systemic immunity, as well as improving nutritional and microbial balance in the intestinal tract" (Naidu et al., 1999). FAO/WHO refined all the definitions and summed up as "live microorganisms, which, when consumed in adequate amounts, confer a health benefit on the host" (FAO/WHO, 2001). Probiotics not only contribute positive effect on the health status but also deliberate resistance against various pathogens, including viruses, bacteria, and parasites. Furthermore, there are numerous health benefits of probiotics, including protection from diarrheal and other infectious diseases, decline the allergies and serum cholesterol level, manage blood pressure, reduction in tumors and carcinogen productions, stimulation in immune system and cytokine production (Patel & Denning, 2013).

Numerous probiotic microorganisms already live in the intestinal microenvironment, but some are supplemented by food and improving the gut health and balancing the gut microbiota. The common type of probiotics is lactic acid bacteria belonging to *Lactobacilli*, *Streptococci*, *Bifidobacteria*, *Enterococcus*, *Pediococcus*, *Leuconostoc* and some yeast like *Saccharomyces boulardii* (George et al., 2018), but the major species, *Lactobacillus casei, Lactobacillus acidophilus,* and *Bifidobacterium* spp., thought to contain probiotics features and have well known health benefits (Ashraf & Shah, 2014; Prescott et al., 2008; Shah, 2007). Probiotics generate several non-toxic and non-pathogenic metabolic byproducts, like organic acids, diacetyl, acetaldehydes, hydrogen peroxide (H_2O_2), ethanol, and bacteriocins, these are resilient to enzymes and have been designated as substitute of antibiotics and interference with the growth of pathogenic microbes in the host (Islam, 2016; Ooi et al., 2015). Probiotics also play an important role in reducing oxidative stress (OS) by generating antioxidants (glutathione). *L. acidophilus*, *L. casei*, *L. bulgaricus*, *L. rhamnosus*, *B. longum*, *B. breve*, and *S. thermophiles* have been reported to significantly increase the glutathione (GSH) level in plasma (Asemi et al., 2013). Some strains of *L. acidophilus* (NCDC14 and NCDC19) reduce OS and lipid peroxidation in pancreatic tissues (Yadav et al., 2008). Other strains of *L. delbrueckii* subsp. *bulgaricus*

A13 and subsp. *Bulgaricus* B3 significantly decrease the OS in colitis (Sengul et al., 2011). Probiotic microflora also ferments non-digestible carbohydrates, including oligofructose, fructooligosaccharide, galactose, xylose-containing oligosaccharides, and inulin from natural food resources in the small intestine (Hutkins et al., 2016). *L. rhamnosus*, *L. acidophilus*, and *L. casei* significantly produce antioxidants, predominantly GSH, help in fermenting inulin and reducing OS (Verma et al., 2014; Kleniewska & Pawliczak, 2017).

5.3 PREBIOTICS

The "prebiotic" term was usually known as food ingredients that indicate advantageous effects on the host by stimulating the growth and/or function of beneficial intestinal microorganisms (Slavin, 2013). There are various types of food components recognized as the prebiotics. Different dietary fibers constituted as carbohydrates (polymers of mono-sugars) including starch, non-starch polysaccharides (PS) (celluloses, hemicelluloses, pectins, gums, and mucilages), inulin, and oligosaccharides (fructooligosaccharides, galactooligosaccharides, and xylooligosaccharides) (Aachary & Prapulla, 2011). Prebiotic fibers could encourage the growth of probiotic bacteria by serving as substrate for them and concurrently maintain homeostasis in the gut environment and good health (Pujari & Banerjee, 2021). The dietary fibers are not digested by the host because it does not contain the enzymes that can degrade the carbohydrate fibers (Gloux et al., 2011). These substances are converted into short-chain fatty acids (SCFAs) along with several other metabolites by probiotic microbiome in the intestinal tract (McNeil et al., 1978). SCFAs represent the anti-inflammatory property and are known to be effective in decreasing colitis (Maslowski et al., 2009; Hamer et al., 2010) and other colonic inflammatory diseases (Huda-Faujan et al., 2010). Butyrate is the most studied SCFA, that is responsible for reduction in inflammation by elevating Treg cells and decreasing production of IFN-γ, in an experiment of inflammatory bowel disease (IBD) (Klampfer et al., 2003). Elevated butyrate stimulated the nuclear transcription factor and peroxisome proliferator activator receptor γ (PPARγ), which suppress the gut inflammation in colitis models by reduction in the pro-inflammatory pathways, including STAT, NF-κB and AP-1 pathways (Su et al., 1999; Luhrs et al., 2002; Schwab et al., 2006). Probiotics-derived metabolites, SCFA perform as immuno-modulatory, would be suggested as prophylactic and therapeutic agent in gut inflammation. Thus, prebiotics along with probiotics, preparing synbiotics, have provided a promising new therapeutic approach for disease treatment (Keservani et al., 2010a, b, 2017, 2020).

5.4 PROBIOTICS GENERATE IMMUNE RESPONSES

The gut microbiome performs a significant role in immunomodulation and regulation of inflammation. The commensal bacteria directly interact with the intestinal epithelial layer modulates the epithelial signaling, improves barrier function, shows fight for intestinal colonization with pathogens, stimulates the activity of immune cells, alters the cytokine production and influencing the immune homeostasis (Keservani et al., 2015; Cani, 2017; Karczewski et al., 2010; Sherman et al., 2009; Wells et al., 2010; Sujino et al., 2016). Both innate and adaptive immune responses are generated by the probiotics and protect from external and internal damages (Donkor et al., 2010). The innate immune system works in a non-specific way and doesn't provide long-lasting immunity to the host. Probiotic microbial components like cell wall structures have the capability to express the antigenicity and to cross the IECs and M cells in Peyer's patches (PP) and generate the immune responses (Galdeano & Perdigon, 2004). Probiotics trigger the cytokine and chemokine production from immune cells, that promote the innate and adaptive response (Azad et al., 2018). *L. acidophilus* NCFB 1748 and *L. paracasei* subsp. *Paracasei* DC412 strain provoke early innate immune response via recruitment of the polymorphonuclear (PMN) cells, phagocytosis, and TNF-α production in animal models (Kourelis et al., 2010). Probiotic microbial cell wall structures (LPS and lipoteichoic acids) are recognized by innate pattern recognition receptors (PRRs), including toll like receptors (TLRs), NOD-like receptors (NLRs) and C-type lectin receptors (CLRs) that initiate the cascade of immunological defense mechanisms, such as pro- and anti-inflammatory cytokine and chemokine responses (Meijerink & Wells, 2010; Anderson, 2000). These signals further instigate the effector and regulatory T-cells of adaptive immunity (Meijerink & Wells, 2010). TLRs are majorly expressed on dendritic cells (DCs), macrophages, B-cells and epithelial cells (Pasare & Medzhitov, 2005).

5.5 PROBIOTICS AND CYTOKINE PRODUCTION

Probiotics modulate the innate and adaptive immunity by regulating the production of different cytokine, including interleukins (ILs), interferons

(IFNs), tumor necrosis factors (TNFs), and chemokines from different immune cells as lymphocytes, macrophages, granulocytes, DCs, and mast cells (Savan & Sakai, 2006; Foligne et al., 2010). *Lactobacilli* and *Bifidobacteria* cell wall components, like lipoteichoic acid, stimulate the NO synthase, that induces TNF-α production by macrophages and also upregulates phagocytosis receptors (TLR and FcγRIII) (Schwandner et al., 1999; Delcenserie et al., 2008). Probiotics regulate the inflammatory processes by association with DCs, enterocytes, and Treg cells. TLRs recognize the foreign particles and participated in better immune defensive machinery by influencing the secretion of pro- and anti-inflammatory cytokine (Anderson, 2000). Cytokines trigger the numerous cellular cascades and perform the adaptive immune defense mechanism against different pathogens (viral, bacterial, fungal, etc.). There are two definite kind of probiotics on the basis of immunomodulatory activity, first immunostimulatory probiotics stimulates pro-inflammatory cytokine, such as IL-12 production, that participated in activation of NK and Th1 cells and have the capability in action against the cancer and infections, and also against allergies by creating a balance between Th1 and Th2 response. Second, immunoregulatory probiotics have shown the ability to modulate the production of anti-inflammatory products, IL-10 and Treg cells, which reduce the inflammatory processes, including autoimmune diseases, allergy, and IBD (Chiba et al., 2010). Two *Lactobacillus* spp. shown different inflammatory effect in Caco-2 cells in an *in vitro* experimental study, as *Lactobacillus sakei* stimulated pro-inflammatory cytokines (IL-8, IL-1β, and TNF-α, etc.). In contrast, *Lactobacillus johnsonii* participated to generate anti-inflammatory effect by producing TGF-β (Haller et al., 2000). A combination of *L. paracasei* and *L. reuteri* was given to IL-10 deficient mice infected with *Helicobacter hepaticus*, decreased the progress of colitis by reducing the pro-inflammatory cytokines (Pena et al., 2005). Another mixture of *Lactobacillus* and *Bifidobacterium* spp. probiotics modulated the inflammatory response by downregulating the pro-inflammatory cytokines (IL-6 and TNF-α) and upregulating the anti-inflammatory cytokine (IL-10) in rat model (Karamese et al., 2016). Various strains of *Lactobacillus* and *Bifidobacterium* instigate IL-6, IL-10, and TGF-β expression by epithelial cell, that further induce IgA production (Resendiz-Albor et al., 2010). IL-6 enhances the clonal expansion of B-lymphocytes that produces IgA, and also induces another antibodies production (IgG and IgM) and reduces the IgE secretion (Galdeano et al., 2007). Probiotics in different combination, species specific and strain specific manner influence the immunomodulation of cytokines, which further associated with treatment of inflammation-associated diseases and GIT inflammation.

5.6 PROBIOTICS STIMULATE MUCOSAL IMMUNITY

Mucosal (local) type of immunity acts as a part of adaptive immune system like systemic immunity. It provides first-line-of-defense protecting from the invading microbes and decreases the requirement of systemic immunity (Donkor et al., 2010). The mucus, antimicrobial proteins, and immune cells in combination, preventing the entry of the pathogen into intestinal epithelium. Gut-associated lymphoid tissue (GALT) that is the part of mucosa-associated lymphoid tissue (MALT), is a major site of the GI tract where the commensal microorganisms (probiotics) recruit the DC populations and participate in immunomodulation. GALT comprises multi-follicular lymphoid tissues, like isolated lymphoid follicles (ILFs) dispersed along the length of the intestine, the vermiform appendix, and PP in the ileum (Morbe et al., 2021).

In the intestinal epithelial layer, specialized M cells covering the mucosal lymphoid tissues efficiently transport the antigens from the intestinal lumen to the lamina propria (LP) (Miller et al., 2007). DCs are more effective antigen presenting cells (APCs) and considerably generate T-cell dependent IgA responses (Macpherson & Uhr, 2004). IgA indicates first-line-of-defense in mucosal immunity via IgA1 in the small intestine and IgA2 in the colonic mucosa (Otten & Egmond, 2004). Specific mucosal immunity motivated by live probiotic microorganisms. *Bifidobacterium* and *Lactobacillus* species efficiently provoked the IgA production and participate to raise immunity in infants (Yasui et al., 1992; Majamaa et al., 1995; Xiao et al., 2017). Fermented milk comprising *L. casei* DN 114001 efficiently enhance the gut mucosal immunity in different extends of non-specific, innate, and adaptive response in mice model, and augment the count of macrophages, T-cells, IgA^+ B-cells and goblet cells (from non-specific barrier), induce the intestinal cells to release IL-6 and encourage the peripheral immunoglobulins (Igs) (Galdeano et al., 2009; Senok et al., 2005). Probiotic fermented milk also stimulates the expression of the enzyme calcineurin that activate the transcription factor NFAT (nuclear factor of activated T-cells) (Galdeano et al., 2009). Excitation of NFAT regulates different immune processes, including apoptosis, anergy, T-cell progression, aging of the immune system (Masuda et al., 1998). Activated T-cells perform effector function, and maintain the integrity of the mucosal barrier and gut homeostasis. Secretory IgA shows proteolytic resistance, hinder the penetration and colonization of the pathogens to epithelial cells by immune escape of foreign antigens (Erickson & Hubbard, 2000; Ouwehand et al., 2002). Other *L. casei* CRL 431 strain modulated effectively immune cells in *ex-vivo*, encouraged IL-6 production significantly, and mainly influenced the innate immune system (Dogi et al., 2008). *L. paracasei* DC412 strains strongly interacted with GALT via activation of TLR2/TLR4-mediated signaling leads to secretion of cytokines IFN-γ, TNF-α, IL-6, and IL-10 in animal models and primarily affected immunoregulatory activity (Kourelis et al., 2010). *L. acidophilus*, *L. casei*, and their mixture remarkably influenced the phagocytic and lymphocytic activity in Swiss mice model. Heat-killed and live preparation of *L. rhamnosus* HN001 strain enhance the phagocytic activity in a dose-dependent way (Gill & Rutherfurd, 2001a). Probiotics modulate the host resistance through interaction with immune and IECs, which regulate the intracellular signaling pathways and alter the cytokine secretion (Ezendam et al., 2008). *L. plantarum* Lp6 administration in mice, increased different gene expression involving in immune responses, cell signaling, cell differentiation, cell adhesion, and signal transcription and transduction (Chang et al., 2009). *L. acidophilus* treatment encouraged the overexpression of several genes participated in pro-inflammatory signaling pathways, such as IL-18, MyD88, STAT2, STAT4, IFNα, and IFNγ (Brisbin et al., 2008).

5.7 PROBIOTICS INDUCE MUCIN PRODUCTION

Mucins are glycosylated proteins comprising tandem repeats of threonine, proline, and serine residues, where O-glycosidic linkages present between the protein core and *N*-acetylgalactosamine (GalNAc) termini of oligosaccharides (Hanisch, 2001; Dekker et al., 2002). These are produced by intestinal epithelial layer and characterize viscoelastic physical properties. They may be membrane-bound (MUC1, MUC3A, MUC3B, MUC4, MUC12, MUC13, and MUC17) involving in signaling pathways or secreted (MUC2, MUC5AC, MUC5B, and MUC6) type that polymerize to the matrix for preparation of the mucus layer (Cornick et al., 2015). The mucus layer protects the host from toxins, pathogens, dehydration, and other foreign activities (Laboisse et al., 1996). The composition of the microbiome is the major driver of mucus alterations. Moreover, our diet contents, including high/low fibrous diet, food additives, prebiotics, and probiotics also affect the mucus variability. Some probiotics, *L. plantarum* 299v and *L. rhamnosus* GG actively participated in high expression of mucin production, MUC2 and MUC3 by IECs, which prevent the adherence of pathogens (Mack et al., 1999). These probiotics not only increase the transcription and translation of MUC3 gene but also induce extracellular secretion of MUC3 mucin, leads to reduce enteropathogenic *E. coli* E2348/69 adherence (Mack et al., 2003). In another experiment, *L. casei* GG increases the MUC2 production in human enterocyte Caco-2 cells that inhibit the bacterial translocation (Mattar et al., 2002). Additionally, *L. acidophilus* A4 provoked MUC2 expression in HT29 cells (Kim et al., 2008). MUC2 is the major constituent of the intestinal mucus. It is actively participating in the formation of mucus skeleton and maintaining gut homoeostasis and oral tolerance (Paone & Cani, 2020).

5.8 PROBIOTIC MICROORGANISMS PARTICIPATE IN THE IMMUNOMODULATION MECHANISM

Probiotic microorganisms not only maintain gut microflora but also induce immune responses and promote the intestinal immunological barrier (Isolauri et al., 1993; Kaila et al., 1992). They compete for nutrients for their growth and proliferation, and for niche to the pathogens (Wilson & Perini, 1988). Probiotic bacteria produce nonspecific as well as specific host's immune response (Isolauri et al., 2001). They can generate the immune responses in both species and strain-specific manner (Meijerink & Wells, 2010; Hill et al., 2014). Immune response is also influenced by altering probiotic microbiota composition and activity of the host microbiota (Bermudez-Brito et al., 2012). The large intestine has a large number of endogenous microbiota than the small intestine, where probiotic consumption modifies the microbiome status and effects the host immune responses (Troost et al., 2008; van Baarlen et al., 2013). The small intestine supremely shows an imperative role in immune stimulation due to thin mucosal layer. The gram-positive bacteria are predominantly identified probiotic microorganisms. Their cell wall structures, such as lipoteichoic acid stimulate monocytes and macrophages and induce the TNF-α and IL-10 production in a dose-dependent fashion (Dogi et al., 2010). Different type of gram-positive bacteria excites differentially monocyte-derived DCs. *Streptococcus pyrogenes* induces high expression of CD80, CD83, and CD86, costimulatory molecules present on DC and generate Th1 response. Although, *L. rhamnosus* affects the expression of costimulatory molecules present on DC at moderate level and lower the cytokine and chemokine levels (Veckman et al., 2004). As differential bacterial species effect, various cell wall structures affect different levels of cytokine response. LPS was more efficiently generate cytokine response than peptidoglycan, and peptidoglycan was more proficient than lipoteichoic acids (Erickson & Hub-Bard, 2000; Hessle et al., 2005). Bacterial metabolites like SCFAs also recognized for immune stimulation (Venegas et al., 2019).

5.8.1 NONSPECIFIC IMMUNE RESPONSE

Lactobacilli can provoke nonspecific immune response against various pathogens. Their several strains exhibited non-specific immunity by enhancing pro-inflammatory cytokine levels (TNF-α and IL-6) (Miettinen et al., 1996). *L. casei*, *L. bulgaricus*, and *L. acidophilus* stimulate phagocytosis by activating macrophages (Perdigon et al., 1986, 1998; Schiffrin et al., 1994). Before antibody production, initially phagocytosis is accountable for the generation of the inflammatory response by accommodating the immunocompetent cells. Proinflammatory cytokines such as IL-1, IFN, and TNF significantly influence the inflammation status. Fermented milk products that contain *Lactobacilli* was denoted to induce the IFN production by PBMCs (De Simone et al., 1986; Halpern et al., 1986), that encourages the intake of antigens in PP (Sutas et al., 1997). Probiotics commensals provide protective effect by maintaining the immunologic barrier of gut mucosa through IFN generation and lowering proinflammatory cytokines (TNF). Integrity of the immunologic barrier may be disrupted by deviant IFN production (Madara & Stafford, 1989; Zhang & Michael, 1990).

5.8.2 SPECIFIC IMMUNE RESPONSE

The most known probiotic cultures, *Lactobacillus casei* Shirota, *Lactobacillus rhamnosus* GG, *Lactobacillus johnsonii* La1, *Bifidobacterium lactis* DR10, *Bifidobacterium animalis* Bb-12 and *Saccharomyces cerevisiaeboulardii*, participated in modulation of immune system (Ashraf & Shah, 2014). Probiotics can specifically modulate the host's immune responses against potential antigens. *Bifidobacterium bifidum* has been responsible to activate antibody response for ovalbumin (Moreau et al., 1990) and *Bifidobacterium breve* was indicated to generate IgA response against cholera toxin in mice model (Yasui et al., 1992). *L. rhamnosus* GG participated to enhance humoral immune response during acute rotavirus infection, that generated specific IgA antibody for rotavirus (Kaila et al., 1992; Majamaa et al., 1995). Lactobacilli oral administration to suckling rats, who were sensitized with cow milk, was responsible to the enrichment of the β-lactoglobulin secreting antibodies (Isolauri et al., 1993). Probiotic microflora participates in the processing of food antigens into low-molecular-weight peptides for activation of immune responses. It can also change the structure of potentially harmful antigens for alteration in immunogenicity. In healthy individuals, probiotic generated proteases can degrade cow milk casein into peptides to suppress lymphocyte proliferation (Sutas et al., 1996a). In another study, probiotics degraded casein regulates the cytokine production and anti-CD3 antibody-induced peripheral blood mononuclear cells in atopic infants who have cow milk allergy (Sutas et al., 1996b). *L. rhamnosus* strain GG and *L. plantarum* have the ability to prevent adherence of enteropathogenic *E. coli* to intestinal epithelial layer (Wilson & Perini, 1988).

5.8.3 STRAIN SPECIFIC MANNER IMMUNE RESPONSE

Probiotic microorganisms provoke the cytokine's secretion in a strain-specific way from intestinal immune cells and influence the immune response. *Lactobacillus plantarum* strains, BFE 5759 and BRE 1685, and *L. johnsonii* strains, BFE 6128, stimulate secretion of IL-8 from HT29 intestinal cells in an *ex-vivo* study (West et al., 2009). *L. paracasei* subsp. *paracasei* strain DC412 induced IFN-γ, TNF-α, IL-6, and IL-10 production (Kourelis et al., 2010). The same cytokines with IL-12β were stimulated by *L. acidophilus* NCFB 1748. These responses were possibly facilitated by TLR2 and TLR9. *L. plantarum* 299V strain was responsible for the production of high level of IL-12. Although, the strains of *Streptococcus thermophilus* and *Leuconostoc* were strongly influenced the production of Thl type cytokine's level, IL-12 and IFN-γ, as compared to *Lactobacillus* strains (Kekkonen et al., 2008). Similarly, *B. adolescentis* and *B. longum* were more potentially induced the pro-inflammatory cytokine (IL-12 and TNF-α) secretion than the other *Bifidobacterium spp.*, *B. bifidum*, *B. breve*, and *B. infantis* (He et al., 2002). IL-12 stimulates cell mediated immunity by activating NK cells, cytotoxic T-cells and Th1 cytokine response (IFN-γ and IL-2). In an *in-vitro* study, *L. casei* Shirota strain was shown to induce secretion of IL-12 and IFN-γ by X-ray-irradiated murine splenocytes (Kato et al., 1999). In another study, the same strain was orally intake by mice induced Th1 cytokine production (IL-12) and inhibited the incidence of tumors and IgE production (Shida et al., 1998; Yasui et al., 1999). In contrast, Th2 cytokine (IL-4, IL-5, IL-6, and IL-10) response was lowered in splenocytes (Shida et al., 1998; Yasui et al., 1999). The strains of *Streptococcus thermophilus* and *Leuconostoc mesenteroides* subsp. *cremoris* (DSM18892) were the best inducers of Thl type cytokine (IFN-γ and IL-12) response (Kekkonen et al., 2008). *L. acidophilus* TMC 0356 significantly increased the amount of IL-6, IL-10, IL-12, and TNF-α (Morita et al., 2002). However, *L. paracasei* strain NCC2461 and

L. casei strain GG stimulated the largest IL-10 production (Von der Weid et al., 2001). Although, *L. paracasei* strain NCC2461 induced the high level of both mutually antagonistic cytokines IL-10 and IL-12 in splenocytes of BALB/c mice (Von der Weid et al., 2001). *Propionibacterium* and *Bifidobacterium* strains also provoked high levels of IL-10 as compared to other studied bacteria (Kekkonen et al., 2008). The strains of *L. rhamnosus* (E509 and E522) and *L. delbrueckii* subsp. *bulgaricus* E585 have been exhibited to encourage Th1 type cytokine response IL-12, IL-18, and IFN-γ in human PBMC (Miettinen et al., 1998).

5.9 PROBIOTICS GENERATE THE IMMUNE RESPONSE DURING THE DISEASES

Probiotics have been shown to be diverse applications in the fields of livestock, poultry, aquaculture, animals, and in humans for the prevention and management of various diseases and disorders, either infectious (e.g., viral, bacterial, fungal, and parasitic diseases) and non-infectious diseases (e.g., cancer, allergies, obesities, metabolic disorders and neurological disabilities), and also managing preoperative and postoperative circumstances (Raheem et al., 2021). During recent years, specific probiotic cultures, *Lactobacillus, Bacillus, Enterococcus, Pediococcus, Streptococcus, Aspergillus*, and *Saccharomyces*, etc., have been used in the poultry and livestock industries (Dhama et al., 2011). Numerous probiotic's species such as *Lactococcus, Lactobacillus*, *Bifidobacterium, Propionibacterium, Pediococcus, Leuconostoc, Enterococcus, Vagococcus, Bacillus*, and some yeasts are known to be used against infectious diseases, to beneficial for human health and to encourage the aquaculture production (Hoseinifar et al., 2018; Zorriehzahra et al., 2016). Probiotics not only improve human health but also protect them from different diseases by showing several properties, like immune stimulation ability, produce adhesive surface molecules (e.g., glyceraldehyde-3-phosphate, enolases, and pyruvate dehydrogenase), to colonize the gut epithelium, protect the adherence of bad pathogens to intestinal lining, produce antimicrobial peptides (AMPs) against the pathogens, illustrates antipathogenic capability, show resistance to gut adverse conditions (pH and bile salts), help to ferment the indigestible foods, produce vitamins (K and B), antioxidant function, cytotoxic effect on cancer cells, and maintain the gut homeostasis (Sagheddu et al., 2019; Thakur et al., 2016; Mazloom et al., 2019; Fukuda, 2017; Kline et al., 2009).

5.9.1 PROBIOTICS INDUCE IMMUNE RESPONSE DURING CANCERS

The probiotics have shown its significant role in treating and preventing cancers. Many *in vitro* studies have been documented and stated that probiotics have beneficial properties in regulating proliferation and apoptosis of cancer cells. For example, it has been demonstrated that *Lactobacillus rhamnosus* GG strain inhibits proliferation and induces apoptosis in colon cancer cell lines of mouse (HGC-27), and of human (Caco-2, DLD-1, and HT-29), also lowering the level of IL-8 (Lu et al., 2021). The patients who suffer from cirrhosis, generally they are marked by leaky gut and dysbiosis reflected by the increased occurrence of gut-derived bacterial infections such as spontaneous bacterial peritonitis. Also, increased bacterial translocation and dysbiosis promote inflammation, fibrogenesis, and the progression to cirrhosis and HCC (Schwabe & Greten, 2020). In the liver, inflammatory responses mediated through interaction between microbe- or pathogen-associated molecular patterns (MAMPs) and PRRs like TLRs (Takeuchi & Akira, 2010). TLR4 presented by hepatic cells, including hepatic stellate cells (HSCs), hepatocytes, Kupffer cells, and endothelial cells (ECs), and is associated with progression of fibrogenesis and hepatocarcinogenesis (Dapito et al., 2012). TLR4 stimulation activates nuclear factor kappa-B (NF-kB) that leads to upregulation of the epidermal growth factor, hepatomitogen epiregulin (EPR), which is shown to be a potent mitogenic properties on hepatocytes (Dapito et al., 2012; Toyada et al., 1995). Further, the probiotic yeast; *Saccharomyces boulardii* is one of the best probiotics for treating dysbiosis, and has demonstrated its anti-microbial effects in many clinical trials (Terciolo et al., 2019). In colorectal cancer (CRC), probiotics are helpful in the prevention of cancer and provide support during treatment. These bacteria exert their role via modifying the intestinal microbiota, improvement of colonic physicochemical conditions, and production of anti-cancerogenic and antioxidant metabolites against carcinogenesis, a decrease in intestinal inflammation, and reduces the production of harmful enzymes. In addition, *Lactobacillus casei* Shirota can induce the production of IL-12 (interleukin 12) through phagocytes, which decreases the activity of the fecal enzymes, and protects against mutagens from food, which strengthen the immune defense of the host. Also, it has an inhibitory role in the progression of CRC and bladder cancer (Molska & Reguła, 2019). Moreover, another study suggested that combined treatment with IL-2 and *Akkermansia muciniphila* (AKK) showed a stronger antitumor efficacy in CRC tissue through activating toll-like receptor 2 (TLR2) signaling pathway (Shi et al., 2020). Several studies show that heavy metals (HMs) exposure is a great concern due to its potency to cause various lethal diseases including cancer. The progression of HMs induced diseases is partly due to perturbations of the gut microbiota, which are the first lines of defense against the HMs toxicity. Thus, HM exposure alters the composition and metabolic profile of the gut microbiota at the functional level, and in turn, the gut microbiota alters the uptake and metabolism of HMs by acting as a physical barrier to HM absorption and by altering the pH, oxidative balance, and concentrations of detoxification enzymes or proteins involved in HM metabolism. Probiotics have been shown to reduce the absorption of HMs in the intestinal tract via the enhancement of intestinal HM sequestration, detoxification of HMs in the gut, changing the expression of metal transporter proteins, and maintaining the gut barrier function (Duan et al., 2020).

5.9.2 PROBIOTICS AND IMMUNE RESPONSE IN VIRAL INFECTIONS

Probiotic microorganisms are aided to lower the symptoms of viral infections, to inhibit the viral growth and to prevent adhesion of viral particles to the cell surface by performing the competition (Khani et al., 2012). When virus entered in the body, stimulates innate immunity through enhancing the phagocytic function, the microbicidal activity of neutrophils and leukocytes, and regulating the action of inflammasomes those destroy the pathogens (Sundararaman et al., 2020; Antushevich, 2020). In addition, there are two types of first-line of defense (innate immunity) engendered against the viral infection by probiotics, first, the production of type I IFNs generates an antiviral effect, and second, production of cytokines, including ILs, promote inflammatory processes and immune responses against viral effect (Antushevich, 2020). Secreting metabolites of probiotics, mainly SCFAs participated in the reduction of pattern recognition receptor (PRR) induction via activation of NF-kB and TNFα (Zolnikova et al., 2018). Most of the studies articulate that the probiotics can enhance the CD4+ T-cells count and regulate the production of IL-6, IL-8, IL-10, IL-12, and TNFα (Ou et al., 2019). Probiotics modulate the innate immune response via PRRs, especially toll like receptors (TLRs), retinoic acid-inducible gene I (RIG-I)-like receptors (RLRs), nucleotide-binding oligomerization domain (NOD)-like receptors (NLRs), CLRs and other signaling pathways, and regulating the inflammatory process (Dhar & Mohanty, 2020; Newton et al., 2016). TLRs recognition leads to the activation of NF-κB transcription factor and IFN regulatory factors (IRFs), including IRF 3, IRF 5, and IRF 7, inspiring the production of pro-inflammatory cytokines and type I IFNs, IFN-α and IFN-β (Kennedy et al., 2012; Kreijtz et al., 2011; Totura et al., 2012). During the viral infection, type III IFN, IFN-λ are further secreted by epithelial cells. RLRs recognize RNA structures, like ssRNA and melanoma differentiation associated 5 (MDA-5) dsRNA, resulting production of type I IFN and type III IFN by epithelial cells through mitochondrial antiviral signaling protein (MAVS). Type I and III IFNs participate in activating the antiviral effect and in delaying the spread of infection (Kennedy et al., 2012; Kuiken et al., 2012). The stimulation of these receptors (TLRs and RLRs) by viruses, leads to further production of NOD-, pyrin domain-containing protein 3 (NLRP3)-, and leucine-rich repeat (LRR) inflammasome (Lupfer et al., 2015). NLRP3 inflammasome modulates IL-18 cytokine and caspase 1-dependent IL-1β expression and inflammatory programed cell death (pyroptosis) (Swanson et al., 2019).

Probiotics have been demonstrated the preventive effect and lower the disease symptoms against the common viruses, specifically respiratory tract viruses, digestive infections viruses including rotavirus, coronavirus, or norovirus, and other viruses such as herpes simplex virus (HSV), human immunodeficiency virus (HIV), hepatitis virus, human papillomavirus (HPV) (Lopez-Santamarina et al., 2021). *Lactobacilli* specific strains modulate both innate and acquired immune responses, protect from respiratory tract infections, and inhibit flu-like respiratory viruses (Lehtoranta et al., 2014; Kawase et al., 2010). *L. plantarum* DK119 generated innate immune response against influenza virus through activation of macrophages and DCs and inactivate influenza A viruses (H1N1 and H3N2) in mouse models (Park et al., 2013). Gut microbiota composition are affected during respiratory infections. In mice study, infection of influenza virus influenced the gut microbiome by enhancing the *Enterobacteriaceae,* and reducing the *Lactobacillus* and *Lactococcus* species count (Looft & Allen, 2012). In addition, antibiotic treatment eliminates a few good bacterial species in the intestine and leads to enhance the risk of influenza infection (Looft & Allen, 2012). Probiotics have represented the preventive role in management of COVID-19 (Stavropoulou & Bezirtzoglou, 2020; Baud et al., 2020). HSP27-inducible polyphosphate (poly P) is represented by the *Lactobacillus* species, that provide the strength to epithelial barrier and maintain intestinal homeostasis via integrin-p38 MAPK pathway (Segawa et al., 2011). The p38 MAPK inhibitor increased the survival rate in SARS-CoV infected mice (Lau et al., 2007). *Lactobacillus* species provide protection from enteric viruses, inducing production of antibodies (IgA) and cytokine, depleting reactive oxygen species (ROS), enhancing mucosal secretion, stimulating IFN production (Albarracin et al., 2017; Vitetta et al., 2018). *Lactobacillus* spp., *Enterococcus*, and yeast (*Saccharomyces*) were advocated the disease symptoms during other viral infections, including HCV, HIV, HPV, and HSV type I (Brun et al., 2017; Ou et al., 2016; Palma et al., 2018; Scheri et al., 2017; Abdolalipour et al., 2020; Khani et al., 2012). Preclinical studies of the probiotics suggested that *L. rhamnosus*, *L. acidophilus*, and *B. lactis* have the ability to induce antiviral immune responses and associated with lowering the risk of respiratory tract infections (RTIs) (Lehtoranta et al., 2020).

5.9.3 PROBIOTICS PERSUADES IMMUNE RESPONSE AGAINST BACTERIAL INFECTIONS

Probiotics not only characterize the preventive effect against the bacterial pathogens but also provide a better alternative option for antibiotics. Probiotics extensively surface the intestinal epithelial lining and interact with DCs, that work as a linkage between the innate and adaptive immune responses. Innate immune response is generated by pathogen-associated molecular patterns (PAMPs), presented on the majority of bacteria (Bermudez-Brito et al., 2012). PAMPs interact with PRRs and starts further immune cell signaling. Probiotics are participated to control the DCs function and polarization (Christensen et al., 2002), and activate the appropriate adaptive immune response (T_H cell to Treg cell differentiation) (Fink et al., 2007), which usually prevents Th1-, Th2-, and Th17-mediated inflammatory response, and lower the TLRs expression by inhibiting NF-κB signaling (Gómez-Llorente et al., 2010; Rhayat et al., 2019). Probiotics also modulate the different cytokine production during the bacterial infection. As immunomodulatory properties of probiotics, during the extreme inflammatory conditions, like IBD, and autoimmune disorders, are resulted in decreased production of Treg cells and IL-10 (Azad et al., 2018; Chiba et al., 2010). *Lactobacillus fermentum* and *L. gasseri* inhibit the aggregation of *Campylobacter jejuni* and *Helicobacter pylori* by expressing collagen-binding protein genes and sortase A gene, respectively (Nishiyama et al., 2015; Zuo et al., 2019). *L. plantarum* WCFS1 defend disruption of tight junction by toxins, pathogens, and cytokines through enhancing occludin and ZO-1 in tight junction via TLR2 dependent pathway (Karczewski et al., 2010).

5.9.4 PROBIOTICS INFLUENCE IMMUNE RESPONSE DURING NEURODEGENERATIVE DISEASES

The compositional changes in the gut microbiota have been strongly associated with the aging process and thus with neurodegenerative disorders such as Alzheimer's disease (AD), Parkinson's disease (PD), autism spectrum disorder (ASD) and Huntington's disease (Suganya & Koo, 2020). The susceptibility of gut microbial alterations is high at the two extremes of life span which are very crucial for the onset of neurodegeneration in old age. Ageing is also resulted into alterations in the composition of microbiome and gradual decrease in gut microbial diversity as the age progress (Claesson et al., 2011). There is an increase in harmful bacterial spp. (*Proteobacteria, Enterobacteriace, Escherichia spp. Clostridium spp. Bacteroids* spp.) known as pathobionts with simultaneous decrease in beneficial bacteria like *Bifidobacteria spp., Ruminococcus spp*. Recently some of interesting reports demonstrated that Probiotic species *Lactobacillus* and *Bifidobacterium* spp. are helpful in reducing the cytokine upregulation and OS related with the neurodegeneration in PD patients (Magistrelli et al., 2019). The upregulated expression of TLR4 and pro-inflammatory mediators such as IL-6, TNF-α, and IL1β is significantly associated with the neurodegenerative diseases. During the aging, inflammation is associated with numerous factors such as amplified polymorphic heterogeneity in promoter of pro-inflammatory cytokines genes, hyper-activated innate immune response (activated microglia and astrocytes in brain), alterations in gut microbiome physiology and ultimately the augmented permeability through the intestinal barrier. The main features of immunosenscence at the time of aging is thymus atrophy, decrease in immune cells like neutrophils and natural killer (NK) cells and reduced antibody (B-cell) production against the antigen. Hence, probiotic therapy or combination of pro, pre, and or synbiotics can be the most effective intervention of preventing measure in young or middle-aged individuals if gut microbiome alteration proved as the causative factor of neurodegenerative diseases (Wang et al., 2016).

5.9.5 PROBIOTICS GENERATE IMMUNITY DURING THE OTHER DISEASES

Probiotics also stimulate the immune response during the other diseases or metabolic disorders, and help to prevent the disease. *Lactobacillus fermentum*

NCIMB5221 shows an anti-proliferative effect, which is potentially able to modulate hyperinsulinemia, insulin resistance, hypercholesterolemia, and hypertriglyceridemia. *Weissella cibaria* JW15 strengthens the function of the immune system by increasing the activity of NK cells (natural killer cells). *Saccharomyces cerevisiae var. boulardii* has anti-inflammatory and antibacterial effects. It increases the secretion of immunoglobulin A (IgA) and maintains the integrity of the epithelial barrier. It helps in the treatment of travelers' diarrhea (Molska & Reguła, 2019). In addition, *Lactobacillus acidophilus, L. plantarum, Bifidobacterium lactis* and *Saccharomyces boulardii* significantly decreased the risk of postoperative complications, namely mechanical ventilation, infections, and anastomotic leakage. Also, reduced the rate of postoperative pneumonia and of surgical site infections (Kotzampassi et al., 2015).

5.10 CONCLUSION AND FUTURE DIRECTION

Probiotics are indispensable elements of our life which exhibit several clinical applications as well as beneficial effects on human health. Prebiotics in addition to probiotics synergistically enhance the health benefits to the host. Probiotics have been reported to regulate the various immune signaling pathways, including NF-κB, PI3K/Akt, MAPKs, and transcriptional regulators, which modulate the immune responses as well as inflammatory cytokine responses. By inducing multiple immune pathways, probiotic commensal bacteria generate significant effect on systemic and mucosal immunity and maintain gut homeostasis. These applications flourish the importance of advanced cell or gene therapy-based approaches for the development of genetically engineered probiotics for better human health. Preventive and therapeutic ability of the probiotics against the diseases aggravate the curiosity to be explored in other human diseases and also at clinical trial levels in the future.

KEYWORDS

- **adaptive immunity**
- **cytokine**
- **innate immunity**
- **mucosal immunity**
- **probiotics**

REFERENCES

Aachary, A. A., & Prapulla, S. G., (2011). Xylooligosaccharides (XOS) as an emerging prebiotic, microbial synthesis, utilization, structural characterization, bioactive properties, and applications. *Comprehensive Reviews in Food Science and Food Safety, 10*, 2–16.

Abdolalipour, E., Mahooti, M., Salehzadeh, A., Torabi, A., Mohebbi, S. R., Gorji, A., & Ghaemi, A., (2020). Evaluation of the antitumor immune responses of probiotic *Bifidobacterium bifidum* in human papillomavirus-induced tumor model. *Microb. Pathog., 145*, 104207.

Albarracin, L., Kobayashi, H., Iida, H., Sato, N., Nochi, T., Aso, H., Salva, S., Alvarez, S., Kitazawa, H., & Villena, J., (2017). Transcriptomic analysis of the innate antiviral immune response in porcine intestinal epithelial cells, Influence of immunobiotic lactobacilli. *Front. Immunol., 8*, 57.

Ambalam, P., Raman, M., Purama, R. K., & Doble, M., (2016). Probiotics, prebiotics and colorectal cancer prevention. *Best Pract. Res. Clin. Gastroenterol., 30*(1), 119–131.

Anderson, K. V., (2000). Toll signaling pathways in the innate immune response. *Curr. Opin. Immunol., 12*, 13–19.

Antushevich, H., (2020). Interplays between inflammasomes and viruses, bacteria (pathogenic and probiotic), yeasts and parasites. *Immunol. Lett., 228*, 1–14.

Arun, K. B., Madhavan, A., Sindhu, R., Emmanual, S., Binod, P., Pugazhendhi, A., Sirohi, R., et al., (2021). Probiotics and gut microbiome – Prospects and challenges in remediating heavy metal toxicity. *Journal of Hazardous Materials, 420*, 126676.

Asemi, Z., Zare, Z., Shakeri, H., Sabihi, S. S., & Esmaillzadeh, A., (2013). Effect of multispecies probiotic supplements on metabolic profiles, hs-CRP, and oxidative stress in patients with type 2 diabetes. *Annals of Nutrition and Metabolism, 63*(1, 2), 1–9.

Ashraf, R., & Shah, N. P., (2014). Immune system stimulation by probiotic microorganisms. *Crit. Rev. Food. Sci. Nutr., 54*(7), 938–956.

Azad, M. A. K., Sarker, M., & Wan, D., (2018). Immunomodulatory effects of probiotics on cytokine profiles. *Biomed. Res. Int.*, 8063647.

Bashir, A., Miskeen, A. Y., Bhat, A., Fazili, K. M., & Ganai, B. A., (2015). *Fusobacterium nucleatum*, an emerging bug in colorectal 20 tumorigenesis. *Eur. J. Cancer Prev., 24*, 373–385.

Baud, D., Dimopoulou, A. V., Gibson, G. R., Reid, G., & Gionnoni, E., (2020). Using probiotics to attend the curve of coronavirus disease COVID-2019 pandemic. *Front Pub. Health., 8*, 186.

Bermudez-Brito, M., Plaza-Díaz, J., Muñoz-Quezada, S., Gómez-Llorente, C., & Gil, A., (2012). Probiotic mechanisms of action. *Ann. Nutr. Metab., 61*, 160–174.

Brisbin, J. T., Zhou, H., Gong, J., Sabour, P., Akbari, M. R., Haghighi, H. R., et al., (2008). Gene expression profiling of chicken lymphoid cells after treatment with *Lactobacillus acidophilus* cellular components. *Dev. Comp. Immunol., 32*, 563–574.

Bron, P. A., Van, B. P., & Kleerebezem, M., (2011). Emerging molecular insights into the interaction between probiotics and the host intestinal mucosa. *Nature Rev. Microbiol., 10*, 66–78.

Brun, P., Scarpa, M., Marchiori, C., Sarasin, G., Caputi, V., Porzionato, A., Giron, M. C., Palù, G., & Castagliuolo, I., (2017). *Saccharomyces boulardii* CNCM I-745 supplementation reduces gastrointestinal dysfunction in an animal model of IBS. *PLoS One, 12*, e0181863.

Cani, P. D., (2017). Gut cell metabolism shapes the microbiome. *Science, 357*(6351), 548, 549.

Castelli, V., D'ANGELO, M., Quintiliani, M., Benedetti, E., Cifone, M. G., & Cimini, A., (2021). The emerging role of probiotics in neurodegenerative diseases: New hope for Parkinson's disease?. *Neural Regeneration Research, 16*(4), 628–634.

Chang, G., Shi, Y., Le, G., Xu, Z., Sun, J., & Li, J., (2009). Effects of *Lactobacillus* plant arum on genes expression pattern in mice jejunal Peyer's patches. *Cell Immunol.*, 2581–2588.

Chen, X., Kokkotou, E. G., Mustafa, N., Bhaskar, K. R., Sougioultzis, S., O'Brien, M., et al., (2006). *Saccharomyces boulardii* inhibits ERK1/2 mitogen-activated protein kinase activation both *in vitro* and *in vivo* and protects against *Clostridium difficile* toxin A-induced enteritis. *J. Biol. Chem., 281*, 24449–24454.

Chiba, Y., Shida, K., Nagata, S., Wada, M., Bian, L., Wang, C., et al., (2010). Well-controlled proinflammatory cytokine responses of Peyer's patch cells to probiotic *Lactobacillus casei. Immunology, 130*, 352–362.

Christensen, H. R., Frøkiær, H., & Pestka, J. J., (2002). Lactobacilli differentially modulate expression of cytokines and maturation surface markers in murine dendritic cells. *J. Immunol., 168*, 171.

Claesson, M. J., Cusack, S., O'Sullivan, O., Greene-Diniz, R., De Weerd, H., Flannery, E., Marchesi, J. R., et al., (2011). Composition, variability, and temporal stability of the intestinal microbiota of the elderly. *Proc. Natl. Acad. Sci. USA., 15*(108), 4586–4591.

Cornick, S., Tawiah, A., & Chadee, K., (2015). Roles and regulation of the mucus barrier in the gut. *Tissue Barriers., 3*(1–2),e982426.

Dapito, D. H., Mencin, A., Gwak, G. Y., Pradere, J. P., Jang, M. K., Mederacke, I., Caviglia, J. M., et al., (2012). Promotion of hepatocellular carcinoma by the intestinal microbiota and TLR4. *Cancer Cell, 21*, 504–516.

De Simone, C., Salvadori, B. B., Negri, R., Ferrazzi, M., Baldinelli, L., & Vesely, R., (1986). The adjuvant effect of yogurt on production of gamma – interferon by ConA-stimulated human peripheral blood lymphocytes. *Nutr. Rep. Int., 33*, 419–433.

Dekker, J., Rossen, J. W., Büller, H. A., & Einerhand, A. W., (2002). The MUC family, an obituary. *Trends Biochem. Sci., 27*(3), 126–131.

Delcenserie, V., Martel, D., Lamoureux, M., Amiot, J., Boutin, Y., & Roy, D., (2008). Immunomodulatory effects of probiotics in the intestinal tract. *Curr. Issues Mol. Biol., 10*(1, 2), 37–54.

Dhama, K., Verma, V., Sawant, P., Tiwari, R., Vaid, R., & Chauhan, R., (2011). Applications of probiotics in poultry, enhancing immunity and beneficial effects on production performances and health: A review. *J. Immunol. Immunopathol., 13*, 1–19.

Dhar, D., & Mohanty, A., (2020). Gut microbiota and COVID-19–possible link and implications. *Virus Res., 285*, 198018.

Dogi, C. A., Galdeano, C. M., & Perdigon, G., (2008). Gut immune stimulation by non-pathogenic gram(+) and gram(−) bacteria. Comparison with a probiotic strain. *Cytokine., 41*, 223–231.

Dogi, C. A., Weill, F., & Perdigo´n, G., (2010). Immune response of non – pathogenic gram(+) and gram(−) bacteria in inductive sites of the intestinal mucosa. Study of the pathway of signaling involved. *Immunobiology, 215*, 60–69.

Donkor, O. N., Shah, N. P., Apostolopoulos, V., & Vasiljevic, T., (2010). Development of allergic responses related to microorganisms exposure in early life. *Int. Dairy J., 20*, 373–385.

Duan, H., Leilei, Y. L., Tian, F., Zhai, Q., Fan, L., & Chen, W., (2020). Gut microbiota, A target for heavy metal toxicity and a probiotic protective strategy. *Science of the Total Environment, 742*, 140429.

Erickson, K. L., & Hubbard, N. E., (2000). Probiotic immunomodulation in health and disease. *J. Nutr., 130*, 403S–409S.

Ezendam, J., De Klerk, A., Gremmer, E. R., & Van, L. H., (2008). Effects of *Bifidobacterium animalis* administered during lactation on allergic and autoimmune responses in rodents. *Clin. Exp. Immunol., 154*, 424–431.

FAO/WHO, (2001). *Health and Nutritional Properties of Probiotics in Food Including Powder Milk with Live Lactic Acid Bacteria*. Basel, Switzerland, World Health Organization.

Fink, L. N., Zeuthen, L. H., Ferlazzo, G., & Frøkiær, H., (2007). Human antigen-presenting cells respond differently to gut-derived probiotic bacteria but mediate similar strain-dependent NK and T cell activation. *FEMS Immunol. Med. Microbiol., 51*, 535–546.

Fitzgibbon, G., & Mills, K. H. G., (2020). The microbiota and immune-mediated diseases, opportunities for therapeutic intervention. *Eur. J. Immunol., 50*, 326–337.

Foligne, B., Dewulf, J., Breton, J., Claisse, O., Lonvaud-Funel, A., & Pot, B., (2010). Probiotic properties of non-conventional lactic acid bacteria: Immunomodulation *by Oenococcus oeni. International Journal of Food Microbiology, 140*(2, 3), 136–145.

Fukuda, K., (2017). Is it feasible to control pathogen infection by competitive binding of probiotics to the host? *Virulence, 8*, 1502–1505.

Fuller, R., (1989). Probiotics in man and animals. *J. Appl. Bacteriol., 66*(5), 365–378.

Galdeano, C. M., De Moreno De, L. A., Vinderola, G., Bibas, B. M. E., & Perdigo´n, G., (2007). Proposed model: Mechanisms of immunomodulation induced by probiotic bacteria. *Clinical and Vaccine Immunology, 14*(5), 485–492.

Galdeano, M., & Perdigon, G., (2004). Role of viability of probiotic strains in their persistence in the gut and in mucosal immune stimulation. *Journal of Applied Microbiology, 97*(4), 673–681.

Galdeano, M., De Moreno, A., De Leblanc, E., Carmuega, R., Weill, & Perdigo´n, G., (2009). Mechanisms involved in the immunostimulation by probiotic fermented milk. *Journal of Dairy Research, 76*(4), 446–454.

Garrett, W. S., Gordon, J. I., & Glimcher, L. H., (2010). Homeostasis and inflammation in the intestine. *Cell, 140*, 859–870.

George, K. R., Patra, J. K., Gouda, S., et al., (2018). Benefaction of probiotics for human health, a review. *J. Food. Drug. Anal., 26*, 927–939.

Gill, H. S., & Rutherfurd, K. J., (2001a). Probiotic supplementation to enhance natural immunity in the elderly, Effects of a newly characterized immunostimulatory strain *Lactobacillus rhamnosus* HN001 (DR20TM) on leucocyte phagocytosis. *Nutr. Res., 21*, 183–189.

Gloux, K., Berteau, O., El Oumami, H., Beguet, F., Leclerc, M., & Dore, J., (2011). A metagenomic beta-glucuronidase uncovers a core adaptive function of the human intestinal microbiome. *Proc. Natl. Acad. Sci. U. S. A., 108*(Supp L1), 4539–4546.

Gómez-Llorente, C., Munoz, S., & Gil, A., (2010). Role of toll-like receptors in the development of immunotolerance mediated by probiotics. *Proc. Nutr. Soc., 69*, 381–389.

Haller, D., Bode, C., Hammes, W. P., Pfeifer, A. M. A., Schirin, E. J., & Blum, S., (2000). Non-pathogenic bacteria elicit a differential cytokine response by intestinal epithelial cell/leucocyte co-cultures. *Gut., 47*(1), 79–87.

Halpern, G. M., Vruwink, K. G., Van De, W. J., Keen, C. L., & Gershwin, M. E., (1991). Influence of long-term yoghurt consumption in young adults. *Int. J. Immunother., 7*, 205–210.

Hamer, H. M., Jonkers, D. M., Vanhoutvin, S. A., Troost, F. J., Rijkers, G., De Bruine, A., et al., (2010). Effect of butyrate enemas on inflammation and antioxidant status in the colonic mucosa of patients with ulcerative colitis in remission. *Clin. Nutr., 29*, 738–744.

Hanisch, F. G., (2001). O-glycosylation of the mucin type. *Biol. Chem., 382*(2), 143–149.

He, F., Morita, H., Ouwehand, A. C., Hosoda, M., Hiramatsu, M., Kurisaki, J. I., Isolauri, E., et al., (2002). Stimulation of the secretion of pro-inflammatory cytokines by *Bifidobacterium* strains. *Microbiol. Immunol., 46*, 781–785.

Hessle, C. C., Andersson, B., & Wold, A. E., (2005). Gram-positive and gram–negative bacteria elicit different patterns of pro-inflammatory cytokines in human monocytes. *Cytokine, 30*, 311–318.

Hill, C., Guarner, F., Reid, G., Gibson, G. R., Merenstein, D. J., Pot, B., Morelli, L., Canani, R. B., Flint, H. J., Salminen, S., et al., (2014). Expert consensus document. The international scientific association for probiotics and prebiotics consensus statement on the scope and appropriate use of the term probiotic. *Nat. Rev. Gastroenterol. Hepatol., 11*, 506–514.

Hoseinifar, S. H., Sun, Y. Z., Wang, A., & Zhou, Z., (2018). Probiotics as means of diseases control in aquaculture, a review of current knowledge and future perspectives. *Front. Microbiol., 9*, 2429.

Huda-Faujan, N., Abdulamir, A. S., Fatimah, A. B., Anas, O. M., Shuhaimi, M., Yazid, A. M., et al., (2010). The impact of the level of the intestinal short chain fatty acids in inflammatory bowel disease patients versus healthy subjects. *Open Biochem. J., 4*, 53–58.

Hutkins, R. W., Krumbeck, J. A., Bindels, L. B., et al., (2016). Prebiotics, why definitions matter. *Current Opinion in Biotechnology, 37*, 1–7.

Islam, S. U., (2016). Clinical uses of probiotics. *Medicine, 95*(5), e2658.

Isolauri, E., Majamaa, I. F., Arvola, T., Rantala, I., Virtanen, E., & Arvilommi, H., (1993). *Lactobacillus casei* strain GG reverses increased intestinal permeability induced by cow milk in suckling rats. *Gastroenterology, 105*, 1643–1650.

Isolauri, E., Sütas, Y., Kankaanpää, P., Arvilommi, H., & Salminen, S., (2001). Probiotics, effects on immunity. *Am. J. Clin. Nutr., 73*(2 Suppl), 444S–450S.

Kaila, M., Isolauri, E., Soppi, E., Virtanen, E., Laine, S., & Arvilommi, H., (1992). Enhancement of the circulating antibody secreting cell response in human diarrhea by a human *Lactobacillus* strain. *Pediatr. Res., 32*, 141–144.

Karamese, M., Aydin, H., Sengul, E., et al., (2016). The immunostimulatory effect of lactic acid bacteria in a rat model. *Iranian Journal of Immunology: IJI, 13*(3), 220–228.

Karczewski, J., Troost, F. J., Konings, I., Dekker, J., Kleerebezem, M., Brummer, R. J., & Wells, J. M., (2010). Regulation of human epithelial tight junction proteins by *Lactobacillus plantarum in vivo* and protective effects on the epithelial barrier. *American Journal of Physiology – Gastrointestinal and Liver Physiology, 298*, G851–885.

Kato, I., Tanaka, K., & Yokokura, T., (1999). Lactic acid bacterium potently induces the production of interleukin-12 and interferon-γ by mouse splenocytes. *Int. J. Immunopharmacol., 21*, 121–131.

Kawase, M., He, F., Kubota, A., Harata, G., & Hiramatsu, M., (2010). Oral administration of lactobacilli from the human intestinal tract protects mice against influenza virus infection. *Lett. Appl. Microbiol., 51*, 6–10.

Kekkonen, R. A., Lummela, N., Karjalainen, H., Latvala, S., Tynkkynen, S., Jaïrvenpaä, S., Kautiainen, H., et al., (2008). Probiotic intervention has strain-specific anti-inflammatory effects in healthy adults. *World J. Gastroentero., 14*, 2029–2036.

Kennedy, J. L., Turner, R. B., Braciale, T., Heymann, P. W., & Borish, L., (2012). Pathogenesis of rhinovirus infection. *Curr. Opin. Virol., 2*, 287–293.

Keservani, R. K., Kesharwani, R. K., Sharma, A. K., & Jarouliya, U., (2015). Dietary supplements, nutraceutical and functional foods in immune response (immunomodulators). In: Debasis, B., Harry, G. P., & Anand, S., (eds.), *Nutraceutical and Functional Foods in Human Life and Disease Prevention* (pp. 343–358). Boca Raton, Fl. CRC Press, Taylor and Francis, Chapter 20. ISBN: 9781482237214.

Keservani, R. K., Kesharwani, R. K., Sharma, A. K., Vyas, N., & Chadoker, A., (2010b). Nutritional supplements: An overview. *International Journal of Current Pharmaceutical Review and Research, 1*(1), 59–75.

Keservani, R. K., Kesharwani, R. K., Vyas, N., Jain, S., Raghuvanshi, R., & Sharma, A. K., (2010a). Nutraceutical and functional food as future food: A review. *Der Pharmacia Lettre, 2*(1), 106–116.

Keservani, R. K., Sharma, A. K., & Kesharwani, R. K., (2017). An overview and therapeutic applications of nutraceutical and functional foods. *Recent Advances in Drug Delivery Technology*, 160–201. ISBN:9781522507543.

Keservani, R. K., Sharma, A. K., & Kesharwani, R. K., (2020). *Nutraceuticals and Dietary Supplements: Applications in Health Improvement and Disease Management*. CRC Press. ISBN: 9781771888738.

Khani, S., Motamedifar, M., Golmoghaddam, H., Hosseini, H. M., & Hashemizadeh, Z., (2012). *In vitro* study of the effect of a probiotic bacterium *Lactobacillus rhamnosus* against herpes simplex virus type 1. *Braz. J. Infect. Dis., 16*, 129–135.

Kim, Y., Kim, S. H., Whang, K. Y., et al., (2008). Inhibition of *Escherichia coli* O157,H7 attachment by interactions between lactic acid bacteria and intestinal epithelial cells. *J. Microbiol. Biotechnol., 18*, 1278–1285.

Klampfer, L., Huang, J., Sasazuki, T., Shirasawa, S., & Augenlicht, L., (2003). Inhibition of interferon gamma signaling by the short chain fatty acid butyrate. *Mol. Cancer Res., 1*, 855–862.

Kleniewska, P., & Pawliczak, R., (2017). Influence of synbiotics on selected oxidative stress parameters. *Oxidative Medicine and Cellular Longevity*.

Kline, K. A., Fälker, S., Dahlberg, S., Normark, S., & Henriques-Normark, B., (2009). Bacterial adhesins in host-microbe interactions. *Cell Host. Microbe., 5*, 580–592.

Kotzampassi, K., Stavrou, G., Damoraki, G., Georgitsi, M., Basdanis, G., Tsaousi, G., & Giamarellos-Bourboulis, E. J., (2015). A four-probiotics regimen reduces postoperative complications after colorectal surgery, a randomized, double-blind, placebo-controlled study. *World J. Surg., 39*, 2776–2783.

Kourelis, A., Zinonos, I., Kakagianni, M., Christidou, A., Christoglou, N., Yian-naki, E., Testa, T., et al., (2010). Validation of the dorsal air pouch model to predict and examine immunostimulatory responses in the gut. *J. Appl. Microbiol., 108*, 274–284.

Kreijtz, J. H., Fouchier, R. A., & Rimmelzwaan, G. F., (2011). Immune responses to influenza virus infection. *Virus Res., 162*, 19–30.

Kuiken, T., Riteau, B., Fouchier, R. A., & Rimmelzwaan, G. F., (2012). Pathogenesis of influenza virus infections, The good, the bad and the ugly. *Curr. Opin. Virol., 2*, 276–286.

Laboisse, C., Jarry, A., Branka, J. E., Merlin, D., Bou-Hanna, C., & Vallette, G., (1996). Recent aspects of the regulation of intestinal mucus secretion. *Proc. Nutr. Soc., 55*, 259–264.

Lau, J. M., Jin, X., Ren, J., Avery, J., De Bosch, B. J., Treskov, I., et al., (2007). The 14–3-3tau phosphoserine-binding protein is required for cardiomyocyte survival. *Mol. Cell Biol., 27*, 1455–1466.

Lehtoranta, L., Kalima, K., He, L., Lappalainen, M., Roivainen, M., Närkiö, M., Mäkelä, M., et al., (2014). Specific probiotics and virological findings in symptomatic conscripts attending military service in Finland. *J. Clin. Virol., 60*, 276–281.

Lehtoranta, L., Latvala, S., & Lehtinen, M. J., (2020). Role of probiotics in stimulating the immune system in viral respiratory tract infections, a narrative review. *Nutrients, 12*(10), 3163.

Lilly, D. M., & Stillwell, R. H., (1965). Probiotics, growth-promoting factors produced by microorganisms. *Science, 147*(3659), 747–748.

Looft, T., & Allen, H. K., (2012). Collateral effects of antibiotics on mammalian gut microbiomes. *Gut Microbes, 3*, 463–467.

Lopez-Santamarina, A., Lamas, A., Del Carmen, M. A., Cardelle-Cobas, A., Regal, P., Rodriguez-Avila, J. A., Miranda, J. M., et al., (2021). Probiotic effects against virus infections, new weapons for an old war. *Foods, 10*, 130.

Lu, K., Dong, S., Wu, X., Jin, R., & Chen, H., (2021). Probiotics in cancer. *Front. Oncol., 11*.

Luhrs, H., Gerke, T., Muller, J. G., Melcher, R., Schauber, J., Boxberge, F., et al., (2002). Butyrate inhibits NF-kappaB activation in lamina propria macrophages of patients with ulcerative colitis. *Scand. J. Gastroenterol., 37*, 458–466.

Lupfer, C., Malik, A., & Kanneganti, T. D., (2015). Inflammasome control of viral infection. *Curr. Opin. Virol., 12*, 38–46.

Mack, D. R., Ahrne, S., Hyde, L., Wei, S., & Hollingsworth, M. A., (2003). Extracellular MUC3 mucin secretion follows adherence of *Lactobacillus* strains to intestinal epithelial cells *in vitro*. *Gut., 52*, 827–833.

Mack, D. R., Michail, S., Wei, S., McDougall, L., & Hollingsworth, M. A., (1999). Probiotics inhibit enteropathogenic *E. coli* adherence *in vitro* by inducing intestinal mucin gene expression. *Am. J. Physiol. Gastrointest. Liver Physiol., 276*, G941–G950.

Macpherson, A. J., & Uhr, T., (2004). Induction of protective IgA by intestinal dendritic cells carrying commensal bacteria. *Science, 303*, 1662–1665.

Madara, J. L., & Stafford, J., (1989). Interferon – directly affects barrier function of cultured intestinal epithelial monolayers. *J. Clin. Invest., 83*, 724–727.

Magistrelli, L., Amoruso, A., Mogna, L., Graziano, T., Cantello, R., Pane, M., & Comi, C., (2019). Probiotics may have beneficial effects in parkinson's disease: *In vitro* evidence. *Front. Immunol., 10*, 969.

Majamaa, H., Isolauri, E., Saxelin, M., & Vesikari, T., (1995). Lactic acid bacteria in the treatment of acute rotavirus gastroenteritis. *J. Pediatr. Gastroenterol. Nutr., 20*, 333–338.

Maldonado, G. C., Cazorla, S. I., Lemme, D. J. M., Vélez, E., & Perdigón, G., (2019). Beneficial effects of probiotic consumption on the immune system. *Ann. Nutr. Metab., 74*, 115–124.

Maslowski, K. M., Vieira, A. T., Ng, A., Kranich, J., Sierro, F., Yu, D., et al., (2009). Regulation of inflammatory responses by gut microbiota and chemoattractant receptor GPR43. *Nature, 461*, 1282–1286.

Masuda, E. S., Imamura, R., Amasaki, Y., Arai, K., & Arai, N., (1998). Signaling into the T-Cell nucleus, NFAT regulation. *Cell. Signal., 10*, 599–611.

Mattar, A. F., Teitelbaum, D. H., Drongowski, R. A., et al., (2002). Probiotics up-regulate MUC-2 mucin gene expression in a Caco-2 cell-culture model. *Pediatr. Surg. Int., 18*, 586–590.

Mazloom, K., Siddiqi, I., & Covasa, M., (2019). Probiotics, how effective are they in the fight against obesity? *Nutrients, 11*, 258.

McNeil, N. I., Cummings, J. H., & James, W. P., (1978). Short chain fatty acid absorption by the human large intestine. *Gut., 19*, 819–822.

Meijerink, M., & Wells, J. M., (2010). Probiotic modulation of dendritic cells and T cell responses in the intestine. *Benef. Microbes., 1*(4), 317–326.

Miettinen, M., Matikainen, S., Vuopio-Varkila, J., Pirhonen, J., Varkila, K., Kurimoto, M., & Julkunen, I., (1998). Lactobacilli and streptococci induce interleukin-12 (IL-12), IL-18, and gamma interferon production in human peripheral blood mononuclear cells. *Infect. Immun., 66*, 6058–6062.

Miettinen, M., Vuopio-Varkila, J., & Varkila, K., (1996). Production of human tumor necrosis factor alpha, interleukin-6, and interleukin-10 is induced by lactic acid bacteria. *Infect. Immun., 64*, 5403–5405.

Miller, H., Zhang, J., Kuolee, R., Patel, G. B., & Chen, W., (2007). Intestinal M cells, the fallible sentinels? *World Journal of Gastroenterology, 13*, 1477–1486.

Molska, M., & Reguła, J., (2019). Potential mechanisms of probiotics action in the prevention and treatment of colorectal cancer. *Nutrient, 11*, 2453.

Mörbe, U. M., Jørgensen, P. B., Fenton, T. M., Von Burg, N., Riis, L. B., Spencer, J., & Agace, W. W., (2021). Human gut-associated lymphoid tissues (GALT); diversity, structure, and function. *Mucosal Immunol., 14*(4), 793–802.

Moreau, M. C., Hudault, S., & Bridonneau, C., (1990). Systemic antibody response to ovalbumin in gnotobiotic C3H/HeJ mice with *Bifidobacterium bifidum* or *Escherichia coli*. *Microecol. Ther., 20*, 309–312.

Morita, H., He, F., Fuse, T., Ouwehand, A. C., Hashimoto, H., Hosoda, M., Mizumachi, K., & Kurisaki, J. I., (1999). Cytokine production by the murine macrophage cell line J774.1 after exposure to lactobacilli. *Biosci. Biotechnol. Biochem., 66*, 1963–1966.

Naidu, A., Bidlack, W., & Clemens, R., (1999). Probiotic spectra of lactic acid bacteria (LAB). *Crit. Rev. Food. Sci. Nutr., 39*(1), 13–126.

Newton, A. H., Cardani, A., & Braciale, T. J., (2016). The host immune response in respiratory virus infection, Balancing virus clearance and immunopathology. *Semin. Immunopathol., 38*, 471–482.

Nishiyama, K., Nakazato, A., Ueno, S., Seto, Y., Kakuda, T., Takai, S., et al., (2015). Cell surface-associated aggregation-promoting factor from *Lactobacillus gasseri* SBT 2055 facilitates host colonization and competitive exclusion of *Campylobacter jejuni*. *Mol. Microbiol., 98*, 712–726.

Oo, K. M., Ayelwin, A., Kyaw, Y. Y., Tun, W. M., Fukada, K., Goshima, A., Shimada, T., & Okada, S., (2016). Safety and long-term effect of the probiotic FK-23 in patients with hepatitis C virus infection. *Biosci. Microbiota Food Health, 35*, 2015–2024.

Ooi, M. F., Mazlan, N., Foo, H. L., et al., (2015). Effects of carbon and nitrogen sources on bacteriocin-inhibitory activity of postbiotic metabolites produced by *Lactobacillus plantarum* I-UL4. *Malaysian Journal of Microbiology, 11*(2), 176–184.

Otten, M. A., & Van, E. M., (2004). The Fc receptor for IgA (FcαRI, CD89). *Immunology Letters, 92*(1, 2), 23–31.

Ou, Y. C., Fu, H. C., Tseng, C. W., Wu, C. H., Tsai, C. C., & Lin, H., (2019). The influence of probiotics on genital high-risk human papilloma virus clearance and quality of cervical smear, a randomized placebo-controlled trial. *BMC Women's Health, 19*, 103.

Ouwehand, A. C., Salminen, S., & Isolauri, E., (2002). Probiotics, An overview of beneficial effects. *Antonie Van Leeuwenhoek., 82*, 279–289.

Palma, E., Recine, N., Domenici, L., Giorgini, M., Pierangeli, A., & Panici, P. B., (2018). Long-term *Lactobacillus rhamnosus* BMX 54 application to restore a balanced vaginal ecosystem, A promising solution against HPV-infection. *BMC Infect. Dis., 18*, 13.

Paone, P., & Cani, P. D., (2020). Mucus barrier, mucins and gut microbiota, the expected slimy partners? G*ut, 69*, 2232–2243.

Park, M. K., Ngo, V., Kwon, Y. M., Lee, Y. Y., Yoo, S., Cho, Y. H., Hong, S. M., Hwang, H. S., Ko, E. J., Jung, Y. J., et al., (2013). *Lactobacillus plantarum* DK119 as a probiotic confers protection against influenza virus by modulating innate immunity. *PLoS One, 8*, e75368.

Parker, R., (1974). Probiotics, the other half of the antibiotic story. *Anim. Nutr. Health, 29*, 4–8.

Pasare, C., & Medzhitov, R., (2005). Toll-like receptors, Linking innate and adaptive immunity. *Advances in Experimental Medicine and Biology, 560*, 11–18.

Patel, R. M., & Denning, P. W., (2013). Therapeutic use of prebiotics, probiotics, and postbiotics to prevent necrotizing enterocolitis. what is the current evidence? *Clinics in Perinatology, 40*(1), 11–25.

Pena, J. A., Rogers, A. B., Ge, Z., et al., (2005). Probiotic *Lactobacillus* spp. diminish *Helicobacter hepaticus*-induced inflammatory bowel disease in interleukin-10-deffcient mice. *Infection and Immunity, 73*(2), 912–920.

Perdigón, G., De Macías, M. E., Alvarez, S., Oliver, G., & De Ruiz, H. A. A., (1986). Effect of perorally administered lactobacilli on macrophage activation in mice. *Infect. Immun., 53*, 404–410.

Perdigón, G., De Macías, M. E., Alvarez, S., Oliver, G., & De Ruiz, H. A. P., (2013). Systemic augmentation of the immune response in mice by feeding fermented milks with *Lactobacillus casei* and *Lactobacillus acidophilus*. *Immunology, 63*, 17–23.

Pot, B., Foligné, B., Daniel, C., & Grangette, C., (2013). Understanding immunomodulatory effects of probiotics. *Nestle Nutr. Inst. Workshop Ser., 77*, 75–90.

Prescott, S. L., Wickens, K., Westcott, L., Jung, W., Currie, H., Black, P. N., Stanley, T. V., et al., (2008). Supplementation with *Lactobacillus rhamnosus* or *Bifidobacterium* lactis probiotics in pregnancy increases cord blood interferon-γ and breast milk transforming growth factor-β and immunoglobin A detection. *Clin. Exp. Allergy, 38*, 1606–1614.

Pujari, R., & Banerjee, G., (2021). Impact of prebiotics on immune response, from the bench to the clinic. *Immunol. Cell Biol., 99*, 255–273.

Raheem, A., Liang, L., Zhang, G., & Cui, S., (2021). Modulatory effects of probiotics during pathogenic infections with emphasis on immune regulation. *Front. Immunol., 12*, 616713.

Resendiz-Albor, A., Reina-Gar as, H., Rojas-Herna ́ndez, S., et al., (2010). Regionalization of pIgR expression in the mucosa of mouse small intestine. *Immunology Letters, 128*(1), 59–67.

Rhayat, L., Maresca, M., Nicoletti, C., Perrier, J., Brinch, K. S., Christian, S., et al., (2019). Effect of *Bacillus subtilis* strains on intestinal barrier function and inflammatory response. *Front Immunol., 10*, 564.

Rubinstein, M. R., Wang, X., Liu, W., Hao, Y., Cai, G., & Han, Y. W., (2013). *Fusobacterium nucleatum* promotes colorectal carcinogenesis by modulating E-cadherin/β-catenin signaling via its FadA adhesin. *Cell 1 Host Microbe., 14*, 195–206.

Sagheddu, V., Guidesi, E., Galletti, S., & Elli, M., (2019). Original paper selection and characterization criteria of probiotics intended for human use from the past to the future. *Food Sci. Nutr., 3.*

Savan, R., & Sakai, M., (2006). Genomics of fish cytokines, comparative biochemistry and physiology part D. *Genomics and Proteomics, 1*(1), 89–101.

Scheri, G. C., Fard, S. N., Schietroma, I., Mastrangelo, A., Pinacchio, C., Giustini, N., Serafino, S., De Girolamo, G., Cavallari, E. N., Statzu, M., et al., (2017). Modulation of tryptophan/serotonin pathway by probiotic supplementation in human immunodeficiency virus-positive patients, Preliminary results of a new study approach. *Int. J. Trytophan Res., 10*, 1178646917710668.

Schiffrin, E. J., Rochat, F., Link-Amster, H., Aeschlimann, J. M., & Donnet-Hughes, A., (1994). Immunomodulation of human blood cells following the ingestion of lactic acid bacteria. *J. Dairy Sci., 78*, 491–497.

Schlee, M., Harder, J., Koten, B., Stange, E. F., Wehkamp, J., & Fellermann, K., (2008). Probiotic lactobacilli and VSL#3 induce enterocyte β-defensin 2. *Clin. Exp. Immunol., 151*, 528–535.

Schwab, M., Reynders, V., Ulrich, S., Zahn, N., Stein, J., & Schroder, O., (2006). PPAR gamma is a key target of butyrate-induced caspase-3 activation in the colorectal cancer cell line Caco-2. *Apoptosis, 11*, 1801–1811.

Schwabe, R. F., & Greten, T. F., (2020). Gut microbiome in HCC – mechanisms, diagnosis and therapy. *J. Hepat., 72*, 230–238.

Schwandner, R., Dziarski, R., Wesche, H., Rothe, M., & Kirschning, C. J., (1999). Peptidoglycan – and lipoteichoic acid-induced cell activation is mediated by toll-like receptor 2. *The Journal of Biological Chemistry, 274*(25), 17406–17409.

Segawa, S., Fujiya, M., Konishi, H., Ueno, N., Kobayashi, N., Shigyo, T., et al., (2011). Probiotic-derived polyphosphate enhances the epithelial barrier function and maintains intestinal homeostasis through integrin-p38 MAPK pathway. *PLoS One, 6*, e23278.

Sengul, N., Isik, S., Aslim, B., Ucar, G., & Demirbag, A. E., (2011). Effect of exopolysaccharide-producing probiotic strains on gut oxidative damage in experimental colitis. *Digestive Diseases and Sciences, 56*(3), 707–714.

Senok, A. C., Ismaeel, A. Y., & Botta, G. A., (2005). Probiotics, Facts and myths. *Clin. Microbiol. Infect., 11*, 958–966.

Seth, A., Yan, F., Polk, D. B., & Rao, R. K., (2008). Probiotics ameliorate the hydrogen peroxide-induced epithelial barrier disruption by a PKC – and MAP kinase-dependent mechanism. *Am. J. Physiol. Gastrointest. Liver. Physiol., 294*, 1060–1069.

Shah, N. P., (2007). Functional cultures and health benefits. *Int. Dairy J., 17*, 1262–1277.

Sherman, P. M., Ossa, J. C., & Johnson-Henry, K., (2009). Unraveling mechanisms of action of probiotics. *Nutrition in Clinical Practice, 24*, 10–14.

Shi, L., Sheng, J., Chen, G., Zhu, P., Shi, C., Li, B., Park, C., Wang, J., Zhang, B., Liu, Z., & Yang, X., (2020). Combining IL-2-based immunotherapy with commensal probiotics produces enhanced antitumor immune response and tumor clearance. *J. Immunother. Cancer,* 8.

Shida, K., Makino, K., Morishita, A., Takamizawa, K., Hachimura, S., Ametani, A., Sato, T., et al., (1998). *Lactobacillus casei* inhibits antigen-induced IgE secretion through regulation of cytokine production in murine splenocyte cultures. *Int. Arch. Allergy Immunol., 115*, 278–287.

Slavin, J., (2013). Fiber and prebiotics, mechanisms and health benefits. *Nutrients, 5*(4), 1417–1435.

Stavropoulou, E., & Bezirtzoglou, E., (1999). Probiotics as a weapon in the fight against COVID-19. *Front. Nutr., 7*, 614986.

Su, C. G., Wen, X., Bailey, S. T., Jiang, W., Rangwala, S. M., Keilbaugh, S. A., et al., (1999). A novel therapy for colitis utilizing PPAR-gamma ligands to inhibit the epithelial inflammatory response. *J. Clin. Invest., 104*, 383–389.

Suganya, K., & Koo, B. S., (2020). Gut–brain axis: Role of gut microbiota on neurological disorders and how probiotics/prebiotics beneficially modulate microbial and immune pathways to improve brain functions. *Int. J. Mol. Sci., 21*, 7551.

Sujino, T., London, M., Van, K. D. P. H., Rendon, T., Buch, T., Silva, H. M., et al., (2016). Tissue adaptation of regulatory and intraepithelial $CD4^+$ T cells controls gut inflammation. *Science, 352*(6293), 1581–1586.

Sundararaman, A., Ray, M., Ravindra, P. V., & Halami, P. M., (2020). Role of probiotics to combat viral infections with emphasis on COVID-19. *Appl. Microbiol. Biotechnol., 104*, 8089–8104.

Sütas, Y., Autio, S., Rantala, I., & Isolauri, E., (1997). IFN – enhances macromolecular transport across Peyer's patches in suckling rats, implications for natural immune responses to dietary antigens early in life. *J. Pediatr. Gastroenterol. Nutr., 24*, 162–169.

Sütas, Y., Hurme, M., & Isolauri, E., (1996a). Downregulation of antiCD3 antibody – induced IL-4 production by bovine caseins hydrolyzed with *Lactobacillus* GG-derived enzymes. *Scand J. Immunol., 43*, 687–689.

Sütas, Y., Soppi, E., Korhonen, H., et al., (1996b). Suppression of lymphocyte proliferation *in vitro* by bovine caseins hydrolyzed with *Lactobacillus casei* GG-derived enzymes. *J. Allergy Clin. Immunol., 98*, 216–224.

Swanson, K. V., Deng, M., & Ting, J. P., (2019). The NLRP3 inflammasome, Molecular activation and regulation to therapeutics. *Nat. Rev. Immunol., 19*, 477–489.

Takeuchi, O., & Akira, S., (2010). Pattern recognition receptors and inflammation. *Cell, 140*, 805–820.

Terciolo, C., Dapoigny, M., & Andre, F., (2019). Beneficial effects of *Saccharomyces boulardii* CNCM I-745 on clinical disorders associated with intestinal barrier disruption. *Clin. Exp. Gastroenterol., 12*, 67–82.

Thakur, N., Rokana, N., & Panwar, H., (2016). Probiotics, selection criteria, safety and role in health and disease. *J. Innov. Biol., 3*, 259–270.

Totura, A. L., & Baric, R. S., (2012). SARS coronavirus pathogenesis, Host innate immune responses and viral antagonism of interferon. *Curr. Opin. Virol., 2*, 264–275.

Toyoda, H., Komurasaki, T., Uchida, D., Takayama, Y., Isobe, T., Okuyama, T., & Hanada, K., (1995). Epiregulin. A novel epidermal growth factor with mitogenic activity for rat primary hepatocytes. *J. Biol. Chem., 270*, 7495–7500.

Troost, F. J., Van, B. P., Lindsey, P., Kodde, A., De Vos, W. M., Kleerebezem, M., & Brummer, R. J. M., (2008). Identification of the transcriptional response of human intestinal mucosa to *Lactobacillus plantarum* WCFS1 *in vivo*. *BMC Genom., 9*, 374.

Van, B. P., Wells, J. M., & Kleerebezem, M., (2013). Regulation of intestinal homeostasis and immunity with probiotic lactobacilli. *Trends Immunol., 34*, 208–215.

Vancheswaran, G. V., Helmink, B. A., Spencer, C. N., Reuben, A., & Wargo, J. A., (2018). The influence of the gut microbiome on cancer, immunity, and cancer immunotherapy. *Cancer Cell., 33*(4), 570–580.

Veckman, V., Miettinen, M., Pirhonen, J., Sire´n, J., Matikainen, S., & Julkunen, I., (2004). Streptococcus pyogenes and *Lactobacillus rhamnosus* differentially induce maturation and production of Th1-type cytokines and chemokines in human monocyte-derived dendritic cells. *J. Leukoc. Biol., 75*, 764–771.

Venegas, D. P., De La Fuente, M. K., Landskron, G., González, M. J., Quera, R., Dijkstra, G., Harmsen, H. J. M., et al., (2019). Short chain fatty acids (SCFAs)-mediated gut epithelial and immune regulation and its relevance for inflammatory bowel diseases. *Front. Immunol., 10*, 277.

Verma, & Shukla, G., (2014). Synbiotic (*Lactobacillus rhamnosus* + *Lactobacillus acidophilus*+inulin) attenuates oxidative stress and colonic damage in 1, 2 dimethylhydrazine dihydrochloride – induced colon carcinogenesis in Sprague' Dawley rats, A long – term study. *European Journal of Cancer Prevention, 23*(6), 550–559.

Vitetta, L., Vitetta, G., & Hall, S., (2018). Immunological tolerance and function, Associations between intestinal bacteria, probiotics, prebiotics, and phages. *Front. Immunol., 9*, 2240.

Von, D. W. T., Bulliard, C., & Schiffrin, E. J., (2001). Induction by a lactic acid bacterium of a population of $CD4^+$ T cells with low proliferative capacity that produce transforming growth factor β and interleukin-10. *Clin. Diagn. Lab. Immunol., 8*, 695–701.

Wang, H., Lee, I. S., Braun, C., & Enck, P., (2016). Effect of probiotics on central nervous system functions in animals and humans: A systematic review. *J Neurogastroenterol Motil., 22*, 589–605.

Wehkamp, J., Harder, J., Wehkamp, K., Wehkamp-Von, M. B., Schlee, M., Enders, C., et al., (2004). NFκB – and AP-1-mediated induction of human β defensin-2 in intestinal epithelial cells by *Escherichia coli* Nissle 1917: A novel effect of a probiotic bacterium. *Infect. Immun., 72*, 5750–5758.

Wells, J. M., Rossi, O., Meijerink, M., & Van, B. P., (2010). Epithelial crosstalk at the microbiota-mucosal interface. *Proceedings of the National Academy of Sciences of the USA*.

West, N. P., Pyne, D. B., Peake, J. M., & Cripps, A. W., (2009). Probiotics, immunity and exercise, A review. *Exerc. Immunol. Rev., 15*, 107–126.

Wilson, K., & Perini, F., (1988). Role of competition for nutrients in suppression of *Clostridium difficile* by the colonic microflora. *Infect. Immun., 56*, 2610–2614.

Xiao, L., Ding, G., Ding, Y., et al., (2017). Effect of probiotics on digestibility and immunity in infants. *Medicine, 96*(14), e5953.

Yadav, H., Jain, S., & Sinha, P. R., (2008). Oral administration of dahi containing probiotic *Lactobacillus acidophilus* and *Lactobacillus casei* delayed the progression of streptozotocin-induced diabetes in rats. *Journal of Dairy Research, 75*(2), 189–195.

Yasui, H., Nagaoka, N., Mike, A., Hayakawa, K., & Ohwaki, M., (1992). Detection of *Bifidobacterium* strains that induce large quantities of IgA. *Microb. Ecol. Health Dis., 5*, 155–162.

Yasui, H., Shida, K., Matsuzaki, T., & Yokokura, T., (1999). Immunomodulatory function of lactic acid bacteria. *Antonie Van Leeuwenhoek, 76*, 383–389.

Zhang, Z., & Michael, G. I., (1990). Orally inducible immune responsiveness is abrogated by IFN-gamma treatment. *J. Immunol., 144*, 4163–4165.

Zivkovic, R., (1999). Probiotics or microbes against microbes. *Acta. Med. Croatica., 53*(1), 23–28.

Zolnikova, O., Komkova, I., Potskherashvili, N., Trukhmanov, A., & Ivashkin, V., (2018). Application of probiotics for acute respiratory tract infections. *Ital. J. Med., 12*, 32–38.

Zorriehzahra, M. J., Delshad, S. T., Adel, M., Tiwari, R., Karthik, K., Dhama, K., et al., (2016). Probiotics as beneficial microbes in aquaculture, an update on their multiple modes of action, a review. *Vet Q., 36*, 228–241.

Zuo, F., Appaswamy, A., Gebreegziabher, H., & Jonsson, A. B., (2019). Role of sortase A in *Lactobacillus gasseri* Kx110A1 adhesion to gastric epithelial cells and competitive exclusion of *Helicobacter pylori*. *Front Microbiol., 10*, 2770.

Zyrek, A. A., Cichon, C., Helms, S., Enders, C., Sonnenborn, U., & Schmidt, M. A., (2007). Molecular mechanisms underlying the probiotic effects of *Escherichia coli* Nissle 1917 involve ZO-2 and PKCzeta redistribution resulting in tight junction and epithelial barrier repair. *Cell Microbiol., 9*, 804–816.

CHAPTER 6

MINERALS AS GATEKEEPERS OF IMMUNE FUNCTION

R. MD. ANWAR NAWAZ and K. SIVAKUMAR

Department of Biomedical Engineering, Karpaga Vinayaga College of Engineering and Technology, Chengalpattu, Tamil Nadu, India

ABSTRACT

The immune system is made up of interacting lymphoid organs, cells, humoral substances, and cytokines. A sufficient quantity and balance of nutrients are required for the proper functioning of the host defenses. Amino acids, fatty acids, minerals, and vitamins have been discovered as dietary variables that influence human and animal immunological responses. When given at pharmacological doses, several nutrients have been discovered to have specific effects on immunity. On the exterior and inside of the body, the skin and mucus membranes function as physical and chemical barriers against bacteria, viruses, fungi, parasites, pollen, dust, and hazardous chemicals. For optimal performance, the structural and functional integrity of the physical barriers must be maintained. Minerals are an essential component of our diet. They are in charge of several bodily tasks, including bone formation, muscle, and neuron activity, and water balance regulation. They are found in hormones, enzymes, and other physiologically active compounds. Some minerals are also required for appropriate immune system function. Minerals such as magnesium, as well as micronutrients such as selenium, zinc, copper, and others, are required for proper immune function, and those who are lacking in these nutrients may be in danger. Most people's diets are lacking in key minerals and trace elements, which is unfortunate. Inadequate intakes and status of certain minerals, trace

Immune-Boosting Nutraceuticals for Better Human Health: Novel Applications. Urmila Jarouliya, PhD, Raj K. Keservani, MPharm, Rajesh K. Kesharwani, PhD, Virendra K. Patel, PhD, and Adi D. Bharti, MPharm (Eds.)

elements, and vitamins improve immune function and health, whereas low intakes and status of minerals, trace elements, and vitamins damage immunity, leading to increased disease, infection, and a poor quality of life.

6.1 INTRODUCTION

The immune system is a network of lymphoid organs, cells, humoral substances, and cytokines that interact with one another (Kellie & Al-Mansour, 2017). Throughout history, infectious illnesses of viral, bacterial, fungal, and protozoan origin have affected both animals and humans. In humans and higher animals, the immune system is made up of numerous important organs such as the spleen and thymus, as well as a range of bone marrow-derived cells, their secretions and cell surface antigens, and histocompatibility antigens. When cells of the humoral system respond to a foreign antigen or hapten by delaying the manufacture of serum antibodies, this is known as the initial immune response. A secondary immunological response results from a second exposure to antigen or hapten, which raises blood antibody levels. Antibodies can attach to antigens and cause cellular lysis or a variety of other biological consequences when combined with serum complement. As previously established, humoral (B-cell) immunity is involved in such responses. In response to histocompatibility antigens, cellular (or T-cell) immunity is commonly triggered, resulting in organ transplant and tumor rejection, delayed hypersensitivity, and phagocytosis. Helper/inducer and cytotoxic/suppressor T-lymphocytes are members of the T-cell lineage, which are thought to have originated from a single lymphoid cell. Although macrophages and natural killer (NK) cells lack T-cell features, they serve comparable functions to T-cells (Delves et al., 2017). Immune cells become fully formed immune responses after coming into touch with antigens and digesting them. Nutrient imbalance can jeopardize this process at any point. *In vivo* assays, such as delayed-type hypersensitivity evaluated by skin testing, and assessment of humoral immune response by analysis of particular antibodies emerging in response to first or subsequent (booster) vaccination, are therefore necessary for determining immune function.

6.1.1 OUR BODY AND THE IMMUNE SYSTEM

All of the mechanisms and responses that the body utilizes to fight itself against bacteria, viruses, and parasites are referred to as immunity. The correct functioning of the host defenses requires an adequate supply and balance of nutrients. Dietary factors that impact human and animal immune responses have been identified as amino acids, fatty acids, minerals, and vitamins. Certain nutrients have been found to have particular impacts on immunity when administered in pharmacological dosages (Trichet, 2010). Poor nutrition with a lack of minerals, micronutrients, and vitamins (Keservani et al., 2010a, b) are known to reduce immune function. Trace minerals such as magnesium, zinc, copper, and selenium have been studied and are known to have an impact on immune health (Keservani et al., 2017, 2020). Trace element deficiencies and excesses can both affect immune system parameters such as antibody responses, cell-mediated immunity, and NK cell activity (Alpert, 2017).

Innate immunity and adaptive immunity are the two main components of the immune system. Innate immunity, by definition, occurs before antigen encounters and so cannot be altered (adapted). It is found in all multicellular creatures and is phylogenetically conserved. Only a few highly conserved patterns in bacteria are recognized by the innate immune system. The adaptive immune system, on the other hand, is highly selective and grows in both specificity and quantity as antigen exposure grows. Unlike the innate immune system, which recognizes infections based on discrete molecular markers, adaptive immune cells must be taught to distinguish between self and non-self (Hoffmann et al., 1999). The skin and the mucus membranes on the outside and inside of the body operate as physical and chemical barriers to bacteria, viruses, fungus, parasites, pollen, dust, and harmful chemicals. The structural and functional integrity of the physical barriers must be maintained for optimal function. These processes need the use of micronutrients. Hair or hair-like structures cover the physical barriers that produce chemically rich fluids like perspiration, saliva, mucus, bile, and gastric acid, all of which are designed to capture or filter foreign material and wash or transport it out of the body (e.g., via coughing or sneezing) or contain antimicrobials, acids, or enzymes to inhibit the growth or destruction of bacteria. Calcitriol (vitamin D's active form) promotes the generation of antimicrobial peptides (AMPs) in epithelial cells, such as those that lining the lungs, protecting them against infection (Gniadecki et al., 1997; Pálmer et al., 2001). The gastrointestinal tract (GIT) is an important line of defense in which epithelial cells provide a physical and biochemical barrier and collaborate with immune cells and gut microflora (some of which produce bacteriocins) to strengthen the gut barrier, fight pathogens, and limit their direct contact with the epithelium (Takiishi et al., 2017). At birth, the immune system is immature, but it develops and changes throughout life until it reaches senescence. The dietary needs for immunological development change from prenatal to postnatal periods, and they are heavily influenced by childhood and teenage growth and development. In addition, the aging process affects dietary needs, which has a significant impact on immune function.

6.2 NUTRITION AND THE IMMUNE SYSTEM

Nutritional status has a big influence on immune system performance. In the blood, regional lymph nodes, and the specialized GI immune system, dietary proteins, carbs, and lipids, as well as micronutrients (vitamins and minerals), interact with immune cells systemically (Cunningham-Rundles, 1998). The immunomodulatory characteristics of glutamate, arginine, taurine, and sulfur-containing amino acids have been studied. Malnutrition and protein insufficiency have been related to both specific and non-specific immunity. There's also evidence that vitamins A, B6, B12, folic acid, C, D, and E, as well as trace minerals like iron, zinc, selenium, and copper, are linked to immunological dysfunction, as are vitamins A, B6, B12, folic acid, C, D, and E (Keservani et al., 2015; Scrimshaw & SanGiovanni, 1997).

While most macro-and micronutrients may be absorbed from natural foods, dietary supplements and/or augmented meals may be advantageous in stressful situations such as childhood, aging, or illness, as well as extreme conditions such as exercise. In terms of dietary aid and the formation of a strong immune system, three degrees of care are being sought. The main level is the availability of all required micro-and macronutrients to sustain immune cells and processes. The immune system is controlled at the second level to respond appropriately to specific but broad areas of concern. At the tertiary level, nutritional therapy tailored to an individual's immunological disposition and circumstances are reflected, and so constitute part of preventative and individualized nutrition (Kussmann & Fay, 2008).

Infection and malnutrition regularly combine to influence immunological function. Immune cells release cytokines in response to infections. The resulting cytokine network interactions are critical for immune response amplification and differentiation. According to a new study, the sort of T-cell-cytokine pattern that develops in response to infection is strongly associated to host defensive effectiveness. A Type 1 response occurs when T-cells produce interferon (IFN)-gamma and interleukin-2 in response to intracellular viral infections that impact mononuclear cells, such as measles, mycobacterial diseases, and leprosy. Because malnutrition alters T-cell activity and cytokine production, determining the cellular immune response is crucial for identifying mechanistic, causal linkages between virus exposure, host immunological response, and outcome (Cunningham-Rundles, 1998). Poor immunological responses, such as cell-mediated immunity, cytokine generation, secretory antibody response, and affinity, are frequently linked to nutritional deficiencies. Micronutrient absorption and consumption are critical for healthy immune responses (e.g., zinc, iron, selenium, vitamin A, pyridoxine, vitamin E). However, macronutrient balance is also important: animal proteins are usually better than vegetable proteins in terms of immune maintenance. The synthesis of glutamine, the most common intracellular amino acid, increases when skeletal muscle metabolizes branched-chain amino acids during times of stress or sickness. Glutamine is a key source of energy for intestinal enterocytes as well as cells that reproduce quickly, such as immunocytes, which respond to damage and illness (Johnson & Kudsk, 1999; Keusch, 2003; Souba, 1991). The immune system regenerates itself every day by producing millions of immune cells. During infectious illness, immune cell renewal is accelerated, and recovery is dependent on the rate of cell division of the invading microbe and immune cells. The immune system needs both macronutrients and micronutrients to make DNA, RNA, and proteins. As a result, undernutrition has a substantial impact on the immune system in people of all ages, but especially in youngsters and the elderly, when the body's nutritional reserves are exhausted. Undernutrition has a significant influence on development at all stages, leading to immunodeficiency and, as a result, greater infection rates (Lesourd, 2004).

6.3 ROLE OF MINERALS IN THE IMMUNE SYSTEM

Minerals are an important part of our nutrition. They are responsible for a variety of functions in the body, including bone formation, muscle, and neuron activity, and water balance management. Hormones, enzymes, and other physiologically active molecules all contain them. Some minerals are also necessary for the immune system to operate properly. The adaptive immune response, as well as the innate defense system, are both implicated. As a result, mineral availability impacts infection susceptibility as well as the development of chronic diseases (Kim & Choi, 2013; Maggini et al., 2018).

Minerals and trace elements are necessary for maintaining excellent health and well-being. Minerals like magnesium, as well as micronutrients like selenium, zinc, copper, and others, are necessary for healthy immune function, and people who are deficient in these nutrients may be at risk. Unfortunately, most people's diets are deficient in important minerals and trace elements. Furthermore, as people age, their demands for macronutrients and micronutrients increase, and some factors, such as the presence of a chronic disease or the use of certain medicines, can worsen a mineral deficiency. It is critical to utilize the best immune system supplements to maintain overall health. The relevance of the majority of important minerals, as well as their immunological role, are discussed further in subsections.

6.3.1 MAGNESIUM

Magnesium (Mg) is the second-most abundant cation in cellular systems. According to researchers, magnesium plays a role in immune function, both in innate and acquired immune responses (Tam et al., 2003). It is essential for enzyme activation, membrane function, and intracellular signaling. The ion is also a necessary cofactor for many enzymes. It is important in RNA and DNA synthesis and replication, as well as the production of enzymes and hormones (Jahnen-Dechent & Ketteler, 2012). They also play a role in immunoglobulin production, immune cell adherence, antibody-dependent cytolysis, IgM lymphocyte binding, macrophage lymphokine response, T-helper B-cell adhesion, and other reactions (Galland, 1988).

6.3.1.1 ROLE OF MAGNESIUM IN THE IMMUNE SYSTEM

Mg influences the acute phase response and macrophage activity, such as their responsiveness to cytokines. Mg supplementation decreased cytokine production following toll-like receptor (TLR) activation in monocytes. This immunoregulatory role was caused by a rise in IκBα levels, which led to decreased nuclear factor NF-κB (kappa-light-chain-enhancer of activated B-cells) translocation Mg shortage may cause cell activation, including macrophages, neutrophils, and endothelial cells (ECs) (Sugimoto et al., 2012). Mg has a significant impact on lymphocyte growth, differentiation, and proliferation. TRPM7, Mg^{2+} transporter, appears to have a unique function in T-cell development. Developmental inhibition and early cell death occurred in cells missing this transporter and consequently the Mg supply. As a result, the T-cell function appears to be dependent on an appropriate supply of Mg, both at the individual cell level and in terms of an individual's

T-cell pool. Moreover, multiple investigations in lymphocytes have revealed that Mg^{2+} from intracellular/extracellular sources regulates phosphoinositide metabolism. Mg is a key cofactor for several enzymes involved in phosphorylation, such as those involved in the phosphorylation cascades of glycolysis and nucleotide polymerization (Weyh et al., 2022).

Mg shortage also appears to hasten thymus involution (Tam et al., 2003). Animals with induced magnesium shortage have impaired cell-mediated immunity and other responses. Magnesium shortage has also been associated with increased antibody production, allergy susceptibility, and persistent fungal and viral infections in humans.

6.3.2 ZINC

The trace element zinc is vital for numerous physiological processes in humans and is one of the most extensively researched components in nutrition and health. It serves as a basic regulator or coenzyme for over 300 enzymes (Schubert et al., 2015). An adult's total Zn concentration is roughly 2–3 g, with 85% of that concentration dispersed inside muscles and bones. The majority is found in red blood cells during the passage across the bloodstream, where it is preserved by superoxide dismutase (SOD) and carbonic anhydrase (Roohani et al., 2013).

6.3.2.1 ROLE OF ZINC IN IMMUNE SYSTEM

In terms of its influence on the host immune system, it is one of the most extensively examined trace elements. Zinc has been examined for its immunologic qualities, and it is necessary for highly proliferating cells, especially those in the immune system. Both the innate and adaptive immune systems are influenced by them. Zinc ions, in their unbound state, have a direct antiviral effect on rhinovirus replication, and they also enhance skin and mucosal membrane integrity. Zinc supplementation boosts innate immune cellular components, antibody response, and cytotoxic CD8+T cell numbers (Alpert, 2017). Zn is essential for the action of neutrophil granulocyte nicotinamide adenine dinucleotide phosphate (NADPH) oxidase (Rink & Haase, 2007). Zinc deficiency has been linked to spleen and lymph node shrinking, as well as a drop in lymphocytes and gamma globulin levels. Zinc's involvement in a variety of zinc-required enzymes, such as DNA and RNA polymerases, is most likely to blame for these side effects. Zn deficiency can reduce the number of NK cells in the peripheral circulation, resulting in poor function. Virus-infected or tumor cells have shown decreased chemotaxis and lysis in this circumstance (Rajagopalan & Long, 1998).

Zinc shortage has a significant impact on T-cell immunity, as evidenced by thymic involution, which can be corrected over time by restoring zinc nutrition. T-helper cell alterations have been observed in severely zinc-deficient animals, as evidenced by an almost uniformly reported decrease in SRBC plaque-forming cells (Filteau & Woodward, 1987). Hypoplasia of the spleen, lymph nodes, Peyer's patches (PP), and other lymphoid tissues are also linked to zinc deficiency. In lymphocytes from zinc-deficient animals and people, responses to concanavalin A, phytohemagglutinin (PHA), and pokeweed mitogens were reduced (elderly individuals and persons with *Acrodermatitis enteropathica*). Adding zinc or zinc transferrin to the culture medium can help restore lymphocytic proliferation (Spallholz & Stewart, 1989). Zn can alter both the generation and signaling of a wide range of inflammatory cytokines. Chronic inflammation is related to illness processes, which can be exacerbated by a Zn shortage. It has been demonstrated that people with systemic inflammatory illnesses, such as rheumatoid arthritis, and a concurrent Zn deficit can exhibit greater expression of IL-1β, IL-1α, and IL-6 when compared to those with a higher Zn intake (Yazar et al., 2005).

6.3.3 COPPER

Copper (Cu) levels in the body are delicately regulated since Cu is required for numerous physiological functions in tiny amounts on the one hand, but too much Cu may be exceedingly damaging on the other. Copper is required for several functions of the immune system. They are well-known for their role in the prevention of anemia as well as their redox characteristics related to proteins such as ceruloplasmin and SOD. Copper deficiency in animals and people with Menke's disease worsens infections with a variety of pathogens (Spallholz & Stewart, 1989).

6.3.3.1 ROLE OF COPPER IN THE IMMUNE SYSTEM

Immunological competency necessitates a certain quantity of Cu availability. Copper enzymes govern several physiological mechanisms, including energy generation, iron metabolism, connective tissue development, and neurotransmission. Copper is a required cofactor for oxidation-reduction reactions. Although cupric copper (Cu^{2+}) makes up the bulk of copper in the body, it may quickly take and give electrons, which explains its function in oxidation-reduction (redox) processes and free radical scavenging (Alpert, 2017). Mild to moderate deficiency in copper resulted in interleukin 2 and T-cell proliferation. The number of neutrophils in human peripheral blood is reduced in severe deficiency, as is their ability to generate superoxide anion and kill ingested microorganisms (Percival, 2018).

Cu deficiency, on the other hand, results in decreased humoral and cellular immunological activity. Cu-deficient mice had a smaller thymus and an expanded spleen, according to research. There is also neutropenia and a decrease in T-cell count. T-cell proliferation driven by mitogens is hindered, as is the function of B- and NK-cells. As a result, increasing plasma Cu levels may improve both innate and adaptive immunity in people (Stabel & Spears, 1989). Cu is considered in this context as a trace element having antiviral properties that can be used to prevent and treat COVID-19 (Raha et al., 2020).

6.3.4 IRON

Iron (Fe) is a dietary mineral that is necessary for key human activities such as erythropoiesis, cellular energy metabolism, and immune system development and function. Despite its importance, Fe deficiency, which causes anemia, is the world's most common nutritional disorder, affecting 25% of the global population; however, the prevalence is significantly greater in specialized groups such as pregnant and non-pregnant women, as well as children (30–47%). Many biological activities require iron as a cofactor. The bulk of iron entering the circulation is recycled after macrophages kill older red blood cells in the reticuloendothelial system.

6.3.4.1 ROLE OF IRON IN THE IMMUNE SYSTEM

Iron plays a role in immunity by promoting immune cell proliferation and maturation, particularly in lymphocytes, which are involved in the generation of immune responses to infection. Because excess iron can potentially aid in the development of infection and tumor cell invasion, the body can reduce the availability of iron consumed by pathogens by binding with proteins such as transferrin and lactoferrin (Alpert, 2017).

The role of Fe in immune system regulation is intriguingly complicated, involving numerous processes that contrast the role and advantages of healthy Fe storage. Fe deficiency, for example, has been demonstrated to affect B-cell proliferation, T-lymphocyte function, and adaptive antibody responses, all of which appear to be successfully rectified by Fe supplementation (Jiang et al., 2019). Within hours after infection, innate human defense has developed to decrease the amount of Fe accessible to invading bacteria. In such cases, a condition of hypoferremia arises, in which Fe concentrations in extracellular fluid and plasma drop drastically, limiting Fe availability to microbial invaders. The inflammatory cytokine cascade stimulates a rise in circulating levels of the hepatic derived peptide, hepcidin, which works to control and decrease the activity of the ferroportin Fe export channels during host invasion. Degradation of the ferroportin Fe export channels is efficient in limiting Fe availability for pathogen use because it leads to Fe sequestration inside macrophages, enterocytes, and hepatocytes, reducing any extracellular Fe levels that may be used for microorganism development (Ganz, 2009; Ganz & Nemeth, 2015). Selenium metabolism, as well as zinc and copper metabolism, may be affected by iron shortage and high food consumption in other reticuloendothelial cells. To avoid lymphoid tissue atrophy (spleen and thymus) and lymphocyte reduction, appropriate amounts of iron are required (Spallholz & Stewart, 1989). Reduced lymphocyte counts due to iron deficiency may result in changes in helper/suppressor T-cell ratios and several lymphocyte surface antigen markers triggered by pokeweed mitogen. Iron deficiency has been shown to reduce or not alter PHA-induced lymphocyte changes (Beisel, 1982).

6.3.5 SELENIUM

Selenium (Se) is a vital trace element that helps humans and animals maintain equilibrium. Skeletal muscle contains 50% of the selenium in the body. It is also an antioxidant found in the selenoprotein glutathione peroxidase (GPx) (Reeves & Hoffmann, 2009). Selenium was discovered in 1817 by the Swedish scientist Jöns Jakob Berzelius and was thought to be hazardous to people and cattle for over 150 years. The advantage of selenium for humans and other animals were discovered in seminal experiments by Klaus Schwartz and Calvin Foltz in 1957, who found that dietary selenium protected rats against liver necrosis (Schwarz & Foltz, 1957).

6.3.5.1 ROLE OF SELENIUM IN THE IMMUNE SYSTEM

Adequate selenium levels are required for immune system function. Since the outbreak of the pandemic, there has been speculation that selenium may have a role in mitigating the consequences of coronavirus illness. Its effect on the symptoms and outcome of SARS-CoV-2 when Se is lacking is of special relevance. The majority of research on this aspect of Se has focused on its nutritive effects in the context of it being one among several components that play roles in immune function, such as vitamins and free fatty acids (FFA) (Broome et al., 2004). According to Zhang et al., surviving COVID-19 patients have higher Se status than non-surviving COVID-19 patients. They also described a linear relationship between the patient's cure rate and their regional Chinese status (Zhang et al., 2020).

Both the innate and acquired immune systems are affected by selenium. It is required for redox control as well as antioxidant activity. For AIDS patients, it is the most critical nutrient. It also protects against the heart-damaging effects of the cytomegalovirus. The antioxidant GSH, which eliminates excess harmful radicals created during oxidative stress (OS), is responsible for selenium's protective benefits (Alpert, 2017). Various aspects of human biologies, such as the central nervous system (Keservani et al., 2016b), male reproductive biology, the endocrine system, muscle function, the cardiovascular system (Keservani et al., 2016a), and immunity, all require adequate quantities of bioavailable selenium (Rayman, 2012).

Selenium deficiency can lead to immune malfunction, making you more susceptible to infections and even cancer. There is some evidence that selenium can affect the pathophysiology of chronic inflammatory diseases of the gut and liver, as well as inflammatory cancers. Increased levels of inflammatory cytokines have been associated with selenium deficiency and decreased selenoprotein expression in a variety of tissues, including the GI system, the uterus, mammary gland tissues, and others. However, when selenium consumption increases from insufficient to sufficient levels, several inflammatory processes increase (Avery & Hoffmann, 2018). Moreover, increasing selenium status by dietary sodium selenite administration create demand and translation of mRNAs encoding stress-related selenoproteins as well as genes implicated in inflammation and IFN responses (Tsuji et al., 2015).

6.3.6 GOLD

Gold (Au) is a trace element, which plays an important role in the treatment of various health disorders and has a big role in the body's immune system. For more than 40 years, antibody-labeled gold nanoconjugates have been utilized in immunohistochemistry (Faulk & Taylor, 1971). From a chemical standpoint, gold compounds demand special attention due to gold's unique location in the periodic table, which eventually leads to the highest electronegativity, electron affinity, and redox potential when compared to other metals. Gold compounds cause cytotoxicity by blocking thiol-containing enzymes, particularly TrxR, and causing mitochondrial and DNA damage, which may all contribute to their clinical anticancer efficacy. Many researchers have recently discovered that auranofin, a gold molecule commonly used in antirheumatic treatment, also has anticancer, antimicrobial, and other characteristics (Yue et al., 2020).

6.3.6.1 ROLE OF GOLD IN THE IMMUNE SYSTEM

The use of gold medicines is also frequently associated with unfavorable immunological responses such as dermatitis, glomerulonephritis, cytopenias, hepatitis, and pneumonitis (Havarinasab et al., 2007). Rheumatoid arthritis is a chronic, systemic, inflammatory disease that affects the articulating joints and is marked by immune system defects such as rheumatoid factor synthesis (immunoglobulin to native IgG), inflammation, and swelling, as well as the invasion of activated lymphocytes, PMNs, monocytes, and macrophages. Biological Significance Gold salts, gold thiomalate (myochrysine), gold thioglucose (aurofin), and penicillamine are used to treat rheumatoid arthritis. To minimize inflammation, gold salts are injected intramuscularly in mg increments over several weeks. Gold's mechanism of action in the treatment of rheumatoid arthritis is unclear, and its application has been primarily dependent on trial and error. Lysozymes from macrophages and PMNs accumulate gold, which inhibits lymphocyte mitogenesis and DNA replication. Gold salts limit lymphocyte mitogenesis and DNA replication in the same manner as deficiencies in selenium, zinc, magnesium, and copper do. Not only does gold therapy affect selenium redistribution *in vivo*, but it also affects zinc and copper redistribution. These minerals, selenium, zinc, and copper, are anti-oxidant enzymes GPx and SODs' prosthetic moieties. As a result, gold therapy is likely to have an impact not just on selenium but also on other minerals (Dillard & Tappel, 1986; Takahashi et al., 1994).

6.3.7 CALCIUM

Calcium (Ca) is the most prevalent mineral in vertebrates, making up more than half of the mineral content of humans. Calcium is almost entirely contained in hydroxyapatite, which is found in bone and teeth. The remaining calcium is ionized and binds strongly to a variety of calcium-binding proteins in extracellular fluids. Calcium has gotten minimal experimental attention into what function(s) it may have on the immune system and reticuloendothelial system cells while contributing to various critical physiological and enzymatic processes, which is probably unsurprising.

6.3.7.1 ROLE OF CALCIUM IN THE IMMUNE SYSTEM

Calcium (Ca^{2+}) signaling's importance in immunity cannot be emphasized. In lymphocytes, increased cytosolic and organellar Ca^{2+} concentrations regulate complex and critical effector actions such as metabolism, proliferation, differentiation, antibody, and cytokine production, and cytotoxicity. A change in Ca^{2+} regulation in lymphocytes causes a variety of autoimmune, inflammatory, and immunodeficiency disorders. T-cells have a variety of Ca^{2+}-permeable plasma membrane and organellar channels. They provide functional specificity by providing extremely limited spatial and temporal Ca^{2+} micro-regions. While the molecular details of these Ca^{2+} microdomains are still being uncovered, it is obvious that they use crosstalk, synergy, and feedback mechanisms to fine-tune T-cell signaling to meet complex immunological responses (Trebak & Kinet, 2019). Calcium-binding proteins, which are located in cell membranes, control Ca^{++} ion channel passage in and out of cells. *In vitro*, PHA mitogenesis raises the intracellular calcium concentration of lymphocytes, whereas lipoproteins that inhibit PHA mitogenesis lower it. EDTA, citrate, and calcium chelators reduce lymphocyte PHA responsiveness, which can be restored by calcium (Hui & Harmony, 1980). In several cell types, including lymphocytes, calcium works as a second messenger. Resting lymphocytes have a low Ca^{2+} concentration. However, activation of antigen receptors causes calcium influx from the extracellular environment via a variety of mechanisms. Ca^{2+} entrance into lymphocytes is primarily mediated by store-operated calcium (SOC) channels. The discovery of two essential molecular components of SOC channels, CRACM1 (the pore-forming subunit) and STIM1 (the stored calcium sensor), has enabled genetic and molecular modification of the SOC entry route (Vig & Kinet, 2009).

6.3.8 SODIUM

Sodium (Na) is one of the most common trace elements present in common salt. Although sodium is required for basic bodily processes, it can be harmful if consumed more than dietary needs. As long as high salt consumption prevails across the world, the repercussions of excessive salt consumption are becoming increasingly obvious.

6.3.8.1 ROLE OF SODIUM IN THE IMMUNE SYSTEM

Salt has long been suspected of promoting hypertension and cardiovascular illness, but it is now being recognized as a possible modulator of inflammatory and autoimmune disorders through its direct and indirect effects on immune cells. The discovery that additional organs, such as the skin, regulate sodium levels in the body in addition to the kidneys led to novel theories, including the idea that skin-resident macrophages may engage in tissue sodium regulation through their interactions with lymphatic vessels (Wilck et al., 2019). In sodium-rich interstitial microenvironments, immune cells such as macrophages and numerous T-cell subsets can be located, where sodium levels alter their activity. It was over two decades ago that the effects of salt on macrophage function were discovered. Osmotic stress generated by hypertonic salt solutions, such as NaCl, boosted cytokine production and proliferation in human and rabbit peripheral blood mononuclear cells (PBMCs) and the human macrophage THP-1 cell line, according to these early findings (Shapiro & Dinarello, 1995). Excess dietary salt alters the gut bacterial flora, which is another important axis via which salt regulates immune cell activity indirectly. Sodium may contribute to protective immunity (for example, by boosting host responses against cutaneous infections) or it may contribute to immunological dysregulation and accelerate the development of cardiovascular and autoimmune illnesses, depending on the inflammatory setting. Excessive salt consumption promotes the development of essential hypertension, according to epidemiological and experimental research. Because hypertension and its organ sequelae are characterized by a significant inflammatory component, it's worth looking at how immune-mediated activities controlled by extracellular salt contribute to hypertension. Hypertension immunology is complicated, including cells from both the innate and adaptive immune systems (Wilck et al., 2019).

6.3.9 PHOSPHORUS

Phosphorus (P) is an important nutrient for both microorganisms and their hosts, since it plays a role in bone formation, energy metabolism, cellular signaling, and cell membrane stabilization, among other things. P is crucial in agriculture and the food industry as a mineral incorporated in materials used for livestock feeding as well as mineral fertilizers, with about 90% of the world's mined phosphate rock being utilized in agriculture and the food industry. Phytate, the major storage form of P in plant seeds, is only available to non-ruminant animals in restricted amounts. The hydrolysis of phytate in non-ruminant animals is inadequate because the proximal GIT lacks sufficient enzymes, such as endogenous mucosalphytase and phosphatase (Heyer et al., 2015).

6.4 CONCLUSION

Humans are protected from infectious germs in the environment as well as unpleasant irritants by their immune systems. When any of these events occur, the body utilizes two strategies to protect itself: innate and adaptive; both cause blood-borne chemicals (such as complement, antibodies, and cytokines) and immune response cells to upregulate (such as macrophages, polymorphonuclear (PMN) cells, and lymphocytes). In the general population, there is interindividual variation in immune functions; these variations are impacted by age, gender, and degree of activity, smoking, alcohol usage, stress, menstrual cycle, obesity, food, and other factors. Immune cells in the blood, lymph nodes, and the specialized immune system of the GIT interact with nutrients derived from dietary proteins, carbohydrates, and fats, as well as micronutrients, vitamins, and minerals. The immune system is immature at birth, matures and changes during life, and eventually reaches senescence. Nutrient needs for immune development differ between the prenatal and postnatal periods and are heavily influenced by childhood and teenage growth and development processes. Furthermore, the aging process impacts dietary needs, which has a significant impact on immune function. Nutrition has a significant impact on the immune system's growth and decline. As a result, nutritional immune regulation has arisen as a major axis in nutrition and health research, with the goal of better "programming" newborn immunity, maintaining immunological homeostasis throughout life, and increasing immunity in the elderly. Customers are followed by current immune-modulating diets at various periods of their lives and activities.

Micronutrients are an important aspect of the immune system, and the body needs adequate amounts to operate properly. Overt micronutrient deficiencies have long been known to harm the immune system and make people more susceptible to infections. Although the effect may be less dramatic than that seen with overt deficiencies, marginal deficits are likely linked to an increased risk of infection. Worldwide, particularly in industrialized nations, dietary consumption of many micronutrients is insufficient, raising the risk of illness. Furthermore, accumulating research shows that consuming more micronutrients than the RDA might assist boost immune function and consequently infection resistance (Gombart et al., 2020). The necessity of the indicated minerals for the regular operation of the immune system is stressed in scientific literature. Although mineral shortages are infrequent, a few at-risk groups should be watched to ensure they receive adequate supplies. A well-balanced diet is a great place to start. Supplementing should only be used to repair a deficiency in extreme conditions; nevertheless, it should be noted that some mineral supplements can be over-consumed, which might impair the immune system. As a result, any rehabilitative nutritional supplements should always be discussed with a doctor and administered only in the amounts advised. Furthermore, a healthy vitamin and mineral balance may provide some protection against infections and inflammatory illnesses. The best method to maintain a healthy balance of macro-and micronutrients is to consume a variety of meals. Supplements are a more convenient method to get the nutrients you need, but they can also have harmful side effects.

ACKNOWLEDGMENT

The authors of this chapter thank the management of their institution for providing support to complete this chapter.

KEYWORDS

- **antibody**
- **antigen**
- **deficiency**
- **immunoglobulin**
- **lymphocytes**
- **trace elements**

REFERENCES

Alpert, P. T., (2017). The role of vitamins and minerals on the immune system. *Home Health Care Management and Practice, 29*(3), 199–202. https://doi.org/10.1177/1084822317713300.

Avery, J. C., & Hoffmann, P. R., (2018). Selenium, selenoproteins, and immunity. *Nutrients, 10*(9). https://doi.org/10.3390/nu10091203.

Beisel, W. R., (1982). Single nutrients and immunity. *The American Journal of Clinical Nutrition, 35*(2), 417–468.

Broome, C. S., McArdle, F., Kyle, J. A. M., Andrews, F., Lowe, N. M., Hart, C. A., Arthur, J. R., & Jackson, M. J., (2004). An increase in selenium intake improves immune function and poliovirus handling in adults with marginal selenium status. *American Journal of Clinical Nutrition, 80*(1), 154–162. https://doi.org/10.1093/ajcn/80.1.154.

Cunningham-Rundles, S., (1998). Analytical methods for evaluation of immune response in nutrient intervention. *Nutrition Reviews, 56*(1 II). https://doi.org/10.1111/j.1753-4887.1998.tb01641.x.

Delves, P. J., Martin, S. J., Burton, D. R., & Roitt, I. M., (2017). *Roitt's Essential Immunology.* pp. 188–204

Dillard, C. J., & Tappel, A. L., (1986). Mercury, silver, and gold inhibition of selenium-accelerated cysteine oxidation. *Journal of Inorganic Biochemistry, 28*(1), 13–20. https://doi.org/10.1016/0162-0134(86)80019-8.

Faulk, W., & Taylor, G., (1971). An immunocolloid method for the electron microscope. *Immunochemistry, 8*(11), 1081–1083.

Filteau, S., & Woodward, B., (1987). The effect of triiodothyronine on the primary antibody response to sheep red blood cells in moderately undernourished weanling mice. *Nutrition Research, 7*(c), 471–479.

Galland, L., (1988). Magnesium and immune function: An overview. *Magnesium, 7*(5, 6), 290–299.

Ganz, T., & Nemeth, E., (2015). Iron homeostasis in host defense and inflammation. *Nature Reviews Immunology, 15*(8), 500–510. https://doi.org/10.1038/nri3863.

Ganz, T., (2009). Iron in innate immunity: Starve the invaders. *Current Opinion in Immunology, 21*(1), 63–67. https://doi.org/10.1016/j.coi.2009.01.011.

Gniadecki, R., Gajkowska, B., & Hansen, M., (1997). 1,25-dihydroxyvitamin D3 stimulates the assembly of adherens junctions in keratinocytes: Involvement of protein kinase C. *Endocrinology, 138*(6), 2241–2248. https://doi.org/10.1210/endo.138.6.5156.

Gombart, A., Pierre, A., & Maggini, S., (2020). A review of micronutrients and the immune system-working in harmony to reduce the risk of infection. *Nutrients, 12*(1), 236. https://doi.org/doi: 10.3390/nu12010236.

Havarinasab, S., Johansson, U., Pollard, K. M., & Hultman, P., (2007). Gold causes genetically determined autoimmune and immunostimulatory responses in mice. *Clinical and Experimental Immunology, 150*(1), 179–188. https://doi.org/10.1111/j.1365-2249.2007.03469.x.

Heyer, C. M. E., Weiss, E., Schmucker, S., Rodehutscord, M., Hoelzle, L. E., Mosenthin, R., & Stefanski, V., (2015). The impact of phosphorus on the immune system and the intestinal microbiota with special focus on the pig. *Nutrition Research Reviews, 28*(1), 67–82. https://doi.org/10.1017/S0954422415000049.

Hoffmann, J. A., Kafatos, F. C., Janeway, C. A., & Ezekowitz, R. A., (1999). Phylogenetic perspectives in innate immunity. *Science, 284*(5418), 1313–1318.

Hui, D. Y., & Harmony, J. A., (1980). Inhibition of Ca^{2+} accumulation in mitogen-activated lymphocytes: Role of membrane-bound plasma lipoproteins. *Proceedings of the National Academy of Sciences of the United States of America, 77*(8), 4764–4768. https://doi.org/10.1073/pnas.77.8.4764.

Jahnen-Dechent, W., & Ketteler, M., (2012). Magnesium basics. *CKJ: Clinical Kidney Journal, 5*(SUPPL. 1). https://doi.org/10.1093/ndtplus/sfr163.

Jiang, Y., Li, C., Wu, Q., An, P., Huang, L., Wang, J., Chen, C., et al., (2019). Iron-dependent histone 3 lysine 9 demethylation controls B cell proliferation and humoral immune responses. *Nature Communications, 10*(1). https://doi.org/10.1038/s41467-019-11002-5.

Johnson, C. D., & Kudsk, K. A., (1999). Nutrition and intestinal mucosal immunity. *Clinical Nutrition, 18*(6), 337–344. https://doi.org/10.1016/S0261-5614(99)80012-0.

Kellie, S., & Al-Mansour, Z., (2017). Overview of the immune system. *Micro – and Nanotechnology in Vaccine Development, 357*, 63–81. https://doi.org/10.1016/B978-0-323-39981-4.00004-X.

Keservani, R. K., Kesharwani, R. K., Sharma, A. K., & Jarouliya, U., (2015). Dietary supplements, nutraceutical and functional foods in immune response (immunomodulators). In: Debasis, B., Harry, G. P., & Anand, S., (eds.), *Nutraceutical and Functional Foods in Human Life and Disease Prevention* (pp. 343–358). Boca Raton, Fl. CRC Press, Taylor and Francis, Chapter 20. ISBN: 9781482237214.

Keservani, R. K., Kesharwani, R. K., Sharma, A. K., Vyas, N., & Chadoker, A., (2010b). Nutritional supplements: An overview. *International Journal of Current Pharmaceutical Review and Research, 1*(1), 59–75.

Keservani, R. K., Kesharwani, R. K., Vyas, N., Jain, S., Raghuvanshi, R., & Sharma, A. K., (2010a). Nutraceutical and functional food as future food: A review. *Der Pharmacia Lettre, 2*(1), 106–116.

Keservani, R. K., Sharma, A. K., & Kesharwani, R. K., (2016a). Nutraceutical and functional foods for cardiovascular health. In: *Food Process Engineering* (pp. 257–278, 291–312). Apple Academic Press, CRC Press. ISBN: 9781771884020.

Keservani, R. K., Sharma, A. K., & Kesharwani, R. K., (2016b). Medicinal effect of nutraceutical fruits for the cognition and brain health. *Scientifica*, 10. Article ID 3109254. doi: 10.1155/2016/3109254.

Keservani, R. K., Sharma, A. K., & Kesharwani, R. K., (2017). An overview and therapeutic applications of nutraceutical and functional foods. *Recent Advances in Drug Delivery Technology*, 160–201. ISBN:9781522507543.

Keservani, R. K., Sharma, A. K., & Kesharwani, R. K., (2020). *Nutraceuticals and Dietary Supplements: Applications in Health Improvement and Disease Management*. CRC Press. ISBN: 9781771888738.

Keusch, G. T., (2003). Symposium: Nutrition and infection, prologue and progress since 1968 the history of nutrition: Malnutrition, infection and immunity. *J. Nutr, 133*, 336–340.

Kim, M. H., & Choi, M. K., (2013). Seven dietary minerals (Ca, P, Mg, Fe, Zn, Cu, and Mn) and their relationship with blood pressure and blood lipids in healthy adults with self-selected diet. *Biological Trace Element Research, 153*(1–3), 69–75. https://doi.org/10.1007/s12011-013-9656-1.

Kussmann, M., & Fay, L., (2008). *Nutrigenomics and Personalized Nutrition* (Vol. 5, pp. 447–455). Nestle Research Center. http://hub.harper-adams.ac.uk/moodle/pluginfile.php/129501/mod_resource/content/0/Nutrigenomicsand personalised nutrition.

pdf%0Ahttps://www.futuremedicine.com/doi/10.2217/17410541.5.5.447 (accessed on 20 February 2023).

Lesourd, B., (2004). Nutrition: A major factor influencing immunity in the elderly. *The Journal of Nutrition, Health & Aging, 8*(1), 28–37.

Maggini, S., Pierre, A., & Calder, P. C., (2018). Immune function and micronutrient requirements change over the life course. *Nutrients, 10*(10). https://doi.org/10.3390/nu10101531.

Pálmer, H. G., González-Sancho, J. M., Espada, J., Berciano, M. T., Puig, I., Baulida, J., Quintanilla, M., et al., (2001). Vitamin D3 promotes the differentiation of colon carcinoma cells by the induction of E-cadherin and the inhibition of β-catenin signaling. *Journal of Cell Biology, 154*(2), 369–387. https://doi.org/10.1083/jcb.200102028.

Percival, S. S., (2018). *Copper and Immunity, 67*, 1064–1068.

Raha, S., Mallick, R., Basak, S., & Duttaroy, A. K., (2020). Is copper beneficial for COVID-19 patients? *Medical Hypotheses, 142*, 109814. https://doi.org/10.1016/j.mehy.2020.109814.

Rajagopalan, S., & Long, E. O., (1998). Zinc bound to the killer cell inhibitory receptor modulates the negative signal in human NK cells. *J. Immunol., 161*, 1299–1305.

Rayman, M. P., (2012). Selenium and human health. *The Lancet, 379*(9822), 1256–1268. https://doi.org/10.1016/S0140-6736(11)61452-9.

Reeves, M. A., & Hoffmann, P. R., (2009). The human selenoproteome: Recent insights into functions and regulation. *Cellular and Molecular Life Sciences, 66*(15), 2457–2478. https://doi.org/10.1007/s00018-009-0032-4.

Rink, L., & Haase, H., (2007). Zinc homeostasis and immunity. *Trends in Immunology, 28*(1), 1–4. https://doi.org/10.1016/j.it.2006.11.005.

Roohani, N., Hurrell, R., Kelishadi, R., & Schulin, R., (2013). Zinc and its importance for human health: An integrative review. *J. Res. Med. Sci., 28*(144), 157.

Schubert, C., Guttek, K., Reinhold, A., Grüngreiff, K., & Reinhold, D., (2015). The influence of the trace element zinc on the immune system. *Laboratoriums Medizin*, 8–9.

Schwarz, K., & Foltz, C. M., (1957). Selenium as an integral part of factor 3 against dietary necrotic liver degeneration. *Journal of the American Chemical Society, 79*(12), 3292–3293. https://doi.org/10.1021/ja01569a087.

Scrimshaw, N. S., & SanGiovanni, J. P., (1997). Synergism of nutrition, infection, and immunity: An overview. *American Journal of Clinical Nutrition, 66*(2), 464–477. https://doi.org/10.1093/ajcn/66.2.464S.

Shapiro, L., & Dinarello, C. A., (1995). Osmotic regulation of cytokine synthesis *in vitro*. *Proc. Natl Acad. Sci., 92*, 12230–12234.

Souba, W. W., (1991). Glutamine: A key substrate for the splanchnic bed. *Annual Review of Nutrition, 11*, 285–308. https://doi.org/10.1146/annurev.nu.11.070191.001441.

Spallholz, J. E., & Stewart, J. R., (1989). Advances in role of minerals in immunology. *Biological Trace Element Research, 19*.

Stabel, J. R., & Spears, J. W., (1989). Effect of copper on immune function and disease resistance. *Adv. Exp. Med. Biol., 258*, 243–252.

Sugimoto, J., Romani, A. M., Valentin-Torres, A. M., Luciano, A. A., Ramirez, K. C. M., Funderburg, N., Mesiano, S., & Bernstein, H. B., (2012). Magnesium decreases inflammatory cytokine production: A novel innate immunomodulatory mechanism. *The Journal of Immunology, 188*(12), 6338–6346. https://doi.org/10.4049/jimmunol.1101765.

Takahashi, K., Griem, P., Goebel, C., Gonzalez, J., & Gleichmann, E., (1994). The antirheumatic drug gold, a coin with two faces: AU(I) and AU(III). Desired and undesired effects on the immune system. *Metal-Based Drugs, 1*(5, 6), 483–496. https://doi.org/10.1155/mbd.1994.483.

Takiishi, T., Fenero, C. I. M., & Câmara, N. O. S., (2017). Intestinal barrier and gut microbiota: Shaping our immune responses throughout life. *Tissue Barriers, 5*(4), 1–12. https://doi.org/10.1080/21688370.2017.1373208.

Tam, M., Gómez, S., González-Gross, M., & Marcos, A., (2003). Possible roles of magnesium on the immune system. *European Journal of Clinical Nutrition, 57*(10), 1193–1197. https://doi.org/10.1038/sj.ejcn.1601689.

Trebak, M., & Kinet, J. P., (2019). Calcium signaling in T cells. *Nature Reviews Immunology, 19*(3), 154–169. https://doi.org/10.1038/s41577-018-0110-7.

Trichet, V. V., (2010). Nutrition and immunity: An update. *Aquaculture Research, 41*(3), 356–372. https://doi.org/10.1111/j.1365-2109.2009.02374.x.

Tsuji, P. A., Carlson, B. A., Anderson, C. B., Seifried, H. E., Hatfield, D. L., & Howard, M. T., (2015). Dietary selenium levels affect selenoprotein expression and support the interferon-γ and IL-6 immune response pathways in mice. *Nutrients, 7*(8), 6529–6549. https://doi.org/10.3390/nu7085297.

Vig, M., & Kinet., J. P., (2009). Calcium signaling in immune cells. *Nat Immunol., 10*(1), 21–27.

Weyh, C., Krüger, K., Peeling, P., & Castell, L., (2022). *The Role of Minerals in the Optimal Functioning of the Immune System*, 1–15.

Wilck, N., Balogh, A., Markó, L., Bartolomaeus, H., & Müller, D. N., (2019). The role of sodium in modulating immune cell function. *Nature Reviews Nephrology, 15*(9), 546–558. https://doi.org/10.1038/s41581-019-0167-y.

Yazar, M., Sarban, S., Kocyigit, A., & Isikan, U. E., (2005). Synovial fluid and plasma selenium, copper, zinc, and iron concentrations in patients with rheumatoid arthritis and osteoarthritis. *Biological Trace Element Research, 106*(2), 123–132. https://doi.org/10.1385/BTER:106:2:123.

Yue, S., Luo, M., Liu, H., & Wei, S., (2020). Recent advances of gold compounds in anticancer immunity. *Frontiers in Chemistry, 8*. https://doi.org/10.3389/fchem.2020.00543.

Zhang, J., Saad, R., Taylor, E. W., & Rayman, M. P., (2020). Selenium and selenoproteins in viral infection with potential relevance to COVID-19. *Redox Biology, 37*, 101715. https://doi.org/10.1016/j.redox.2020.101715.

CHAPTER 7

ROLE OF BIOACTIVE COMPOUND OF FRUITS IN BOOSTING IMMUNITY

SWATI GUPTA,[1] SHANTHY SUNDARAM,[1] and
MD. AKHLAQUR RAHMAN[2]

[1]*Center of Biotechnology, Nehru Science Center, University of Allahabad, Prayagraj, Uttar Pradesh, India*

[2]*Department of Biotechnology, S.S. Khanna Girls' Degree College, Prayagraj, Uttar Pradesh, India*

ABSTRACT

The aim of this study was to determine the phytochemical substances which are present in different citrus fruit peels such as orange (*citrus sinensis*), kinnow (*Citrus reticulata*), and mosambi (*Citrus limetta)* grapefruit which are rich source of immune boosting substances. The citrus fruit peels are a good source of antioxidant, phenolic, and bioactive compounds. Organic and aqueous extracts of citrus fruit such as orange (*Citrus sinensis*), kinnow (*Citrus reticulata*), and mosambi (*Citrus limetta*), grapefruit (*Citrus paradise*), etc., having the phytochemical compound which boosts the immunity and provide the protection against many diseases in human beings. Citrus fruit showed the antioxidant activity, which also helps to improve immunity in humans. So, it is used as a therapeutic agent. Citrus fruit shown the pharmacological activity and exhibited the good result against the pathogenic bacteria. Citrus fruit having the phenolic and pectin compound, secondary compound, bioactive compound, which shown the antiviral, antifungal, anti-inflammatory, anti-typhoid, anticancer, anti-obesity, anti-diabetic activity. So citrus fruit plays an important role as an immunity booster compound.

Immune-Boosting Nutraceuticals for Better Human Health: Novel Applications. Urmila Jarouliya, PhD, Raj K. Keservani, MPharm, Rajesh K. Kesharwani, PhD, Virendra K. Patel, PhD, and Adi D. Bharti, MPharm (Eds.)

7.1 INTRODUCTION

Citrus fruit belongs to the family, Rutaceae, and it is a flowering plant. Citrus fruit mainly originated from the South-east Asia or China. Different varieties of the citrus fruit originate from the different countries of the whole world. China and India cultivate different varieties of citrus fruit such as orange, sweet oranges, mandarin in a large scale in every year, whereas the Malaysia originated the acid citrus in a large amount. However, different varieties of the citrus fruit such as mosambi, kinnow, grapefruit, lime, lemon, mandarin, etc., are originated by north and south Asia. Citrus fruit has been cultivated in an ever-widening area since ancient times; the best-known examples are the oranges, lemons, grapefruit, and limes. Citrus fruits are an excellent source of the vitamin C, minerals and other biological compounds including different phytochemical compound such as alkaloids, tannin, pectin's, terpenoid, ascorbic acid, flavonoids, and steroids that are very beneficial for human health (Chowdhury et al., 2008; Zheng et al., 2001). The main flavonoids present in citrus plant are Hesperidin, Narirutin, and Naringin. Citrus fruit also contain a rich source of secondary compound. Citrus fruit is an important economic fruit crop which may have produced 90 million tons of citrus fruit in an all over world (Negi et al., 2012; Kim et al., 2010; Singh et al., 2020). The Citrus fruit is a good source of bioactive compounds and many polyethoxylated flavones (PMFs) compounds, which are very rarely present in other plants (Etebu & Nwauzoma, 2014).

Pharmacological industries have produced a new antibiotics drug which resistance to microorganisms. In general, some bacteria have the genetic ability to transmit and resistance to drugs, which are also used as a therapeutic agent. Citrus fruit contains the anticancer, antiviral, antidiabetic, and antifungal properties, so, it is having the great potential for the treatments of therapeutic (Babbar et al., 2011). Numerous citrus plants have been used because of their antibacterial, antifungal, and antiviral trait. Citrus plants can be used as an essential oil, which are more effective in many different diseases (Goulas et al., 2012). Hence different organic extracts of different fruit and essential oils (EOs) of different fruit have contained the many beneficial properties which effective on the microorganism. Citrus fruit minimize the infectious and treatment of different kind of diseases (Kim et al., 2010).

Citrus fruit is a medicinal plant which produce many bioactive compounds and contained anticancer, anti-inflammatory, antidiabetic properties and is very effective against the numerous bacterial diseases. Organic and aqueous extracts obtained from different parts of citrus fruits have been reported against clinically significant bacterial strains. Citrus fruit contains diverse secondary metabolic compounds like as flavonoids, alkaloids, tannin, terpenoids, steroids, etc., which showed biological activity against the bacteria, fungi, virus, and other pathogenic microorganism (Negi et al., 2012; Babber et al., 2011). Flavonoids and alkaloids showed direct antioxidants activity, which has the capacity to increase the enzymatic activities and inhibit cell proliferation (Moo-Huchin et al., 2015). In plants, they inhibit the pathogens, such as bacteria, fungi, and viruses (Negi & Jayaprakasha, 2001). Crude extract of Citrus fruits is a good source of alkaloids, flavonoids, tannin, steroids, glycosides, coumarins. Citrus fruits contained the many polymethoxylated flavones which shows several important bioactivities (Zou et al., 2016). Citrus plant provides numerous compounds of widely differing structures and is constitutes a source of new chemical compounds, so it is having the great potential for the medicinally and physiologically uses. Citrus plant is a traditional and ethnomedicinal plant which, having the good pharmacological activity against some disease and also acts as analgesic, anti-helminthic, anti-typhoid, anti-inflammatory agent. "Phytotherapy" focuses on medicinal plants with well-characterized extracts that are used to treat specific illnesses. An important aspect of phytotherapy is the standardization of plant extracts and methods for the biological and pharmacological testing of such extracts, which are considered to be effective because of the combination of active ingredients rather than a single constituent (Babber et al., 2011; Espiard et al., 2002).

As Citrus plant are rich source of flavonoids, vitamin, minerals, dietary fiber, sugar, ascorbic acid, polyphenols, and other beneficially nutrient, they are applicable to minimize the infection of the various chronic disease such as cardiovascular disease (CVD) or heart disease (Etebu & Nwauzoma, 2014; Keservani et al., 2016a). Different fruits of citrus include the phenolic acids, gallic acid and having the 60 flavonoids, which divided in to the three groups of flavones, flavanones, and flavanols and it can be prevent the stress of the oxidative and reduced the serious illness in the bodies. Hesperidin, naringenin, and aglycone compound have been showed the therapeutic activity and reduced the activity of some enzyme such as cyclooxygenase, phosphodiesterase, protein kinase, which are very important for the activation and transduction of the cell signals. Citrus fruit have the hemicelluloses fiber, ash, fat, and protein which can be used in the nutraceutical and pharmaceutical industry (Keservani et al., 2010a, b, 2017, 2020). Bioactive compounds present in Citrus fruits has tendency to inhibit the oxidative stress (OS) and also helped to prevent the tissue damage, which caused by the harmful free radical in the cell. Different extracts of citrus fruits such as mandarin, kinnow, mosambi, grapefruit, and orange also helped to prevent the disease of *Mycobacterium tuberculosis* (TB) (Etebu & Nwauzoma., 2014; Rajendran et al., 2014). Citrus fruits (Keservani et al., 2016b) are used in the pharma industry, cosmetic industry, and other food industry, where the different items are made by the citrus fruits such as jams, Jellies, juice, and other food supplement in a very reasonable price. Citrus fruits contain many beneficial secondary compounds, which are rarely to present in the other plants (Zou et al., 2016; Babber et al., 2011; Espiard et al., 2002). Phytochemicals also known as secondary metabolites, which is used as a drug and they provide the characteristics of plant parts such as flavor and color. Phytochemical estimation is very useful to produce some active biological compounds in medicinal plants and vegetables. The presence of phytochemicals compounds explains the use of fruits and vegetables in ethnomedicine for the management of various elements. Secondary metabolites obtained from citrus fruits showed various pharmacological and biochemical activities in animals. Citrus plants contained the medicinal and chemical properties and perform pharmacological action against the pathogens, which cause the serious illness in the human body. Extracts of different parts (pulp, peel, root, juice, seed, and stem) of kinnow, orange, mosambi, and grape fruits are very effective against the various pathogenic bacteria (Singh et al., 2020; Kulkarni et al., 2012). Citrus fruits exhibited the good biological activity such as antiviral, anti-obesity (Jarouliya & Keservani, 2019a), antibacterial, antifungal, anti-diabetic, anti-proliferative, and anti-typhoid activities due to presence of many secondary compound. All stated citrus fruits are generally of sour and sweet flavor (Figure 7.1).

FIGURE 7.1 Different citrus fruit.

Citrus fruits are mainly used for the extraction of phenolic, organic phytochemical compounds and essential oil, which are very important for the human health and have anticancer, anti-cardiovascular, antioxidant, and other properties, which is beneficial for the health.

Citrus fruits are cultivated in all over the world. Different varieties of citrus fruits are grown in different countries of the northern and southern hemispheres. Some citrus fruits are cultivated in all seasons approximately and less susceptible to harm. Total production of citrus fruits is much higher than the other fruit crops in all over the world. Brazil produces a large volume of the different varieties of citrus fruit in the world such as oranges (65%), kinnow (26%), mosambi (19%), lemons (14%), limes (12%) grapefruit (8%). Other countries produce orange fruit in different capacities, such as the China cultivate (26%), India (25%), United States (12%), Egypt (10%), Spain (10%). And Japan (6% each), while China (55%), India (23%), Mexico (15%), Argentina (14%), Spain (9%), Brazil, and Japan (5%) produces the kinnow fruits. In the other way India (17%), China (14%), Spain (9%) and Japan (5% each) and India (17%) produces the lemons in a large amount. Further, United States (18%) Mexico (15%), Argentina (14%), Spain (8%) and South Africa (7%) was produced grape fruit in all over world (Ladaniya et al., 2008; Berk et al., 2016) (Figure 7.2). Citrus fruit cultivated in new areas, where it produces a large number of different varieties of fruits, which is not affected in a harsh condition in any season. As the costs of land, labor, and fuel rise, new technologies will also help to increase the productivity of citrus fruits in the world and increase the productivity of land for fruits in the world market.

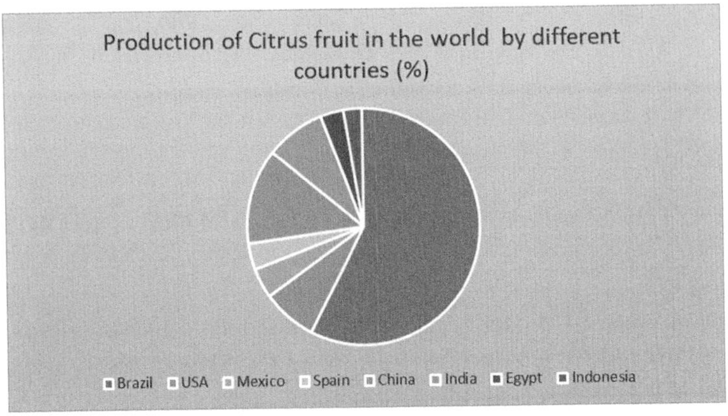

FIGURE 7.2 Total production of citrus fruit in the world by different countries.

A group of the sweet orange contains navel oranges and blood orange. Citrus fruits contain different compounds including quercetin, myricetin, fisetin, pectin, dietary compound, fiber, essential oil, etc. Citrus fruits commonly used in pharmaceutical, cosmetic, and food industries. Citrus fruits are important for economical and used in all countries in a large scale, since it is an excellent source of energy Citrus waste are also used in pharmaceuticals and other industries, where it produces many beneficial products in a marginal cost. Some other industries produce the food supplement such as juice, jams, and other product in a small scale. Moreover, some other large industries are also involved in the production of the EOs and different kind of fruit products due to rich source of vitamin C, good taste, pleasant aroma, and other phytochemical compounds (Sing et al., 2020). Utilization of different citrus fruit are given in Figure 7.3.

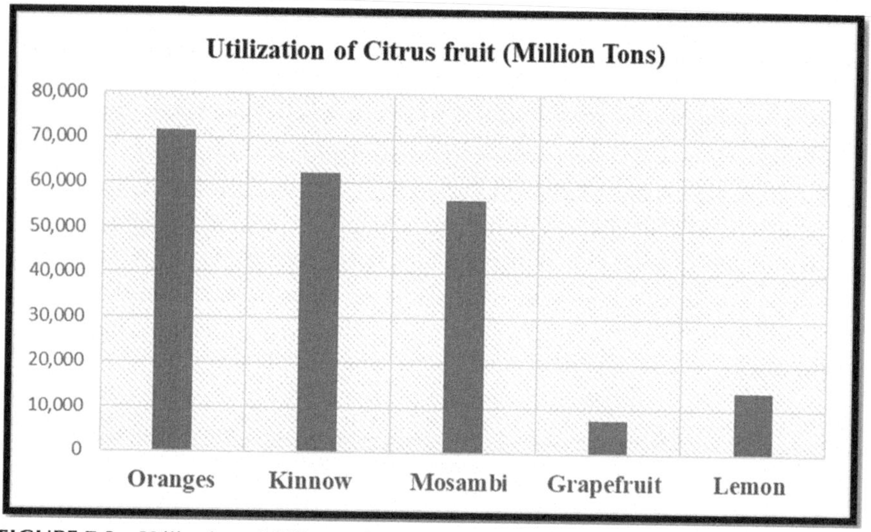

FIGURE 7.3 Utilization of different citrus fruit in the world.

7.2 PHYTOCHEMICAL COMPOUND OF DIFFERENT CITRUS FRUIT

The various type of bioactive chemical components like alkaloids, terpenoids, coumarins, phenylpropanoids, tannin's, flavonoids (Keservani & Sharma, 2014), polysaccharide, carotenoids, seed oil, steroids, phenolics, pectin's have been isolated and identified from various parts of citrus fruit

(Goulas et al., 2012). Different phytochemical compounds are isolated from citrus plant parts such as flowers, seeds, leaves, roots, and fruit are discussed in subsections.

7.2.1 ALKALOIDS

Alkaloids are secondary compounds, contains the nitrogen atom and are found naturally. Alkaloids showed physiological activity in human's body. Alkaloids are very useful in the pharmaceuticals industry and also used as a supplement. Alkaloids are effective against the bacteria, fungi, and virus, which provides the protection against the many pathogenic disease (Gang et al., 2021). Citrus fruits contained the various type of alkaloids, which have been reported as morphine, quinine, ephedrine, nicotine, cocaine, vincristine, vinblastine, berberine, L-hyoscyamine, L-phenylalanine, L-histidine, L-arginine, L-ornithine, ethyl cinnamide, aegelenine, O-methyl halfordinine, isopentenyl halfordinol (Negi et al., 2012) are given in Figure 7.4.

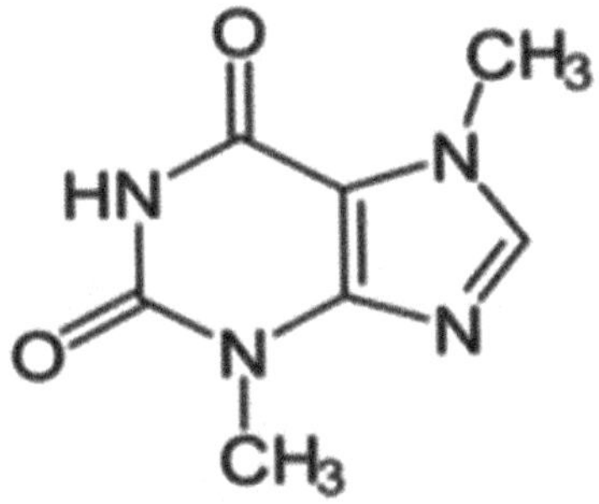

FIGURE 7.4 Chemical structure of alkaloids.

7.2.2 COUMARINS

Several coumarins such as aegeline, aegelenine, marmelin, furocoumarins, O-isopentenyl halfordinal, marmelosine, scoporane, apiaceae, oleaceae, clusiaceae, guttifeae, umbelliferone, and psoralon. Beside these Methoxsalen (8-methoxypsoralen) coumarins also has been reported. Esculetin, fraxinus, esculin compounds are chemical constituent, which are present in the leaves, wood, and root of citrus fruit (Amutha et al., 2017) (Figure 7.5).

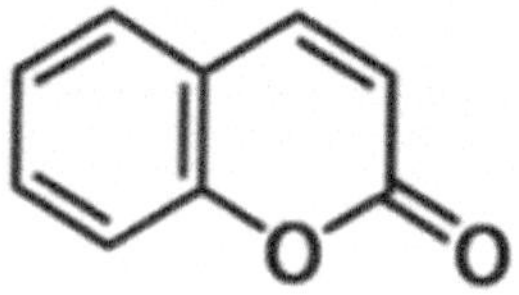

FIGURE 7.5 Chemical structure of phenylpropanoids.

7.2.3 TERPENOIDS

The essential oil of citrus fruits is very important from a research point of view in India by various researchers since 1950. The ἀ-pinene, d-Limonene, is a common compound, which found in different parts of citrus fruits (Negi et al., 2012). The limonene, squalene, geosmin are the main chemical compounds of citrus fruit, which are given in Figure 7.6.

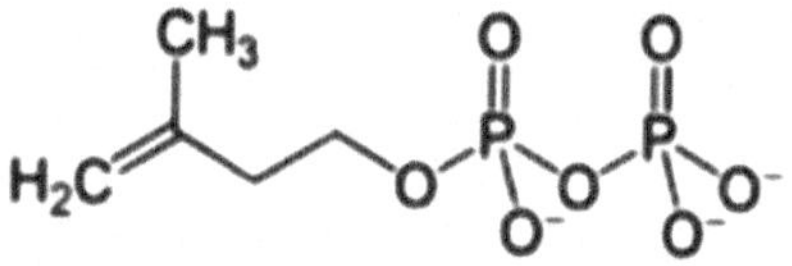

FIGURE 7.6 Chemical structure of terpenoids.

7.2.4 PHENYLPROPANOIDS

Phenylpropanoids are secondary compounds, which contains 3 carbon compounds and attached with the 6 carbon rings inside the chain of aromatic ring. The phenylpropanoids also included hydroxyl, coumarins, lignans, and phenylpropanes (Parle & Chaturvedi, 2012) (Figure 7.7).

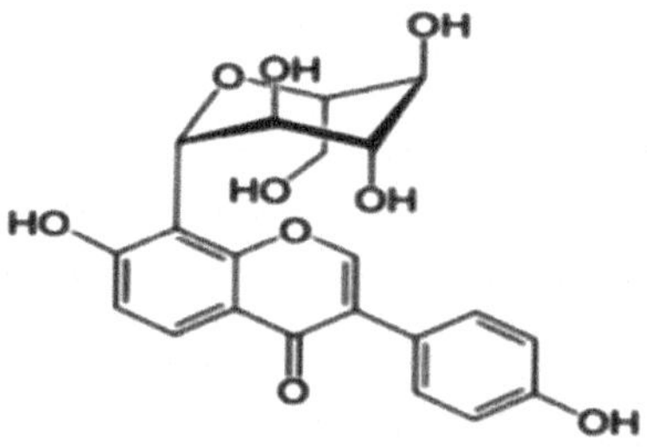

FIGURE 7.7 Chemical structure of phenylpropanoids.

7.2.5 TANNIN

Tannin is one of the most important components, which found in different parts of citrus fruits. Tannin provides the protection against the pathogens, and it is present in large quantities in the pulp and peels of fruits. Gallotannins, ellagitannins compounds are also present in the peel and fruit of citrus (Singh et al., 2020) (Figure 7.8).

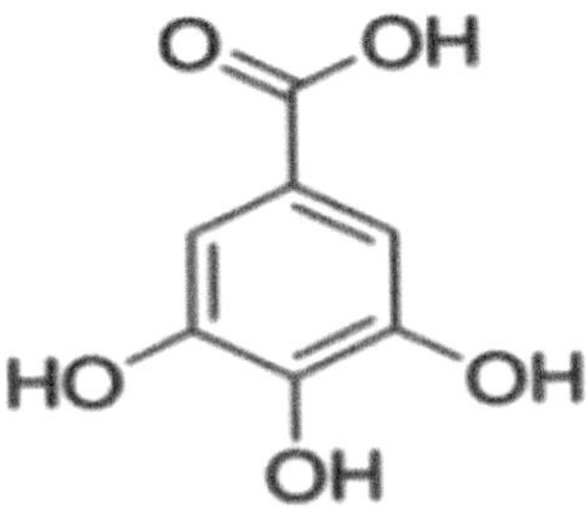

FIGURE 7.8 Chemical structure of tannin.

7.2.6 FLAVONOIDS

Flavonoids a chemical compound, which is including naringenin, hesperidin, kaempferol, luteolin, aglyconerutins, flavone, fisetin, quercetin, myricetin compounds and inhibits the infection of harmful diseases (Kim et al., 2010; Keservani & Sharma, 2014). More than founded 6,000 types of Flavonoids founded in citrus fruits. One of the flavonoids is quercetin which is given in Figure 7.9.

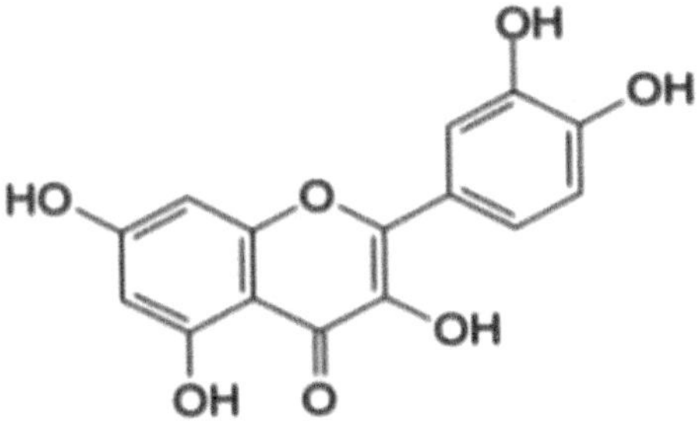

FIGURE 7.9 Chemical structure of flavonoid.

7.2.7 CAROTENOIDS

Carotenoids are responsible for the color in fruit and vegetables. Zeaxanthin, lycopene, xanthophyll are active therapeutic compounds, which are found in citrus plants. The minimum constituents such as crude fibers, ascorbic acid, sitosterol, tannin, ὰ-amyrin and crude proteins are also present in citrus fruits. The xanthotoxin, compounds are present in the root of the citrus tree (Negi et al., 2012; Kim et al., 2010; Singh et al., 2020). The compounds such as betulumic acid, trans-cinnamic acid, montanin, and praealtin-D present in citrus fruit and it is given in Figure 7.10.

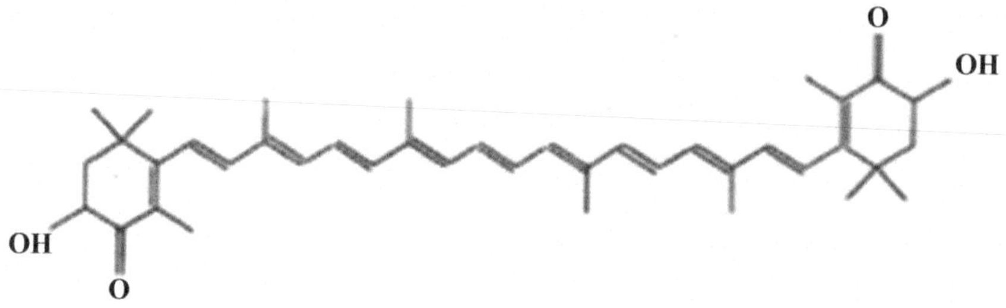

FIGURE 7.10 Chemical structure of carotenoids.

7.3 ANTIOXIDANT COMPOUND OF CITRUS FRUIT

7.3.1 PHENOLIC COMPOUND

Phenolic compounds are antioxidant compounds, which are synthesized by different plants of the citrus family, and it is present in the vacuoles of tissues (Chowdhury et al., 2008). The phenolic compounds help in protecting the human health against pathogenic microorganism. Phenolic compounds shown most important activity for the antioxidant, where it contains free radical scavenging activities and many reactive oxygen species (ROS), such as superoxide radical and hydroxyl radical are produced major harmful agents of pathogenic microorganisms in human (Hayata et al., 2009). The phyto-phenolic compound helps in hindering the free radical, which produced the diseases in human body. For example, heart diseases and different types of cancer are also generated by free radicals and ROS. Many research have shown that the free radicals are present in the human and generate the oxidative degradation of bio-molecules such as protein (Singh et al., 2017b; Jarouliya & Keservani, 2019b) and bioactive compound produces the degenerative disorders (Bocco et al., 1998). Many other phytochemical compounds are present in orange, kinnow, and other citrus peels. Hydrogen-donating mechanisms, able to inhibit the free radicals by removal of the ROS. ROS and RNS originator by quenching chain initiator catalyst. During last decade, citrus fruits such as orange mosambi, kinnow, grapefruit, lemon, etc., contain very effective biological compounds which named 4′geranyloxyferulic (GOFA) and boropinic acid (Zou et al., 2016; Kim et al., 2010; Singh et al., 2020). It is a pharmacological drug, which provides protection against neuro and cancer diseases and shown the anti-inflammatory properties. The oxidation of lipid modifies the properties of nutritional foods and change the color and flavor. Therefore, synthetic of antioxidants is used for prevent the oxidation of lipid for a long time. Demerit of anti-oxidants, synthetic replaced by extract of *Citrus reticulata* (kinnow), *Citrus sinensis* (orange), and *Citrus limetta* (mosambi). From previous study it is shown that the extract of kinnow, orange are good sources of phenolic compound, which exhibited the free radical scavenging activities (Table 7.1). Citrus fruits, peels, and leaves are rich sources of antioxidants (Singh et al., 2017a). The phenolic compounds of citrus fruit also contain the capacity to act as phytoalexins, which supply to the sensory and organoleptic properties to the citrus plants. Therefore, many studies have confirmed that the presence of bioactive, antioxidant, phytochemical, and other biological active compounds make a citrus fruit, which is very beneficial for health (Zheng et al., 2001).

TABLE 7.1 Phenolic Compound of Different Citrus Fruit (μg/g)

Phenolic Compound (μg/g)	*C. sinensis*	C. reticulata	C. limetta	C. limon	C. paradisi
Caffeic acid	14.1±1.18	4.07±0.04	8.24±0.76	81.0±3.74	74.3±1.14
Chlorogenic acid	301±8.68	322±16.30	340±5.01	178±4.66	198±0.75
Ferulic acid	46±0.41	151±5.90	146±2.42	60.1±0.94	87.8±4.56
Sinapic acid	45.2±1.63	95.3±1.23	179±6.61	59.7±0.48	68.9±0.54
q-Coumaric acid	230±1.22	347±3.46	184±6.61	265±4.20	198±3.25

7.3.2 PHENOLIC ACIDS

Citrus fruits are rich sources of phenolic acids and showed the different levels of antioxidant activity such as free radical scavenging and reducing power activity. Antioxidant properties of phenolic acids of citrus show the strong dehydrogenation of hydroxyl groups and exhibited the effect of ortho-substitution on ring of benzene (Liew et al., 2018; Ismaill et al., 2010). Total content of phenolic acids of citrus fruits contained the 5,565 mg/g DW in peel of Ponkan (*Citrus poonensis Hort.* ex Tanaka) (Wang et al., 2008; Hang et al., 2018).

7.3.3 VITAMINS

Citrus fruits contained 180 antioxidants, including many different types of secondary compound such as phenolic compounds, pectin, phenolic acids, vitamins, mineral, terpenoids, qucertin, and rutin (Zou et al., 2016). Organic substances are vitamins, which is very vital for the body function and very important for our life. Vitamin A, vitamin B, vitamin C, vitamin B2, vitamin E and vitamin B3 has been reported in the literature of citrus fruits (Singh et al., 2017a; Wang et al., 2008). Antioxidant activities were evaluated by vitamins A, vitamin B, vitamin C and vitamin E in citrus fruits (Etebu & Nwauzoma, 2014) (Table 7.2).

7.3.3.1 VITAMIN A

Vitamin A is an organic compound, which is soluble in the fat and includes the several (b-carotene) and retinoic acid. Fruits of Citrus such as orange, mosambi, kinnow, grapefruit are rich source of vitamin A, retinol, and carotene compounds. Total content of vitamin A in orange and kinnow fruits (*Citrus sinensis*) is 0.30 mg/kg and 3.68 mg/kg in (*Citrus reticulate*). According to the reports, free radicals are inhibited by the vitamin A and shown the antioxidant activities (Amutha et al., 2017).

7.3.3.2 VITAMIN C

Vitamin C that is ascorbic acid is soluble in the water. Citrus fruit is a major source of vitamin C, and it is found in the pulp and peel of fruits. Extraction of vitamin C in Pomelo (*Citrus maxima*), orange (*C. sinensis*) and Kinnow (*C. reticulata*) 40–50 mg/g, 69 mg/g and 34 mg/g, respectively. Vitamin C is a free radical scavenger, which reduces the ROS and sulfur radicals (Negi et al., 2012).

7.3.3.3 VITAMIN E

Vitamin E is a fat-soluble compound, which includes tocopherols. Vitamin E is mainly present in the peels, pulps, and seeds of citrus fruits and its total content of vit E in orange (*C. sinensis*), mosambi (*C. limetta*), kinnow (*C. reticulata*) and lemon (*Citrus limon*) can reach 6.50 mg/g, 5.40 mg/g, 7.32 mg/g and 12.40 mg/g respectively. Vitamin E also provides the protection of the Cell membranes and very effective against the lipid peroxidation damages (Negi & Jayaprakasha, 2001).

7.3.4 MINERAL ELEMENTS

Mineral elements calcium (Ca), phosphorus (P), magnesium (Mg), potassium (K), sulfur (S), sodium (Na), iron (Fe), manganese (Mn), nickel (Ni), boron (B), silicon (Si), copper (Cu), zinc (Zn), molybdenum (Mo), selenium (Se), cobalt (Co), chromium (Cr), germanium (Ge) and arsenic (As) are the chemical compounds, which found in the mammalian bodies (Zou, 2016). Plants obtained from the soil for their growth and development with the exception of carbon, hydrogen, and oxygen (Zheng et al., 2001) (Table 7.2). Chemical element of Mn, Fe, Cu, Zn, and Se have been reported to show the antioxidant activities in organisms (Amutha et al., 2017). Mn, Fe, Cu, Zn, Se can protect the tissues against, oxidative damage and destroy the free radicals in bodies. The total content of Se in extraction of the orange (*C. sinensis*), Kinnow (*C. reticulate*) and lemon (*C. limon*) are 0.36 mg/g, 0.524 mg/g and 0.604 mg/g, respectively (Parle & Chaturvedi, 2012).

TABLE 7.2 Nutritional Composition of Different Citrus Fruit

Compound	Orange	Kinnow	Mosambi	Lemon	Grapefruit
Crude protein	0.68 g	0.92 g	0.64 g	0.67 g	0.72 g
Moisture (%)	15.0 g	9.2 g	13.2 g	11.6 g	14.5 g
Crude fat (%)	0.24 g	–	0.10 g	0.15 g	0.18 g
Ash content	7.2 g	5.0 g	4.6 g	6.3 g	7.4 g

TABLE 7.2 *(Continued)*

Compound	Orange	Kinnow	Mosambi	Lemon	Grapefruit
Carbohydrate	1.01 g	0.98 g	0.89 g	0.92 g	0.95 g
Crude fiber	14.5 g	10.8 g	8.9 g	9.7 g	11.6 g
Iron (%)	0.6 mg	0.3 mg	0.2 mg	0.5 m	0.4 mg
Calcium	0.6 mg	38 mg	44 mg	47 mg	46 mg
Vitamin A	1.88 g	0.98	0.82 g	0.78 g	0.81 g
Energy	52 Kcal	48 Kcal	39 Kcal	42 Kcal	40 Kcal

7.3.5 FLAVONOIDS

Flavonoids is a group of chemical compounds, which is divided into different groups such as flavones, flavanol, and flavonones, which having biological and pharmaceutical properties and it is very effective against the many harmful diseases in human body. It showed the anti-inflammatory, anti-diabetic, anti-proliferative, and antiviral activities (Wang et al., 2008; Kim et al., 2006; Montanari et al., 1998). Flavonoids isa low molecular weight compound and it is obtained from plants. On the basis of structure flavonoids are polyphenolic compounds with a phenyl benzopyrene structure. Flavonoids are present in plants as glycosides. The biochemical study of flavonoids in peel of orange, kinnow, mosambi, grapefruit, lemon, etc., demonstrate that, it enhanced antioxidant capacity opposed to lipid peroxidation and suppress the OS (Goulas et al., 2012; Samman et al., 1996). This biochemical study also revealed that the flavonoids possess the advantageous effects as anti-proliferative and anti-inflammation activity (Middleton et al., 1994). The dried form of citrus peel is routinely used as a medicinal drug in China for treatment of many diseases such as dyspepsia, asthma, and cardiac diseases. Hesperidin is the most important flavonoid in orange, kinnow, mosambi, lemon, grapefruit, and peel (Table 7.3) contain inhibitory activity against neuro-disorder (Parle & Chaturvedi, 2012; Kim et al., 2010).

7.3.6 LIMONOIDS

Limonoids are phytochemical compounds, which is present in all Citrus fruits and anti-typhoid, anti-inflammatory, antioxidant, antibacterial, antifungal, antiviral, and anticancer activities shown by the limonoids (Kim et al., 2006). Citrus limonoids enhance the glutathione (GSH) transferase activity and inhibit lung, skin, and colon tumor and reduced the cholesterol effect (Table 7.4). *In vivo* tests Citrus limonoids inhibit the proliferation of estrogenic receptor in human breast cancer cells *in vitro*. Nomilin agent contained the anti-obesity and anti-hyperglycemic activity that are likely to be mediated through the activation of TGR591 (Kim et al., 2010). Limonin and nomilin in plant products are significant indicators of the pest control. Limonoids and nomilin, which are obtained from pulps and peels of citrus fruit show antimicrobial activity against pathogenic bacteria.

7.3.7 COUMARINS

Citrus fruits also contain the coumarins bioactive compounds, which have the health promoting properties. Extract of citrus fruits and oils contained 5-methoxy psoralen and it is the biggest contributor of furocoumarin content, which is being used as a fragranced product (Singh et al., 2020). Essential oil is very effective in UV irradiation. Phototoxicity of citrus fruit is considered as a treatment of *Lentigo maligna* in some cases (Kim et al., 2010).

7.3.8 CAROTENOIDS

Carotenoids are biological compound, and it showed the antioxidant activities including free scavenging activity, reducing power activity and FRAP activity and inhibited the harmful free radicals in the bodies (Table 7.4). Carotenoids also protect the lipids cell membrane from oxidative damage. They maintain the communication signals between cells and receptors on the cell membrane and enhance immunity in human body (Wang et al., 2008).

7.3.9 PECTIN

Pectin found in the cell wall of plants. It is a chemically polysaccharide, consisting of a linear chain of linked galacturonic acid. Pectin showed antioxidant activities and inhibiting the blood lipid level and peroxidative status (Zheng et al., 2001) (Table 7.4). The suppressive effect of pectin inhibits the free radicals and enhance the antioxidant enzymes in tissue, where high pectin content showed more potent in different fruits such as orange, kinnow, mosambi, pomelo, and lemon. Total pectin contents range from 39.0±1.40 mg/g to 92.4±3.36 mg/g obtained from citrus fruits (Wang et al., 2008).

TABLE 7.3 Total Flavonoid and Carotenoid Contents in Citrus Fruit Peel

Scientific Name	Total Flavonoid	Total Carotenoid
C. sinensis	36.6±1.04	0.454±0.009
C. reticulata	50.1±1.34	2.05±0.037
C. limon	33.8±1.07	0.112±0.002
C. maxima	40.8±1.04	1.60±0.012
C. paradisi	48.2±1.05	0.546+0.015

TABLE 7.4 Antioxidant Compound and Antioxidant Activity of Different Fruit of Citrus

Antioxidant Compounds	Antioxidant Activity
Phenolic compound	Reduced oxidation of carotenoid and enhancing their antioxidant activity scavenging
Flavonoids	Improve the oxidant enzyme activity *in vitro* oxidation of anti-lipid
Limonoids	Enhance apoptosis and free radical
Coumarins	Inhibit the production of free radical
Pectin	Enhance antioxidant enzymes
Vitamin	Decrease the sulfur radical
Phenolic acid	Have free radical scavenging effect and regulate the sulfur radicals
Carotenoids	Reduced harmful free radicals
Essential oils	Reduced the antimicrobial activity

7.3.10 ESSENTIAL OILS (EOS)

Essential oils (EOs) are aromatic liquids, which is obtained from the plant materials such as flower, seeds, l eaves, barks, herbs (Sharma et al., 2020; Rane et al., 2020, 2020a–c), woods, fruit, and roots by distillation. Individual components of EOs are obtained by extract of plant materials (Negi et al., 2012; Griffn et al., 2000). Individual components of EOs also contained terpenes, alcohols, acetones, phenols, acids, aldehydes, and esters compounds, which are also used in the food favoring and pharmaceutical industry (Table 7.4). Despite the fact that the majority of (Eos) are GRAS (generally recognized as safe) substances, their use as preservatives in food is frequently restricted due to favor considerations (Parle & Chaturvedi, 2012; Singh et al., 2020). Different studies have demonstrated the effectiveness of

EOs and the active compounds control or inhibit the growth of pathogenic and spoilage microorganisms and reported its dependence on chemical structure and concentration of EOs reactive compounds besides the number and type of microorganisms. The bacterial susceptibility to Eos increases with reauctioning of pH in the food and hydrophobicity of the oil increases, when the pH is low, it is easily enabled to dissolve in the lipid. Further, Eos showed antimicrobial effectiveness against microorganisms in concentration dependent manner. Therefore, EOs are very effective against the pathogenic microorganism and also used in the pharmaceutical industry in a large scale (Negi et al., 2012; Anwar et al., 2008).

7.4 PHARMACOLOGY PROPERTIES OF DIFFERENT CITRUS FRUIT

7.4.1 ANTIMICROBIAL ACTIVITIES

Phytochemical compounds like flavonoids, alkaloids, steroids, glycosides, tannins, and terpenoids are present in citrus fruits, which is effective against many bacteria and other pathogens. Different Pharmacological applications of citrus fruit are mentioned in Figure 7.11. It also contains many biologically active compounds, which shows antibacterial activity. Citrus fruits contain a rich quantity of bioactive compound and therefore it is most important source of antibacterial activity. In an experimental study of antimicrobial activity by agar cup diffusion method, different solvents such as ethanol, methanol, acetone extract of *Citrus sinensis*, *Citrus limetta* and other organic extract of citrus fruits showed wide range of antibacterial activity against pathogenic bacteria, here organic and aqueous extracts of *citrus sinensis* fruits shows antibacterial activity against many harmful pathogenic bacteria such as *B. subtilis*, *S. aureus*, *E. coli*, and *S. typhi* (Mwaurah et al., 2020; R´ıos et al., 2005; Dorta et al., 2015). Fruit extracts have antibacterial activity and its zone of inhibition are 28, 24, 20, and 19 mm (diameter) for methanolic extract and 20, 26, 14, and 18 mm for aqueous extract were effective against gram positive and gram-negative bacteria (Parashar et al., 2014; Negi et al., 2001). It was also exhibited that citrus fruit juice decreases the HIV and another disease (Dhanavade et al., 2012; Negi et al., 2001). In a recent research work, the solvent extract and aqueous extract of orange (*C. sinensis*), mosambi (*C. limetta*) lemon (*C. limon*) and kinnow (*C. reticuleta*) were tested for antibacterial activity against two gram positive and two-gram negative bacterial species (*streptococcus aureus, Bacillus subtilis,*

Escherichia coli and Salmonella Typhi). Result showed that maximum zone of inhibition of *S. aureus* (18 mm) and *B. subtilis* (15 mm) at 1,000 μg/disc concentration, while another two negative bacteria *E. coli* species and *S. typhi* showed zones of inhibition of 17 mm and 22 mm. at 600 μg/disc concentration (Liew et al., 2018).

7.4.2 ANTICANCER ACTIVITY

Recently studies have been reported that the consumption of different citrus fruits also decreases the proliferative activity and prevent the several kinds of cancer disease inhuman body (Wang et al., 2008). Flavonoids, carotenoids, phenolic, pectin, and other purified compounds and essential oil of different citrus fruits are very effective against the different types of cancer disease (Lahlou, 2003).

7.4.3 ANTI-ALLERGIC ACTIVITY

Citrus fruits are vital source of anti-allergic compounds. *Citrus limetta* juice is a beneficial alternative medicine and provides the relief immediately to allergic rhinitis or asthma patients (Parle & Chaturvedi, 2012). It is reported that the gencydo has the power to decrease the degranulation and release the histamine of the IgE. Basophilic cells as well as mast cells activated by IgE and enhance the IL-8, TNF-αCitrus fruits such as *citrus medica* ssp. *Limonum* and *cydonia oblonga* are tested on the immunological pathway, which is present in seasonal allergic rhinitis (SAR) (Al-Snafi, 2016).

7.4.4 HYPERGLYCEMIA ACTIVITY

Citrus fruits such as orange, grapefruit, kinnow, mosambi, and lemon contain secondary metabolites, i.e., flavonoids, alkaloids, etc. That, play an important role in reducing the progression of hyperglycemia, when bind to starch, increasing hepatic glycolysis and the glycogen concentration and decreasing the hepatic gluconeogenesis. Blood glucose level is lowered by the Flavonoids and inhibit the hyperglycemia effect in the human body (Parmar et al., 2008; Yolanda et al., 2010; Sidana et al., 2013).

7.4.5 ANTHELMINTIC ACTIVITY

Methanolic and aqueous extract of various citrus fruits were tested against the Indian variety of earth worm *Pheritima posthuman* (Parle & Chaturvedi, 2012; Parmar et al., 2008). For this anthelminthic test several concentrations of extract were prepared and tested, the results were calculated in terms of time of death of the worm as well as time of paralysis. For this experimental work piperazine citrate (15 mg/ml) was used as standard and distilled water was used as control. The anthelmintic activity was represented by the different extract of citrus fruits. Ethanolic extract of citrus fruits exhibited the anthelmintic activity against the *Ascaris lumbricoides* in human.

7.4.6 ANTIOXIDANT ACTIVITY

Antioxidant activities of different citrus fruits have been studied by the ethanolic and acetonic extracts. In this experiment result shown that both ethanolic and acetone extract have the potency to free radical scavenger and also prevent peroxidation process. From the test, it was detected that ethanolic extract showed maximum antioxidant activity as compared to acetone extract in lipid peroxidation assay (Hayata et al., 2009; Ames et al., 1993; Wong et al., 2006). Different extract of citrus fruit has exhibited the antioxidant activity (Zou et al., 2016; Kim et al., 2006).

7.4.7 ANTI-INFLAMMATORY ACTIVITY

Citrus fruits contain many phytochemical compounds and many regulatory enzymes such as phosphodiesterase, cyclooxygenase, which showed the anti-inflammatory activity (Negi et al., 2012). Citrus fruit decreases the activation of endothelial cells (ECs) by regulating the formation of biological mediators, which is responsible for the inflammation. Bioactive compounds of citrus fruit are also able to regulate the inflammation responses by inhibiting the kinases and phosphodiesterase's activity. It is responsible for the activation of immune response (Etebu & Nwauzoma, 2014; Lin et al., 1999; Sidana et al., 2013). Secondary compounds of citrus fruit also prevent the atherosclerosis and inhibits the formation of atheroma. It is reported that different flavonoids, obtained from citrus fruit contains the therapeutic potential, which shown inhibitory effect on lipopolysaccharide (LPS) in hesperidin of mouse macrophage cells, which induced expression of synthetase nitric oxide (NO) and production of prostaglandin E2. Combining the results, it followed that at 250 mg the orange, kinnow, grapefruit, and mosambi fruit extracts exhibited significant (P<0.01) reduction in paw edema at 6th and 7th hour, which was found to be 1.45±0.01, 1.84±0.09, 1.90±0.02; 1.9±0.05, 1.35±0.08, 1.10±0.05; 1.77±0.03, 1.65±0.05, 1.88±0.05, 1.75±0.05, 1.16±0.01, and 1.11±0.04, While kinnow fruits extract showed significant reduction only at 7th hour. The reduction in paw volume in orange, kinnow, grapefruit, and mosambi fruit was prominent at 500 mg dose recording mean values and percent inhibition of [1.5±0.04 (30.8%), 1.06±0.05 (55.1%), 1.18±0.07 (56.5%)], [1.7±0.04 (36.52%), 0.99±0.08 (58.68%), 1.09±0.05 (57.62%)], [1.69±0.08 (27.36%), 1.39±0.08 (51.56%) and 1.41±0.19 (53.39%)] at 5th and 6th hour respectively. But Orange and mosambi fruit extract elicited significant (P<0.01) reduction of 1.62±0.05 (10.89%), 1.67±0.06 (22.81), 1.28±0.05 (42.11), 0.68±0.04 (65.29), 0.44±0.02 (88.72), 1.52±0.06 (7.84%), 1.69±0.06 (18.78%), 1.8±0.08 (42.85%), 0.82±0.045 (67.26%) and 0.61±0.08 (74.44%) at different time period (Lin et al., 1999; Malleshappa et al., 2018). Anti-inflammatory activity also shows a positive correlation with the time intervals, where it is exhibiting the highest inhibition of paw volume at the 7th hour at all the treatment under study (Wang et al., 2008).

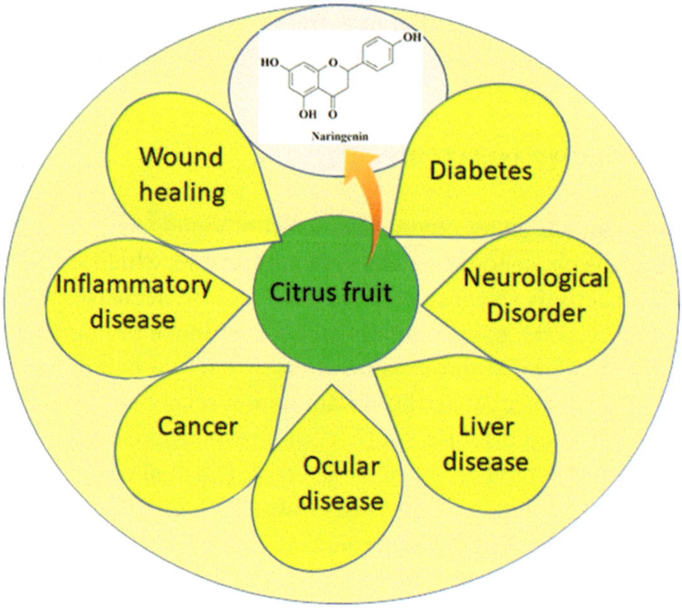

FIGURE 7.11 Pharmacological application of citrus fruit.

7.4.8 ANTI-OBESITY

Citrus fruit is an excellent source of phytochemical compounds, phenolic acids, flavonoids, pectins, phenolic compounds and is very effective against the many harmful diseases in human being. Pectin protects the membrane of mucous, when expose to toxic element directly and bind to chemicals and causes the cancer in colon. Here the pectin compounds have decreased the cholesterol levels in the blood by binding the bile acids in the colon and decrease the re-absorption in the colon (Parle & Chaturvedi, 2012; Parmar et al., 2008). The compound of antioxidant is oxidizing the LDL and reduces OS in a human body.

7.4.9 ANTI-TYPHOID ACTIVITY

Salmonella typhi bacteria causes the typhoid infection, and it causes major health problems in the human body. Ingredients of different fruit of citrus include flavonoids like citacridone, citbrasine, and saponins shows anti-typhoid activities (Parmar et al., 2008).

7.4.10 PROTECTION AGAINST HEART DISEASES

Citrus fruit provides protection against the heart, cancer, typhoid, obesity, diabetes diseases. Orange, kinnow, mosambi, grapefruit, lime, and lemon fruits contained vitamin (Singh et al., 2017a), minerals and dietary fibers and many secondary compounds such as naringenin, hesperidin, qucertin, rutin provides protection against the cardiac disease (Parle & Chaturvedi, 2012; Malleshappa et al., 2018). Although, different citrus fruits such as orange, mosambi, kinnow, grapefruit are contained poly-methoxylated flavones (PMFs), which decrease the cholesterol level in blood and also used as a drug, which is not showing any side effect in the body. PMFs inhibited the synthesis of cholesterol inside the liver (Parmar et al., 2008). Therefore, different citrus fruits can be used in the form of juice, tea, salads, etc., which reduces cholesterol in the blood and provide the benefit for human health (Negi et al., 2012; Sidana et al., 2013).

7.4.11 OVARIAN CANCER

Different citrus fruits suppressed the proliferation of ovarian cancer cell and induces apoptosis, which repair the DNA damage and enhance the cleaved poly(ADP-ribose) polymerase enzyme levels. It decreases the potential of mitochondrial membrane and activate the autophagy in body and always provide contribution in gasdermin E-mediated pyroptosis (Parle & Chaturvedi, 2012).

7.4.12 REDUCED RISK OF KIDNEY STONES

According to the research report, if half a liter of orange juice is taken regularly by a woman, then they are decreasing the risk of stone of calcium oxalate by increasing the pH value of urinary and excretion of citric acid (Gupta et al., 2021).

7.4.13 ANTI-ULCER

Regular Intake the orange juice is minimized the infection of *Helicobacter pylori* (*H. pylori*) and may also help to reduce the development of ulcers (Gupta et al., 2021).

7.5 TRADITIONAL APPLICATION POTENTIAL OF DIFFERENT CITRUS FRUIT FOR HUMAN HEALTH

7.5.1 ANTI-DIABETIC

Citrus fruit juice is valuable for the person who is diabetic. For the treatment of diabetes, it is recommended that, prepare a mixture of 1 teaspoons of citrus juice, 2 teaspoons amla juice and half teaspoon honey and take this in empty stomach early in the morning.

7.5.2 IN THE TREATMENT OF SCURVY

Deficiency of vitamin C causes disease scurvy and is characterized by swollen gums, common cold, frequent bouts of flu and cracked lip corners.

7.5.3 IN ANTI-ANXIETY

As citrus fruit is rich in vitamin C; it is effective in curing anti-anxiety. Tranquilizer is orange oil, which is used in the Aroma-therapists effect. Scientists have found that the oil obtained from different type of citrus fruit, which is an anxiolytic agent.

7.5.4 IMMUNITY BOOSTER

Regularly consumption of citrus fruit juice controls proper blood circulation by enhancing the proper function of the heart. This consequence is a significant source of improved immune system.

7.5.5 WEIGHT REDUCTION

Being low in fat and calories, mosambi juice helps to reduce the body weight. We all can drink a mixture of mosambi juice and honey to burn extra calories.

7.5.6 BENEFICIAL IN PREGNANCY

Citrus fruit juice is rich and healthier source of calcium, which is beneficial for both fetus as well as mother too.

7.5.7 TREATMENT OF URINARY DISORDERS

Juice of mosambi is rich in potassium, which helps to treat urinary disease such as cystitis. Cystitis produces inflammation in the urinary bladder, which is also known as urinary tract infection. Mosambi juice boiled with water and should be taken within a few hours after cooling, it provides immediate relief in cystitis. Potassium facilitates the detoxification of kidneys and bladder by potassium facilitates, which prevent various types of urinary tract infections (UTI).

7.5.8 REMEDY FOR COMMON COLD

Citrus fruit is a rich source of vitamin C, which helps to protect the body to common cold and boost human immunity.

Parashar, S., Sharma, H., & Garg, M., (2014). Antimicrobial and antioxidant activities of fruits and vegetable peels: A review *Journal of Pharmacognosy and Phytochemistry, 3*(1), 160–164.

Parle, M., & Chaturvedi, D., (2012). Orange: Range of benefits. *International Research Journal of Pharmacy, 3*(7), 59–63.

Parmar, H. S., & Kar, A., (2008). Antiperoxidative, anti-thyroidal, antihyperglycemic and cardioprotective role of citrus sinensis peel extract in male mice. Phytotherapy Research: An *International Journal Devoted to Pharmacological and Toxicological Evaluation of Natural Product Derivatves, 22*(6), 791–795.

Rajendran, P., Nandakumar, N., Rengarajan, T., et al., (2014). Antioxidants and human diseases. *Clin. Chim. Acta, 436*, 332–347.

Rane, B. R., & Keservani, R., (2020). Introduction to nutraceuticals and dietary supplements. In: *Nutraceuticals and Dietary Supplements* (pp. 35–56). Apple Academic Press. ISBN: 9781771888738.

Rane, B. R., Bharath, M. S., Patil, R. R., Keservani, R. K., & Jain, A. S., (2020b). Novel approaches in nutraceuticals. In: Rajesh, K. K., Raj, K. K., & Anil, K. S., (eds.), *Enhancing the Therapeutic Efficacy of Herbal Formulations Through Novel Drug Delivery Systems* (pp. 241–266). IGI Global International Publisher, Pennsylvania, USA, Chapter 11. ISBN13, 9781799844532.

Rane, B. R., Patil, A. S., Keservani, R. K., & Jain, A. S., (2020a). Novel approaches in herbal formulation. In: Rajesh, K. K., Raj, K. K., & Anil, K. S., (eds.), *Enhancing the Therapeutic Efficacy of Herbal Formulations Through Novel Drug Delivery Systems* (pp. 43–68). IGI Global International Publisher, Pennsylvania, USA, Chapter 2. ISBN13, 9781799844532.

Rane, B. R., Tadavi, S. A., & Keservani, R. K., (2020c). Naturopathy. In: Anil, K. S., Raj, K. K., & Surya, P. G., (eds.), *Herbal Product Development* (pp. 321–347). Apple Academic Press, CRC Press, Taylor & Francis Group, chapter 12. ISBN: 9781771888776.

Rıos, J. L., & Recio, M. C., (2005). Medicinal plants and antimicrobial activity. *Journal of Enthopharmacology, 100*, 80–84.

Sharma, A. K., Keservani, R. K., & Gautam, S. P., (2020). *Herbal Product Development.* Apple Academic Press, CRC Press, Taylor & Francis Group. ISBN: 9781771888776.

Singh, P., Kesharwani, R. K., & Keservani, R. K., (2017a). Antioxidants and vitamins: Roles in cellular function and metabolism. In: *Sustained Energy for Enhanced Human Functions and Activity* (pp. 385–407). Academic Press. ISBN:9780128054130.

Singh, P., Kesharwani, R. K., & Keservani, R. K., (2017b). Protein, carbohydrates, and fats: Energy metabolism. In: *Sustained Energy for Enhanced Human Functions and Activity* (pp. 103–115). Academic Press. ISBN: 9780128054130.

Ye, Y., Xu, G., & Li, D. L., (2021). Acridone alkaloids and flavones from the leaves of citrus reticulate. *Natural Product Research Formerly Natural Product Letters.*,10.1080/14786419.2021.1876047.

CHAPTER 8

ROLE OF NUTRACEUTICALS IN GUT HEALTH

SUPRIYA RAY and SURESH B. ARAKERA

Department of Applied Genetics, Karnatak University, Dharwad, Karnataka, India

ABSTRACT

Nutraceutical is an amalgam of 'nutrition' and 'pharmaceutical.' In broad, nutraceutical is a part of food that plays a major role in maintaining and modifying the physiological function of human health. The human gut is inhabitant with a group of bacterial species which create metabolic and immune function that has an enormous impact on host health and gut health. Intact gastrointestinal barrier maintains the equilibrium of microbiota, protect pathogen infection, and prevent tissue injury that may lead to many diseases. Disruption to the balance of microbial composition (known as Dysbiosis) could lead to many diseases. Dietary bioactive components such as probiotics, prebiotics, polyphenols, phytoestrogens, polyunsaturated fatty acids (PUFAs) play a beneficial role in maintaining the host and gut health. The beneficial effects of prebiotics rely on the composition of gut microbiota and the fermented products like short chain fatty acids (SCFAs) which have diverse biological roles. The bioavailability of polyphenols depends on the transformation of the gut microbiota composition. Phytoestrogens include isoflavonoids, flavonoids, and anthocyanins get absorbed by gut microbial species that have high antioxidant, anti-inflammatory, antiproliferative, and

Immune-Boosting Nutraceuticals for Better Human Health: Novel Applications. Urmila Jarouliya, PhD, Raj K. Keservani, MPharm, Rajesh K. Kesharwani, PhD, Virendra K. Patel, PhD, and Adi D. Bharti, MPharm (Eds.)

antiestrogenic effects on host and gut health. Moreover, PUFAs (omega-3 and omega-6 fatty acid) influence immune regulation, cholesterol, and triglyceride levels, blood pressure and cardiovascular function in humans and animals. However, probiotics inhibit antibiotic associated diarrhea (AAD) in children and phytoestrogens may also protect against pathogenic bacteria and stimulate the growth of beneficial bacteria. This chapter focuses on the close association and interaction between nutraceuticals and gut microbiota and the after-effects of this interaction in human and gut health.

8.1 INTRODUCTION

The term nutraceutical defined by Broweras, "a food or food part that provides medical or health benefits, including the prevention and/or treatment of a disorder or disease" (1998). Later nutraceutical was defined by Santini et al. as "a food or part of a food that provides benefits to health in addition to its nutritional content" (2017). In recent years, there has been a paradigm shift in the diet health of an individual. People adopting new lifestyle has completely changed the food habits of latter. Improper nutrition due to consumption of junk foods has led to increase a large number of diseases including obesity which has now become a global issue, cardiovascular diseases (CVDs) which remains to be a primary cause of death worldwide followed by cancer, arthritis, osteoporosis, diabetes, and many more (Adelaja & Schilling, 1999). Most of the therapeutic areas are covered by nutraceuticals (Keservani et al., 2010a, b, 2017, 2020) such as prevention of cancers, cholesterol level, diabetes, blood pressure, depression, digestion, cough, and cold (Pandey et al., 2010; Sami Labs, 2002). Nutraceuticals play a key role in maintaining as well as providing protection against chronic diseases (Das et al., 2012; Romano et al., 2012).

Potential nutraceuticals and established nutraceuticals are the classification of nutraceuticals, where majority of the nutraceuticals belong to potential nutraceuticals as more of clinical and preclinical study are in hold to prove their beneficial effect. Few classes such as probiotics, pre-biotics, antioxidants, and omega 3 fatty acids belong to second classes where many studies have confirmed the beneficial effects (Pandey et al., 2010; Alisi et al., 2014; Riva et al., 2017). Other two classifications are based on chemical constituents such as dietary supplements, nutrients, herbs, and Traditional or Non-traditional nutraceuticals (Gupta et al., 2010). Nutraceuticals cannot be referred without connecting them with human microbiota that represents the total microorganisms which colonizes in humans almost all body parts such as gastrointestinal tract (GIT), skin, respiratory tract, vagina, and so on (Thursby & Juge, 2017).

The human intestinal tract comprises of more than 800 different bacterial species which dictates the health and nutritional status of the host. The metabolic activity of the gut microbiota plays a part in the digestion of dietary components, energy salvage and supply of micronutrients to the human system (Nadal et al., 2007; Santacruz et al., 2009). Several endogenous and environmental factors contribute to the composition of gut microbiota. Diet has believed to be the major factor for alteration in the gut microbial diversity, which affects the host functionality (Lee et al., 2008). Diet has direct insinuation on gut health. Certain complications related to gut that results from an imbalance of the intestinal microflora such as irritable bowel syndrome, ulcerative colitis, gluten therapy resistant celiac and Crohn's disease are associated with one's diet. The intestinal microflora remains sterile at birth and colonizes right after birth which varies depending on the factors such as mode of birth, age, diet, use of antibiotics, exercise, mode of feeding and so on (Bronzwaer, 2008).

To simplify a largely symptom-free bowel status, the term "gut health" has been used frequently in the literature. Although the concept "gut health" has not been clear in the literature and is considered to be vary with different culture such as in Asian medicine, gut is reflected as location of soul and in west, gut is considered as an organ for digestion and absorption. Recent studies in medicine showed that the gut has a much broader impact on the overall health of the body, brain signaling, resistance against infections, immune tolerance than nutrition uptake and food processing. Epidemiological studies done on animal findings suggest that "gut health" is now viewed as a factual in medicine not only for the treatment of gastrointestinal disorders but also to resist illness and maintaining the well-being of an individual (Bischoff, 2011). Based on World Health Organization (WHO) definition of health, Bischoff defined "gut health" as "a state of physical and mental well-being in the absence of gastrointestinal complaints that require the consultation of a doctor, in the absence of indications or risks of bowel disease, and in the absence of confirmed bowel disease" (2011).

Gastrointestinal researchers collected evidence over the years on gut health and stated that the health of the gut depends on two major standards such as balanced gut microecology and intact gut barrier. Lower GIT which is a part of the gut comprises of all the large intestines and largely of the small intestines which consists of stratified monolayer of epithelial cells.

This layer of the intestinal epithelial cells (IECs) acts as a site of defense barrier against foreign substances such as allergens, toxins, bacteria, etc., both biochemically and physically. The intestinal barrier also plays a role in intestinal immune responses through secretion of chemokines, cytokines, mucins, antimicrobial peptides (AMPs). The major bacterial phyla exist mostly in the large intestine of around 10^4 trillion commensal bacteria. The major six bacterial phyla which dominates the gut microbiota of an adult healthy individual are: *Bacteroidetes* and *Firmicutes* (90%) followed by *Proteobacteria*, *Actinobacteria*, *Fusobacteria*, and *Verrucomicrobia* (Woting & Blaut, 2016). This large pool of intestinal bacteria diversely recognized as microbiome, microbiota, microecology or microflora. Any microbial imbalance in the gut known as dysbiosis can weaken the intestinal defense system which can further lead to various diseases such as irritable bowel syndrome, obesity, diabetes, CVDs and so on (Dethlefsen et al., 2008; Jeong et al., 2009). Gut microbiota and intestinal barrier are the two factors which affects the gut health. Any intervention in the barrier or to the microbiota composition can lead to gastrointestinal disorders (Sharma et al., 2010). Therefore, gut health governs the overall health of an individual and any malfunction in the gut has a negative impact on an individual's health.

8.2 IMPORTANCE OF GUT MICROBIOTA IN MAINTAINING HOST HEALTH

The human intestine comprises of trillions of microorganisms, and they usually referred to as indigenous microflora or commensal or normal microflora which are parted from internal environment by a continuous single lining of epithelial cells. Microorganism present in our gut are more than 10 times of the total body cells (Tremaroli & Backhed, 2012; Pennisi, 2013). Protection of balanced microbial ecology as an ecosystem barrier to external attacks is essential for gut health. Any disruption in the microbial ecology could lead to dysbiosis, weakens the intestinal homeostasis and causes excessive growth of toxic bacteria which promote colon cancer through chronic inflammation or suppression of local immune system (Wu et al., 2011). Studies showed that the absence of microbes in germ-free rodents resulted in defective tissue formation, molecular, and cellular profiles of immune system are compromised which make the rodents more susceptible to infections. These germ-free rodents have increased calorie intake to maintain their body weight but condensed serum immunoglobulin levels, cytokine production, enzyme activity, muscular turnover, muscle wall thickness, vascularity, lesser intraepithelial cells and Peyer's patches (PP) (Salzman et al., 2007; Lee & Mazmanian, 2010).

The gut microbiota plays an essential role in various functions such as metabolic, protection, and immune functions which have a huge impact on the health and nutritional status of the host. Indigenous microbiota and transient bacteria, including probiotics through interaction with the gut-associated lymphoid tissue (GALT) and epithelium influence the regulation and development of the host's defenses (Sanz & De Palma, 2009). The intestinal epithelium consists of a physical barrier which is strengthened by mucus layer (mucins), other bile, acids, enzyme secretions and synthesis of AMPs. This epithelium regulates the transfer of exogenous substances and damages the entry of luminal antigens. The commensal bacteria are the primary line of defense and take part in the regulation of secretion of AMPs by Paneth cells of the intestine and gene expression by goblet cells. The gut microbiota is crucial for post-natal immune development swaying immunoglobulin A producing B-cells, serum immunoglobulin levels and intraepithelial T-cells (Tlaskalová-Hogenová et al., 2004). Some probiotic strains for their beneficial effects have been recognized for the treatment of acute diarrhea, elimination of *Helicobacter pylori* infection and deterrence of atopic eczema in humans (Sanz & De Palma, 2009; Kalliomäki et al., 2007).

The gut microbiota affects the metabolism of the host by regulating the gene expression encompassed in the utilization of carbohydrates and lipids (Hooper et al., 2002; Gill et al., 2006; Wilson & Nicholson, 2009). Several genomic and physiological studies have shown that intestinal microbiota deliver certain enzymes specific for the utilization of host-derived mucin, non-digestible carbohydrates, biosynthesis of vitamins, reduction of cholesterol, amino acid and xenobiotics metabolism, decarboxylation, and deconjugation of bile acids (Kurokawa et al., 2007; Hooper et al., 2002; Gill et al., 2006). The genes in the microbiome involved in the uptake and metabolism of carbohydrate which indicates that the complex polysaccharides (PS) are the primary source of energy for microbiota in the colon (Kurokawa et al., 2007). Some copious bacterial enzymes (β-glycosidases and β-glucuronidases) plays both harmful and beneficial role, are involved in complex PS and xenobiotics degradation (Dabek et al., 2008). The intestinal microbiota utilizes complex PS to contribute to gather energy from diet which symbolize daily energy supply of about 10% (Flint et al., 2008).

Short chain fatty acids (SCFAs) such as butyrate, acetate, propionate, and gasses such as carbon dioxide and hydrogen are generated by fermentation of dietary PS. These SCFAs gets metabolized by muscle, colonic epithelium and liver. Butyrate is viewed as a healthy metabolite as it promotes cell growth and differentiation which also exerts anti-inflammatory properties (Hamer et al., 2008). Acetate and propionate have an effect on lipid metabolism. Acetate synthesizes cholesterol in the liver and propionate inhibits the influence of acetate. Enzyme β-glycosidases and β-glucuronidases secrete toxins which are glucuronated in the liver and expelled into the gut with the bile, which lead to the accumulation of carcinogenic compounds within the gut, hence escalating the threat of carcinogenesis. Moreover, re-glucuronidation in the liver can increase the retention time of xenobiotic compounds in the body. Study showed that most of the *Bacteroides thetaiotaomicron* and *Bifidobacterium* have β-glucosidase activity. Besides, the enzyme activity level in some species of gut microbiota gets affected by the type of diet consumed, since, the exposure level to glycosides in colon is dependent on the consumption pattern of diet, hence, it affects the functions of gut microbiota (Dabek et al., 2008).

Enzymes of gut microbiota are engaged in cholesterol and bile acid metabolism, where cholesterol is reduced to coprostanol by microbiota, therefore increasing its secretion into feces. Cholesterol synthesizes bile acids in the liver as primary bile acids, chenodeoxycholic acid and cholic acid. The bacteria in the intestine then convert these acids into various secondary bile acids, hence, limiting dietary lipid absorption all over intestine (Ridlon et al., 2006). Nevertheless, all these activities can lead to generate secondary bile acids, which are considered as possible carcinogens such as *Bacteroides fragilis* and *Escherichia coli* (Fukiya et al., 2009). Metabolic activity of gut microbiota is also involved in the supply of amino acid in the human system. The colonic bacteria grow and synthesize protein in presence of carbohydrate substrates which can be a damage amino acid for the host (Cummings et al., 1997).

Nonetheless, the metabolic activity of microbiota is involved in food degradation of nitrogen compounds which can lead to generate possible carcinogenic substances (Rowland, 2009). The gene expression involved in absorption and processing of complex carbohydrates and lipids are regulated by the commensal bacteria which lead to gain in body weight and increase in fat storage (Hooper et al., 2001; Bäckhed et al., 2004). Studies have demonstrated that in germ free mice, colonization results in reduction fasting-induced adipose factor (Fiaf) level, liver level and the skeletal muscle which led to fat storage (Bäckhed et al., 2007). Therefore, the intestinal tract is populated with various microbiota which regulate the utilization and synthesization of nutrients in response to diet pattern consumption. The biochemical and metabolic activity generate healthy as well as harmful compounds from the diet by the microbiota is therefore essential to maintain a healthy status of an individual.

8.3 ROLE OF NUTRACEUTICALS IN MAINTAINING GUT HEALTH

Treating gastrointestinal diseases by taking NSAIDs in some cases instead of drugs could possibly do more harm than any good, therefore dietary intervention with functional foods has unfolded a gap of opportunity for gut health. Functional foods such as probiotics, prebiotics, and polyphenols are considered to be an alternative to drugs where both probiotics and prebiotics modulate the gut microbiome and in contrast polyphenols are well known for its anti-inflammatory properties (Shimizu, 2010; Magrone et al., 2014). Other dietary constituents such as vitamins (Singh et al., 2017a) and minerals and food derivatives such as SCFAs and peptides are rich sources of energy for metabolic signaling (Jandhyala et al., 2015; Woting & Blaut, 2016). The major bacterial phyla present in colon is *Bacteroidetes* that produce acetate and propionate and *Firmicutes* which primarily produce butyrate as metabolic end products (Den Besten et al., 2013; Louis & Flint, 2017). Acetate, butyrate, and propionate plays a significant role in lipogenesis, gluconeogenesis, and regulation of hepatic lipids and glucose homeostasis (Morrison & Preston, 2016). The SCFAs inhibit both histone deacetylase (Hdac2) and G-protein coupled receptors (Keservani et al., 2015; Jarouliya & Keservani, 2019b; Singh et al., 2017b) (GPCRs) ligand which explains the important role played by SCFAs in regulation of nervous system, blood pressure, kidney function and protection against colon cancer (Alex et al., 2013; Pluznick, 2014; Steinmeyer et al., 2015; Joseph et al., 2017).

Studies shown that SCFAs play significant role in prevention as well as treatment of metabolic syndrome, cancer, and certain gastrointestinal diseases such as ulcerative colitis, Crohn's disease, diarrhea (Kim et al., 2014; Den Besten et al., 2013; Weitkunat et al., 2017; Zeng et al., 2017). Studies have also demonstrated the involvement of SCFAs in free fatty acid receptors (FFAR-2 and FFAR-3) activation which functions as signaling molecules in many functional processes (Hara et al., 2011; Alvarez-Curto & Milligan, 2016). Previously stated that exposure of any environmental factors can alter the gut microbiota with a direct effect in the production of SCFAs. Interaction between the gut microbiota and the host immunity has delivered a better understanding in the pathogenesis of various gastrointestinal diseases such as Irritable bowel disease, chronic pancreatitis, irritable bowel syndrome, liver cirrhosis, gastroesophageal reflux disease and other diseases such as neurodegenerative disease, autoimmune disease, metabolic syndrome, asthma, obesity (Jarouliya & Keservani, 2019a), allergy, CVDs and psychiatric disease (Tang & Hazen, 2014; Carding et al., 2015). Although the microbiome changes over time and there is a vast difference in the variations of microbiota between children and elderly (Buford, 2017; Elderman et al., 2017). Based on the most recent survey, there are many dietary supplements that are used to restore the balance of the gut microbiota. The nutraceuticals with beneficial effect on gut microbiota are represented in Table 8.1.

TABLE 8.1 Impact of Nutraceuticals on Gut Health

Nutraceuticals	Beneficial Gut Microbiota's Diversity	Beneficial Effect of Nutraceuticals on Health	References
Probiotics	*Lactobacilli* spp., *Bifidobacterium* sp., *Bacillus* sp., *Leuconostoc cremoris*, *Streptococcus* spp., *Lactococcus* spp., *Clostridium butyricum*, *Saccharomyces* spp., *Akkermansia muciniphila*	Inhibit antibiotic associated diarrhea in children (AAD). Treatment of metabolic endotoxemia (ME) and many metabolic disorders.	Johnston et al. (2011); McFarlin et al. (2017); Zhou (2017)
Prebiotics	GOS and FOS increases *Bifidobacteria* and *Lactobacilli* and decreases *Clostridium*. Inulin and FOS increases *Lactobacillus* and *Bifidobacteria*.	Resemblance with breast fed infants microbiota. Treatment on oncologic patients.	Sangwan et al. (2011); Velasco et al. (2012)
Polyphenols	Quercetin increases *Firmicutes/Bacteroidetes* ratio. Tea contains catechin and epicatechin which reduces the growth of *Salmonella*, *Clostridium*, *Pseudomonas*, *E. coli*, *Bacteroides*. Resveratrol stimulates *Bifidobacterium* and *Lactobacillus* growth, reduces *Enterococcus faecalis* growth and increases *Bacteroidetes* and *Firmicutes* ratio.	Prevents obesity. Antioxidant and anti-inflammatory properties. Prevents the risk of cardiovascular diseases.	Etxeberria et al. (2015); Lee et al. (2006); Oz et al. (2013); Larrosa et al. (2009); Qiao et al. (2014); Bonnefont-Rousselot (2016); Imamura et al. (2017); Wahab et al. (2017)

TABLE 8.1 *(Continued)*

Nutraceuticals	Beneficial Gut Microbiota's Diversity	Beneficial Effect of Nutraceuticals on Health	References
Phytoestrogens	Flavonoids influence *Akkermansia* for intestinal permeability and strengthening barrier function. Anthocyanins increases *Actinobacteria* and *Bacteroidetes*.	High antimicrobial activity, protect against pathogenic bacteria and can be used for treatment and prevention of mild infection. Antioxidant properties. Treatment with anthocyanins to prevent neuro-logical, metabolic, and cardiovascular diseases.	Cassidy & Minihane (2017); Iranshahi et al. (2015); Overall et al. (2017); Chaiyasut et al. (2017)
PUFAs (omega-3 and omega-6 fatty acids)	Decreases *Bacteroidetes* and *Firmicutes* ratio, reduces the growth of *Clostridiales*, *Firmicutes*, *Helicobacter*, *Sphingomonadales*, and *Pseudomonas* sp.	Influence immune regulation, cholesterol, and triglyceride levels, blood pressure and cardiovascular function. Treatment with PUFAs to prevent ulcer and obesity.	Liu et al. (2012); Yu et al. (2014); Ruso et al. (1995); Kawasaki et al. (2000); Straniero et al. (2008); Chen et al. (2009)
Algae (Spirulina)	Inhibit the growth *of E. coli*, *Bacillus subtilis*, *Pseudomonas aeruginosa*, *Staphylococcus aureus*.	Antibacterial and antimicrobial properties stimulate the growth of probiotics.	Beheshtipour et al. (2013); Finamore et al. (2017)
Alkaloids (Berberine)	Increases *Bacteroides* and decreases *Ruminococcus*.	Antimicrobial, antiproliferative, and antioxidant property.	Zhang et al. (2015); Guo et al. (2016)
Curcuminoids (Curcumin)	Increases *Lactobacillus* and decreases *Coriobacteriales*.	Treatment for inflammatory bowel disease.	Ramalingam et al. (2016)

8.3.1 PROBIOTICS

According to WHO, probiotics are defined as "live microorganisms which, when administered in adequate amounts confer a health benefit for the host" (FAO/WHO, 2002). By producing SCFAs, probiotics can reinstate the natural diversity and abundance of microbiota composition. Many studies established that *L. casei, Rhamnosus*, and *acidophilus* which belong to *Lactobacillus* species can inhibit or diminish the severity of AAD (antibiotic associated diarrhea) (Huazano-Garcia et al., 2017; Park et al., 2017) whereas *Bifidobecterium, Escherichia coli* and *Lactobacillus* species can inhibit or treat certain gastrointestinal diseases and metabolic disorders (Woting & Blaut, 2016). To determine whether the administration of probiotic inhibit or reduce AAD in children, Johnston and co-workers implemented a meta-analysis study of 23 subjects. The probiotic treatment involved was *Lactobacilli spp., Bifidobacterium sp., Bacillus sp., Leuconostoc cremoris, Streptococcus spp., Lactococcus spp., Clostridium butyricum, Saccharomyces spp.*, alone or in combination and the results revealed the beneficial effect of probiotics as 8% probiotic vs. 19% control on AAD incidence in children (Johnston et al., 2011).

A mucin degrading bacteria, *Akkermansia muciniphila* discovered in 2004 (Derrien et al., 2004), is commonly found in gut and represents almost 3–5% of human's gut are well known for its probiotic properties and promising benefits in inhibition and treatment of many metabolic disorders (Zhou, 2017). Due to its inhabitant in the mucus layer, this bacterium is involved in the alteration of the gut barrier integrity and immunity, mucus layer thickness (Ottman et al., 2017; Zhou, 2017). *Akkermansia muciniphila* acts as an energy sensor, its copiousness increases with caloric deficiency and decreases with surplus energy since it plays a chief role in adapting the body to cold temperature (Chevalier et al., 2015). Dysbiosis pushes the body to metabolic endotoxemia (ME), therefore, probiotics leading gut microbiome modulation can prevent or protect the body in conditions like ME, but the major issue with probiotics is, it cannot survive the gastric passage to enter in the small or large intestines intact. Nonetheless, there are certain spore probiotics that have the ability to survive the gastric passage and can enter in the intestines viable. Therefore, studies showed, bacterial spores are the only strain that can be used to treat ME (McFarlin et al., 2017; Zhou, 2017). Hence, probiotics can recuperate gut health by stimulating mucin secretion, secretory immunoglobulin A or AMPs secretion, modulation of microbiota (Ohland & MacNaughton, 2009; Wan et al., 2016).

8.3.2 PREBIOTICS

Prebiotic refers to non-digestible dietary carbohydrates that selectively stimulate the growth of the bacteria reside in the colon by altering the composition of the gut microbiota that beneficially affects the hosts (Macfarlane et al., 2006; Gibson & Roberfroid, 1995). Prebiotics are nutraceuticals which endorsed for strengthening the growth of beneficial bacterial species. The microbiota composition and diversity can be altered in a positive way due to the action of SCFAs. The main source of the SCFAs such as acetate, butyrate, and propionate are carbohydrates (Shashkova et al., 2016). These prebiotics are short chain PS that are not digested by humans (Hord, 2008). Most commonly used prebiotics are galacto-oligosaccharides (GOS), fructo-oligosaccharides (FOS), inulin, lactulose, cellulose, xylo-oligosaccharides (XOS), pectins, hemicelluloses, and starches (Dahiya et al., 2017; Espín et al., 2017).

A prebiotic need to fulfill certain criteria such as resistance to certain mammalian enzymes and gastric acidity, fermented by certain colonic bacteria and capability to stimulate growth and activity of beneficial gut microbiota (Gibson & Roberfroid, 1995). The primary function of prebiotics is modulating composition of gut microbiota, regulating immune system, energy production, improving intestinal barrier and increasing absorption of mineral (Bron et al., 2017). Sangwan et al. demonstrated that if a combination of 90% GOS and 10% FOS were fed to preterm infants, there will be an increase in *Bifidobacteria* and *Lactobacilli* gut microbiota composition that will resemble breast fed infants microbiota. The main sources of GOS are lactose from cow's milk and soybeans (2011). Studies showed that GOS alone can only increase *Lactobacillus* while a mixture of GOS and FOS will increase *Bifidobacteria* and will decrease *Clostridium* in the gut (Vandenplas et al., 2017).

Treatment with prebiotics inulin and FOS on 40 oncologic patients treated with pelvic radiotherapy has shown beneficial effect on *Lactobacillus* and *Bifidobacteria* populations of gut (Velasco et al., 2012). Onions are considered to be a good source of carbohydrates FOS and PS with prebiotic effect (Rodríguez et al., 2008). Bactericidal effect of garlic against *Enterococcus faecalis*, *E. coli*, *Neisseria gonorrhoeae*, *Salmonella typhimurium* and *Staphylococcus aureus* has been demonstrated by Santhosha et al. (2013). Aloe vera contains carbohydrate, amino acids, vitamins, and mineral compound and are considered to be a natural source of prebiotics. It was witnessed that due to oxidation, the leaf gel leads to fermentation and regulates bacterial growth (Kim et al., 2014; Cuvas-Limón et al., 2016).

8.3.3 POLYPHENOLS

Polyphenols have beneficial effects on both CVD and cancer. Dark chocolate, cocoa powder, black, and green tea, nuts, berries, vegetables, soy, beans, and wine are the main sources of polyphenols (Manach et al., 2004). The gut microbiota increases the bioavailability of polyphenols, whereas unabsorbed polyphenols maintain the equilibrium of gut microbiota. This will lead to protection against any pathogens and gastrointestinal disorders, increase in mucus secretion, immune function modulation and cytokines modulation (Ozdal et al., 2016). The most studied compound from polyphenols class is quercetin, derived from plants and are found in grapes, apples, berries, onions, nuts, tea, tomatoes, and in medicinal botanicals such as *Ginkgo biloba*, *Hypericum perforatum*, and *Sambucus canadensis* (Li et al., 2016). Studies showed that quercetin enhances *Firmicutes/Bacteroidetes* ratio when overweight rats were fed with high-fat sucrose diet (Etxeberria et al., 2015).

Tea (green and black) is manufactured from *the Camellia sinensis* plant, which is the most commonly consumed beverages in today's world (Cabrera et al., 2003). Tea contains polyphenols and flavanols like catechin and epicatechin and several unabsorbed compounds like caffeic acid of tea have shown to persist in the gut and play a vital role in the intestine by restraining the growth of pathogenic intestinal bacteria such as *Salmonella, Clostridium, Pseudomonas, E. coli, Bacteroides* (Lee et al., 2006). Studies demonstrated that black tea has certain metabolites that has antimicrobial properties such as phenyl propionic acid, phenylacetic acid and benzoic acid (Duynhoven et al., 2013), whereas green tea contains polyphenols that has certain antioxidant and anti-inflammatory properties which induce the activity of cyclooxygenase-2, NF-kB, and IL-2 (Oz et al., 2013). Studies showed that phenols and derivatives from tea affected the commensal bacteria less such as *Bifidobacterium* and *Lactobacillus* than other pathogenic bacteria. Hence, the phenolic compounds and derivatives from tea can modulate the gut microbiota (Lee et al., 2006). Studies demonstrated that resveratrol stimulates the growth of *Bifidobacterium* and *Lactobacillus*, reduces the growth of *Enterococcus faecalis* and increases the ratio of *Bacteroidetes* and *Firmicutes* (Larrosa et al., 2009; Qiao et al., 2014). Resveratrol is a polyphenol and used as a dietary supplement by patients suffering from CVDs (Imamura et al., 2017; Wahab et al., 2017; Bonnefont-Rousselot, 2016).

8.3.4 PHYTOESTROGENS

Phytoestrogens are polyphenols that are plant derived and share similar structure with human estrogens. This natural plant derived compounds are classified into isoflavonoids (Keservani & Sharma, 2014), lignans, flavonoids, ellagitannins. Besides isoflavonoids, ellagitannins, and lignans are absorbed into equol, urolithins, and enterolignans by gut microbial species which have high antioxidant, anti-inflammatory, antiproliferative, and antiestrogenic effects (Bilal et al., 2014; Gaya et al., 2016). Whole grains, fruits, and vegetables are the main source of phytoestrogens (Sirotkin & Harrath, 2014). Also, different parts of a plant contain different quantity of phytoestrogens (Jaro^{2}ová et al., 2015). For human gut microbiota, flavonoids produce a substrate, and in the intestine, these flavonoids get transformed by bacterial species. Flavonoids regulate intestinal permeability and barrier and it was shown that flavonoids have a direct influence on *Akkermansia* rather than on the production of mucin (Cassidy & Minihane, 2017). Flavonoids provide protection against pathogenic bacteria as it has a high antimicrobial effect, therefore, flavonoids can be considered as an alternative option to antibiotics for treatment and prevention of mild infection (Iranshahi et al., 2015).

Anthocyanins belong to a class of flavonoids, play a significant role in prophylaxis of many neurological (Keservani et al., 2016b), metabolic and CVDs (Keservani et al., 2016a) due to its antioxidant properties and can be found in berries, apples, grapes, and many more (Chaiyasut et al., 2017). Overall and co-workers revealed a substantial change in gut microbiota when mice's diet was supplemented with blackcurrants and blueberries. An increase in anaerobic bacteria *Actinobacteria* and *Bacteroidetes* was seen due to the antioxidant properties of blueberries and blackcurrants (Overall et al., 2017). As conjugate glucosides, isoflavones do not remain active in primary form. The intestinal mucosa and gut microbiota convert these glucosides into compounds which are metabolized by intestinal microbial species into other metabolites (Vitale et al., 2013). Isoflavonoids constitute of glycosides like daidzein, genistein, and glycitein. Daidzein has antioxidant and estrogenic property and gets metabolized to equol by gut microbiota and play a vital role in the health benefits of soy (Franke et al., 2014).

8.3.5 POLYUNSATURATED FATTY ACIDS (PUFAS)

PUFAs are essential fatty acids that play an important role in body functions and are taken externally through diet (Escott-Stump & Mahan, 2000). PUFAs can be categorized into omega-3 and omega-6 fatty acids. Omega-3 fatty acids constitute of linolenic, docosahexaenoic, and eicosapentaenoic acids (EPAs) while omega-6 fatty acids include linoleic and arachidonic acids (ARAs). The underlying mechanism through which PUFAs exert its beneficial effect involves production of pro-inflammatory cytokine, formation of thromboxane's and prostacyclin's, induction of acetylcholine release and modulation of anti-inflammatory responses of hypothalamic-pituitary-adrenal gland (Das, 2000). Hence, diet rich in PUFAs influence immune regulation, cholesterol, and triglyceride levels, blood pressure and cardiovascular function in humans and animals (Ruso et al., 1995; Kawasaki et al., 2000; Straniero et al., 2008; Chen et al., 2009).

Omega-3 and omega-6 fatty acids are the most consumed supplements in hypertriglyceridemia. The effect of PUFAs on changes in the composition of gut microbiota was studied on 47 male mice, divided into 3 groups. Different types of diet such as saturated fatty acids (SFAs), omega-3 and omega-6 fatty acids were fed to each group for 14 weeks. Mice fed with saturated fatty acid rich diet showed a significant decrease in the ratio of *Bacteroidetes* and *Firmicutes* compared to diet rich in PUFAs and mice fed with SFAs rich diet obtained a microbial profile similar to obese animals (Liu et al., 2012). Fish oil contains a high amount of omega-3 fatty acids and is one of the most consumed dietary supplements. A study conducted by Yu and co-workers demonstrated the effect of fish oil on gut microbiome. The study was conducted using mice and the mice were divided into three groups. The first group received saline, the second group received 5 mg/kg fish oil and the third group received 10 mg/kg fish oil. The study revealed that fish oil impedes the growth of uncultured bacteria, *Clostridiales*, *Firmicutes*, *Helicobacter*, *Sphingomonadales*, and *Pseudomonas sp. Helicobacter pylori* is known for its role in the pathogenesis of ulcer and *Firmicutes* for its involvement in obesity (2014). Thus, these results could open a window for the possibility of using this supplement in the treatment of ulcers as well as obesity.

8.4 IMPACT OF OTHER DIETARY NUTRACEUTICALS ON GUT HEALTH

8.4.1 SPIRULINA

Spirulina is a blue-green alga mostly consumed as a dietary supplement which inhibits the growth *of E. coli*, *Bacillus subtilis*, *Pseudomonas aeruginosa*, *Staphylococcus aureus*, and many Gram-positive and Gram-negative bacteria. Spirulina has acquired this antibacterial property due to a metabolite produced by this plant. Several compounds found on this plant such as phenolic compounds, carbohydrates, glutamate, and aspartate was attributed by the microbiota's modulatory effect of spirulina which are well-known for its antimicrobial effect that has the ability to stimulate the growth of probiotics (Beheshtipour et al., 2013; Finamore et al., 2017).

8.4.2 BERBERINE

Berberine is an alkaloid, which is isolated from *Berberis sp.* or *Rhizoma coptidis* and has antimicrobial effect on *Bacteroidetes* and *Firmicutes*. It modulates the gut microbiota by increasing the intestinal gene expression of fasting induced adipose factor (Fiaf) in mice (Zhang et al., 2015; Xu et al., 2017). Berberine has both antiproliferative and antioxidant properties. It acts as an antibiotic which decrease *Ruminococcus* and increase *Bacteroides* in the terminal ileum and colon (Guo et al., 2016).

8.4.3 CURCUMIN

Curcumin is a bioactive compound that is isolated from *Curcuma longa* plant. A study of inflammatory bowel disease (IBD) in mice has reported a significant change in the gut microbiota of mice when supplemented curcumin in the diet. A significant increase in the growth of *Lactobacillus* and decrease in *Coriobacteriales* was observed (Ramalingam et al., 2016).

8.5 CONCLUSIONS

This chapter attempted to express the close association and interaction between nutraceuticals and gut microbiota (Figure 8.1). The gut microbiota influences the nutritional and health status of an individual through metabolic and immune function modulation. The active components of nutraceuticals have benefits on human and gut health. The bioactive components also exert substantial effects on modulation of gut microbiota's composition and on the intestinal environment.

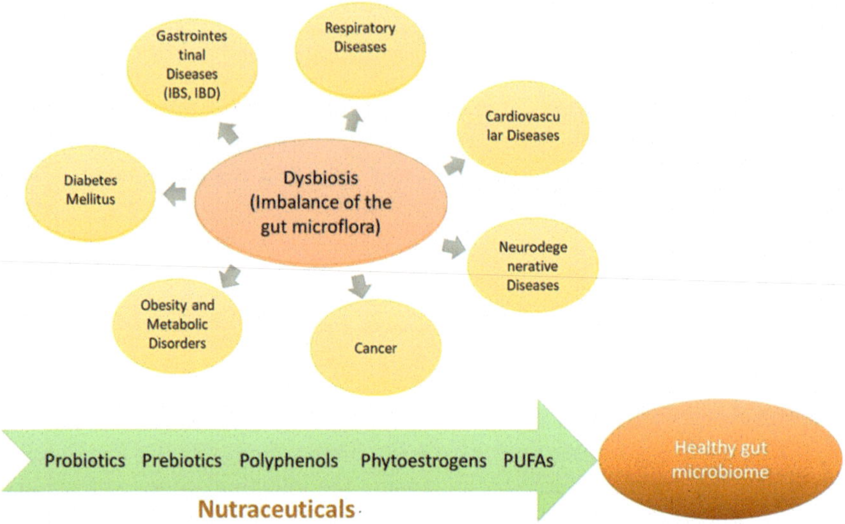

FIGURE 8.1 Link between nutraceuticals and gut microbiota in health and disease.

The intestinal barrier plays as a protective role for maintaining the balance of gut microbiota and preventing from tissue injury that may lead to many diseases. It is well demonstrated that nutraceuticals such as probiotic, prebiotic, PUFAs, polyphenols, phytoestrogens influence the gut health as well as overall human health. These nutraceuticals reinstate the microbial equilibrium, provide protection against pathogenic agent, and induce the inflammation pathway. The changes in the gut microbial diversity are considered as an important constituent in the development of many diseases.

In earlier years, people used to consume nutraceuticals for several diseases without any medical recommendation. People are unaware of both the risk and benefits of nutraceuticals. Therefore, through proper awareness and education, populations can benefit from nutraceuticals. It is evident that imbalance of gut microbiota could lead to many diseases. Thus, critical knowledge needs to be established for use of the correct quantity of nutraceuticals which will affect the composition of microbiota in preventing many diseases. Therefore, it can be concluded that all the above-mentioned nutraceuticals have assuring clinical results and are expected to have beneficial effect on gut and host health. Nevertheless, further studies are needed to fulfill the assessment of nutraceuticals involved in health's benefit as an outcome of alteration in the composition of gut microbial species.

KEYWORDS

- **dysbiosis**
- **gut microbiota**
- **health**
- **nutraceuticals**

REFERENCES

Adelaja, A. O., & Schilling, B. J., (1999). Nutraceutical: Blurring the line between food and drugs in the twenty-first century. *Mag. Food. Farm. Resour. Issues., 14*, 35–40.

Alex, S., Lange, K., Amolo, T., Grinstead, J. S., Haakonsson, A. K., Szalowska, E., Koppen, A., et al., (2013). Short-chain fatty acids stimulate angiopoietin-like 4 synthesis in human colon adenocarcinoma cells by activating peroxisome proliferator-activated receptor. *Mol. Cell. Biol., 33*(7), 1303–1316.

Alisi, A., Bedogni, G., Baviera, G., Giorgio, V., Porro, E., Paris, C., Giammaria, P., et al., (2014). Randomized clinical trial: The beneficial effects of VSL#3 in obese children with non-alcoholic steatohepatitis. *Aliment. Pharmacol. Ther., 39*(11), 1276–1285.

Alvarez-Curto, E., & Milligan, G., (2016). Metabolism meets immunity: The role of free fatty acid receptors in the immune system. *Biochem. Pharmacol., 114*, 3–13.

Bäckhed, F., Ding, H., Wang, T., Hooper, L. V., Koh, G. Y., Nagy, A., Semenkovich, C. F., & Gordon, J. I., (2004). The gut microbiota as an environmental factor that regulates fat storage. *Proc. Natl. Acad. Sci. USA, 101*(44), 15718–15723.

Bäckhed, F., Manchester, J. K., Semenkovich, C. F., & Gordon, J. I., (2007). Mechanisms underlying the resistance to diet-induced obesity in germ-free mice. *Proc. Natl. Acad. Sci. USA, 104*(3), 979–984.

Beheshtipour, H., Mortazavian, A. M., Mohammadi, R., Sohrabvandi, S., & Khosravi-Darani, K., (2013). Supplementation of *Spirulina platensis* and *Chlorella vulgaris* algae into probiotic fermented milks. *Compr. Rev. Food. Sci. Food. Safety, 12*(2), 144–154.

Bilal, I., Chowdhury, A., Davidson, J., & Whitehead, S., (2014). Phytoestrogens and prevention of breast cancer: The contentious debate. *World. J. Clin. Oncol., 5*(4), 705–712.

Bonnefont-Rousselot, D., (2016). Resveratrol and cardiovascular diseases. *Nutrients, 8*(5), 1–24.

Bron, P. A., Kleerebezem, M., Brummer, R. J., Cani, P. D., Mercenier, A., Macdonald, T. T., Garcia-Ródenas, C. L., & Wells, J. M., (2017). Can probiotics modulate human disease by impacting intestinal barrier function? *Br. J. Nutr., 117*(1), 93–107.

Bronzwaer, S., (2008). EFSA scientific forum "from safe food to healthy diets'. EU risk assessment–past, present and future. *Trends. Food. Sci. Technol., 19*(1), S2-S8.

Brower, V., (1989). Nutraceuticals: Poised for a healthy slice of the healthcare market? *Nat. Biotechnol., 16*(8), 728–731.

Buford, T. W., (2017). (Dis)Trust your gut: The gut microbiome in age-related inflammation, health, and disease. *Microbiome., 5*(1), 80.

Cabrera, C., Giménez, R., & López, M. C., (2003). Determination of tea components with antioxidant activity. *J. Agric. Food. Chem., 51*(15), 4427–4435.

Carding, S., Verbeke, K., Vipond, D. T., Corfe, B. M., & Owen, L. J., (2015). Dysbiosis of the gut microbiota in disease. *Microb. Ecol. Health Dis., 26*, 1–9.

Cassidy, A., & Minihane, A. M., (2017). The role of metabolism (and the microbiome) in defining the clinical efficacy of dietary flavonoids. *Am. J. Clin. Nutr., 105*(1), 10–22.

Chaiyasut, C., Pengkumsri, N., Sirilun, S., Peerajan, S., Khongtan, S., & Sivamaruthi, B. S., (2017). Assessment of changes in the content of anthocyanins, phenolic acids, and antioxidant property of *Saccharomyces cerevisiae* mediated fermented black rice bran. *AMB Express, 7*, 114.

Chen, Z. Y., Peng, C., Jiao, R., Wong, Y. M., Yang, N., & Huang, Y., (2009). Anti-hypertensive nutraceuticals and functional foods. *J. Agric. Food. Chem., 57*(11), 4485–4489.

Chevalier, C., Stojanović, O., Colin, D. J., Suarez-Zamorano, N., Tarallo, V., Veyrat-Durebex, C., Rigo, D., et al., (2015). Gut microbiota orchestrates energy homeostasis during cold. *Cell, 163*(6), 1360–1374.

Cummings, J. H., & Macfarlane, G. T., (1997). Role of intestinal bacteria in nutrient metabolism. *JPEN. J. Parenter. Enteral. Nutr., 21*(6), 357–365.

Cuvas-Limón, R., Julio, M., Carlos, C., Mario, C., Mussatto, S., & Ruth, B. C., (2016). *Aloe vera* and probiotics: A new alternative to symbiotic functional foods. *An. Res. Rev. Biol., 9*(2), 1–11.

Dabek, M., McCrae, S. I., Stevens, V. J., Duncan, S. H., & Louis, P., (2008). Distribution of betaglucosidase and beta-glucuronidase activity and of beta-glucuronidase gene Gus in human colonic bacteria. *FEMS. Microbiol. Ecol., 66*(3), 487–495.

Dahiya, D. K., Puniya, M., Shandilya, U. K., & Dhewa, T., (2017). Gut microbiota modulation and its relationship with obesity using prebiotic fibers and probiotics: A review. *Front. Microbiol., 8*, 563.

Das, L., Bhaumik, E., Raychaudhuri, U., & Chakraborty, R., (2012). Role of nutraceuticals in human health. *J. Food. Sci. Technol., 49*(2), 173–183.

Das, U. N., (2000). Beneficial effect(s) of n-3 fatty acids in cardiovascular diseases: But, why and how? *Prostaglandins. Leukot. Essent. Fatty. Acids, 63*(6), 351–362.

Den, B. G., Van, E. K., Groen, A. K., Venema, K., Reijngoud, D. J., & Bakker, B. M., (2013). The role of short-chain fatty acids in the interplay between diet, gut microbiota, and host energy metabolism. *J. Lipid. Res., 54*(9), 2325–2340.

Derrien, M., Vaughan, E. E., Plugge, C. M., & De Vos, W. M., (2004). *Akkermansia muciniphila* gen. Nov., sp. Nov., a human intestinal mucin-degrading bacterium. *Int. J. of Syst. Evol. Microbiol., 54*(Pt 5), 1469–1476.

Dethlefsen, L., Huse, S., Sogin, M. L., & Relman, D. A., (2008). The pervasive effects of an antibiotic on the human gut microbiota, as revealed by deep 16S rRNA sequencing. *PLoS. Biol., 6*(11), e280.

Duynhoven, J. V., Vaughan, E. E., Dorsten, F. V., Gomez-Roldan, V., Vos, R. D., Vervoort, J., Hooft, J. J. J. V. D., et al., (2013). Interactions of black tea polyphenols with human gut microbiota: Implications for gut and cardiovascular health. *Am. J. Clin. Nutr., 98*(6), 1631S-1641S.

Elderman, M., Sovran, B., Hugenholtz, F., Graversen, K., Huijskes, M., Houtsma, E., Belzer, C., et al., (2017). The effect of age on the intestinal mucus thickness, microbiota composition and immunity in relation to sex in mice. *PLoS One, 12*(9), e0184274.

Escott-Stump, E., & Mahan, L. K., (2008). *Krause's Food, Nutrition and Diet Therapy* (12th edn., pp. 553–559). Saunders/Elsevier; Philadelphia.

Espín, J. C., González-Sarrías, A., & Tomás-Barberán, F. A., (2017). The gut microbiota: A key factor in the therapeutic effects of (poly)phenols. *Biochem. Pharmacol., 139*, 82–93.

Etxeberria, U., Arias, N., Boqué, N., Macarulla, M. T., Portillo, M. P., Martínez, J. A., & Milagro, F. I., (2015). Reshaping faecal gut microbiota composition by the intake of trans-resveratrol and quercetin in high-fat sucrose diet-fed rats. *J. Nutr. Biochem., 26*(6), 651–660.

FAO/WHO. (2002). *Guidelines for the Evaluation of Probiotics in Food.* Food and Agriculture Organization of the United Nations and World Health Organization Working Group Report. 1–11.

Finamore, A., Palmery, M., Bensehaila, S., & Peluso, I., (2017). Antioxidant, immunomodulating, and microbial-modulating activities of the sustainable and eco-friendly *Spirulina. Oxid. Med. Cell. Longev.,* 3247528.

Flint, H. J., Bayer, E. A., Rincon, M. T., Lamed, R., & White, B. A., (2008). Polysaccharide utilization by gut bacteria: Potential for new insights from genomic analysis. *Nat. Rev. Microbiol., 6*(2), 121–131.

Franke, A. A., Lai, J. F., & Halm, B. M., (2014). Absorption, distribution, metabolism, and excretion of isoflavonoids after soy intake. *Arch. Biochem. Biophys., 559*, 24–28.

Fukiya, S., Arata, M., Kawashima, H., Yoshida, D., Kaneko, M., Minamida, K., Watanabe, J., et al., (2009). Conversion of cholic acid and chenodeoxycholic acid into their 7-oxo derivatives by *Bacteroides* intestinalis AM-1 isolated from human feces. *FEMS. Microbiol. Lett., 293*(2), 263–270.

Galdón, B. R., Rodríguez, E. M. R., & Romero, C. D., (2008). Flavonoids in onion cultivars (*Allium cepa* L.). *J. Food. Sci., 73*(8), C599–C605.

Gaya, P., Medina, M., Sánchez-Jiménez, A., & Landete, J., (2016). Phytoestrogen metabolism by adult human gut microbiota. *Molecules, 21*(8), 1034.

Gibson, G. R., & Roberfroid, M. B., (1995). Dietary modulation of the human colonic microbiota: Introducing the concept of prebiotics. *J. Nutr., 125*(6), 1401–1412.

Gill, S. R., Pop, M., Deboy, R. T., Eckburg, P. B., Turnbaugh, P. J., Samuel, B. S., Gordon, J. I., et al., (2006). Metagenomic analysis of the human distal gut microbiome. *Science, 312*(5778), 1355–1359.

Guo, Y., Zhang, Y., Huang, W., Selwyn, F. P., & Klaassen, C. D., (2016). Dose-response effect of berberine on bile acid profile and gut microbiota in mice. *BMC. Complement. Altern. Med., 16*(1), 394.

Gupta, S., Chauhan, D., Mehla, K., Sood, P., & Nair, A., (2010). An overview of nutraceuticals: Current scenario. *J. Basic. Clin. Pharm., 1*(2), 55–62.

Hamer, H. M., Jonkers, D., & Venema, K., (2008). The role of butyrate on colonic function. *Aliment. Pharmacol. Ther., 27*(2), 104–119.

Hara, T., Hirasawa, A., Ichimura, A., Kimura, I., & Tsujimoto, G., (2011). Free fatty acid receptors FFAR1 and GPR120 as novel therapeutic targets for metabolic disorders. *J. Pharm. Sci., 100*(9), 3594–3601.

Hooper, L. V., Midtvedt, T., & Gordon, J. I., (2002). How host-microbial interactions shape the nutrient environment of the mammalian intestine. *Annu. Rev. Nutr., 22*, 283–307.

Hooper, L. V., Wong, M. H., Thelin, A., Hansson, L., Falk, P. G., & Gordon, J. I., (2001). Molecular analysis of commensal host-microbial relationships in the intestine. *Science, 291*(5505), 881–884.

Hord, N. G., (2008). Eukaryotic microbiotic crosstalk: Potential mechanisms for health benefits of prebiotics and probiotics. *Annu. Rev. Nutr., 28*, 215–231.

Huazano-Garcia, A., Hakdong, S., & Lopez, G., (2017). Modulation of gut microbiota of overweight mice by agavins and their association with body weight. *Nutrients, 9*(9), 821.

Imamura, H., Yamaguchi, T., Nagayama, D., Saiki, A., Shirai, K., & Tatsuno, I., (2017). Resveratrol ameliorates arterial stiffness assessed by cardio-ankle vascular index in patients with type 2 diabetes mellitus. *Int. Heart. J., 58*(4), 577–583.

Iranshahi, M., Rezaee, R., Parhiz, H., Roohbakhsh, A., & Soltani, F., (2015). Protective effects of flavonoids against microbes and toxins: The cases of hesperidin and hesperetin. *Life. Sci., 137*, 125–132.

Jandhyala, S. M., Talukdar, R., Subramanyam, C., Vuyyuru, H., Sasikala, M., & Reddy, D. N., (2015). Role of the normal gut microbiota. *World. J. Gastroenterol., 21*(29), 8836–8847.

Jarošová, B., Javůrek, J., Adamovský, O., & Hilscherová, K., (2015). Phytoestrogens and mycoestrogens in surface waters--their sources, occurrence, and potential contribution to estrogenic activity. *Env. Int., 81*, 26–44.

Jarouliya, U., & Keservani, R. K., (2019a). Pathways leading to child obesity. In: Debasis, B., (ed.), *Global Perspectives on Childhood Obesity: Current Status, Consequences and Prevention* (2nd edn., pp. 137–146). Academic Press, Elsevier, Chapter 12. ISBN: 9780128128404.

Jarouliya, U., & Keservani, R. K., (2019b). Protein function as cell surface and nuclear receptor in human diseases. In: Sanjay, B., & Debarshi, K. M., (eds.), *Medicinal Chemistry with Pharmaceutical Product Development* (pp. 1–32). Apple Academic Press, CRC Press, Taylor & Francis Group. Chapter 1. ISBN: 9781771887106.

Jeong, S. H., Song, Y. K., & Cho, J. H., (2009). Risk assessment of ciprofloxacin, flavomycin, olaquindox and colistin sulfate based on microbiological impact on human gut biota. *Regul. Toxicol. Pharmacol., 53*(3), 209–216.

Joseph, J., Depp, C., Shih, P. A. B., Cadenhead, K. S., & Schmid-Schönbein, G., (2017). Modified Mediterranean diet for enrichment of short chain fatty acids: Potential adjunctive therapeutic to target immune and metabolic dysfunction in schizophrenia? *Front. Neurosci., 11*(155), 1–16.

Kalliomäki, M., Salminen, S., Poussa, T., & Isolauri, E., (2007). Probiotics during the first 7 years of life: A cumulative risk reduction of eczema in a randomized, placebo-controlled trial. *J. Allergy. Clin. Immunol., 119*(4), 1019–1021.

Kawasaki, T., Seki, E., Osajima, K., Yoshida, M., Asada, K., Matsui, T., & Osajima, Y., (2000). Antihypertensive effect of valyl-tyrosine, a short chain peptide derived from sardine muscle hydrolyzate, on mild hypertensive subjects. *J. Hum. Hypertens., 14*(8), 519–523.

Keservani, R. K., & Sharma, A. K., (2014). Flavonoids: Emerging trends and potential health benefits. *Journal of Chinese Pharmaceutical Sciences, 23*(12), 815.

Keservani, R. K., Kesharwani, R. K., Sharma, A. K., Vyas, N., & Chadoker, A., (2010b). Nutritional supplements: An overview. *International Journal of Current Pharmaceutical Review and Research, 1*(1), 59–75.

Keservani, R. K., Kesharwani, R. K., Vyas, N., Jain, S., Raghuvanshi, R., & Sharma, A. K., (2010a). Nutraceutical and functional food as future food: A review. *Der Pharmacia Lettre, 2*(1), 106–116.

Keservani, R. K., Sharma, A. K., & Jarouliya, U., (2015). Protein and peptide in drug targeting and its therapeutic approach. *Ars Pharmaceutica (Internet), 56*(3), 165–177.

Keservani, R. K., Sharma, A. K., & Kesharwani, R. K., (2016a). Nutraceutical and functional foods for cardiovascular health. In: *Food Process Engineering* (pp. 257–278, 291–312). Apple Academic Press, CRC Press. ISBN: 9781771884020.

Keservani, R. K., Sharma, A. K., & Kesharwani, R. K., (2016b). Medicinal effect of nutraceutical fruits for the cognition and brain health. *Scientifica*, 10. Article ID 3109254. doi: 10.1155/2016/3109254.

Keservani, R. K., Sharma, A. K., & Kesharwani, R. K., (2017). An overview and therapeutic applications of nutraceutical and functional foods. *Recent Advances in Drug Delivery Technology*, 160–201. ISBN:9781522507543.

Keservani, R. K., Sharma, A. K., & Kesharwani, R. K., (2020). *Nutraceuticals and Dietary Supplements: Applications in Health Improvement and Disease Management*. CRC Press. ISBN: 9781771888738.

Kim, C. H., Park, J., & Kim, M., (2014). Gut microbiota-derived short-chain fatty acids, T cells, and inflammation. *Immune. Netw., 14*(6), 277–288.

Kim, Y. W., Jeong, Y. J., Kim, A. Y., Son, H. H., Lee, J. A., Jung, C. H., Kim, C. H., & Kim, J., (2014). *Lactobacillus brevis* strains from fermented *Aloe vera* survive gastroduodenal environment and suppress common food borne enteropathogens. *PLoS One, 9*(3), e90866.

Kurokawa, K., Itoh, T., Kuwahara, T., Oshima, K., Toh, H., Toyoda, A., Takami, H., et al., (2007). Comparative metagenomics revealed commonly enriched gene sets in human gut microbiomes. *DNA. Res., 14*(4), 169–181.

Larrosa, M., Yañéz-Gascón, M. J., Selma, M. V., González-Sarrías, A., Toti, S., Cerón, J. J., Tomás-Barberán, F., et al., (2009). Effect of a low dose of dietary resveratrol on colon microbiota, inflammation and tissue damage in a DSS – induced colitis rat model. *J. Agric. Food. Chem., 57*(6), 2211–2220.

Lee, H. C., Jenner, A. M., Low, C. S., & Lee, Y. K., (2006). Effect of tea phenolics and their aromatic fecal bacterial metabolites on intestinal microbiota. *Res. Microbiol., 157*(9), 876–884.

Lee, Y. K., & Mazmanian, S. K., (2010). Has the microbiota played a critical role in the evolution of the adaptive immune system? *Science, 330*(6012), 1768–1773.

Ley, R. E., Lozupone, C. A., Hamady, M., Knight, R., & Gordon, J. I., (2008). Worlds within worlds: Evolution of the vertebrate gut microbiota. *Nat. Rev. Microbiol., 6*(10), 776–788.

Li, Y., Yao, J., Han, C., Yang, J., Chaudhry, M. T., Wang, S., Liu, H., & Yin, Y., (2016). Quercetin, inflammation and immunity. *Nutrients, 8*(3), 1–14.

Liu, T., Hougen, H., Vollmer, A. C., & Hiebert, S. M., (2012). Gut bacteria profiles of Mus musculus at the phylum and family levels are influenced by saturation of dietary fatty acids. *Anaerobe, 18*(3), 331–337.

Louis, P., & Flint, H. J., (2017). Formation of propionate and butyrate by the human colonic microbiota. *Environ. Microbiol., 19*(1), 29–41.

Macfarlane, S., Macfarlane, G. T., & Cummings, J. H., (2006). Review article: Prebiotics in the gastrointestinal tract. *Aliment. Pharmacol. Ther., 24*(5), 701–714.

Magrone, T., Kumazawa, Y., & Jirillo, E., (2014). Polyphenol-mediated beneficial effects in healthy status and disease with special reference to immune-based mechanisms. In: Watson, R. R., Preedy, V. R., & Zibadi, S., (eds.), *Polyphenols in Human Health and Disease* (Vol. 1, pp. 467–479). Elsevier: USA.

Manach, C., Scalbert, A., Morand, C., Rémésy, C., & Jiménez, L., (2004). Polyphenols: Food sources and bioavailability. *Am. J. Clin. Nutr., 79*(5), 727–747.

McFarlin, B. K., Henning, A. L., Bowman, E. M., Gary, M. A., & Carbajal, K. M., (2017). Oral spore – based probiotic supplementation was associated with reduced incidence of post – prandial dietary endotoxin, triglycerides, and disease risk biomarkers. *World. J. Gastrointest. Pathophysiol., 8*(3), 117–126.

Morrison, D. J., & Preston, T., (2016). Formation of short chain fatty acids by the gut microbiota and their impact on human metabolism. *Gut Microbes., 7*(3), 189–200.

Nadal, I., Donat, E., Ribes-Koninckx, C., Calabuig, M., & Sanz, Y., (2007). Imbalance in the composition of the duodenal microbiota of children with coeliac disease. *J. Med. Microbiol., 56*(Pt 12), 1669–1674.

Ohland, C. L., & Mac, N. W. K., (2009). Probiotic bacteria and intestinal epithelial barrier function. *Am. J. Physiol. Gastrointest. Liver. Physiol., 298*(6), G807-G819.

Ottman, N., Reunanen, J., Meijerink, M., Pietila, T. E., Kainulainen, V., Klievink, J., Huuskonen, L., et al., (2017). Pili-like proteins of *Akkermansia muciniphila* modulate host immune responses and gut barrier function. *PLoS One, 12*(3), e0173004.

Overall, J., Bonney, S. A., Wilson, M., Beermann, A., Grace, M. H., Esposito, D., Lila, M. A., & Komarnytsky, S., (2017). Metabolic effects of berries with structurally diverse anthocyanins. *Int. J. Mol. Sci., 18*(2), 422.

Oz, H. S., Chen, T., & De Villiers, W. J. S., (2013). Green tea polyphenols and sulfasalazine have parallel anti-inflammatory properties in colitis models. *Front. Immunol., 4*, 132.

Ozdal, T., Sela, D. A., Xiao, J., Boyacioglu, D., Chen, F., & Capanoglu, E., (2016). The reciprocal interactions between polyphenols and gut microbiota and effects on bioaccessibility. *Nutrients., 8*(2), 78.

Pandey, M., Verma, R. K., & Saraf, S. A., (2010). Nutraceuticals: New era of medicine and health. *Asian. J. Pharm. Clin. Res., 3*(1), 11–15.

Park, M., Kwon, B., Ku, S., & Ji, G., (2017). The efficacy of *Bifidobacterium longum* BORI and *Lactobacillus acidophilus* AD031 probiotic treatment in infants with rotavirus infection. *Nutrients, 9*(8), 887.

Pennisi, E., (2013). Cancer therapies use a little help from microbial friends. *Science, 342*(6161), 921.

Pluznick, J., (2014). A novel SCFA receptor, the microbiota, and blood pressure regulation. *Gut Microbes, 5*(2), 202–207.

Qiao, Y., Sun, J., Xia, S., Tang, X., Shi, Y., & Le, G., (2014). Effects of resveratrol on gut microbiota and fat storage in a mouse model with high-fat-induced obesity. *Food. Funct., 5*(6), 1241–1249.

Ramalingam, R., Harrison, C. A., Besselsen, D. G., & John, H., (2016). The role of curcumin in modulating colonic microbiota during colitis and colon cancer prevention. *Inflamm. Bowel. Dis., 21*(11), 2483–2494.

Ridlon, J. M., Kang, D. J., & Hylemon, P. B., (2006). Bile salt biotransformations by human intestinal bacteria. *J. Lipid. Res., 47*(2), 241–259.

Riva, A., Togni, S., Giacomelli, L., Franceschi, F., Eggenhoffner, R., Feragalli, B., Belcaro, G., et al., (2017). Effects of a curcumin-based supplementation in asymptomatic subjects with low bone density: A preliminary 24-week supplement study. *Eur. Rev. Med. Pharmaco. Sci., 21*, 1684–1689.

Romano, M., Vitaglione, P., Sellitto, S., & D'Argenio, G., (2012). Nutraceuticals for protection and healing of gastrointestinal mucosa. *Curr. Med. Chem., 19*(1), 109–117.

Rowland, I. R., (2009). The role of the gastrointestinal microbiota in colorectal cancer. *Curr. Pharm. Des., 15*(13), 1524–1527.

Russo, C., Olivieri, O., Girelli, D., Azzini, M., Stanzial, A. M., Guarini, P., Friso, S., et al., (1995). Omega-3 polyunsaturated fatty acid supplements and ambulatory blood pressure monitoring parameters in patients with mild essential hypertension. *J. Hypertens., 13*(12 Pt 2), 1823–1826.

Salzman, N., Underwood, M., & Bevins, C., (2007). Paneth cells, defensins, and the commensal microbiota: A hypothesis on intimate interplay at the intestinal mucosa. *Semin. Immunol., 19*(2), 70–83.

SamiLabs-Pioneer in Nutraceuticals. The Hindu Newspaper. https://www.thehindu.com/news/cities/Hyderabad/nutraceutical-startup-raises-37-crore/article65314683.ece (accessed on 20 February 2023).

Sangwan, V., Tomar, S. K., Singh, R. R. B., Singh, A. K., & Ali, B., (2011). Galactooligosaccharides: Novel components of designer foods. *J. Food. Sci., 76*(4), R103–R111.

Santacruz, A., Marcos, A., Wärnberg, J., Martí, A., Martin-Matillas, M., Campoy, C., Moreno, L. A., et al., (2009). EVASYON study group. Interplay between weight loss and gut microbiota composition in overweight adolescents. *Obesity (Silver Spring), 17*(10), 1906–1915.

Santhosha, S. G., Jamuna, P., & Prabhavathi, S. N., (2013). Bioactive components of garlic and their physiological role in health maintenance: A review. *Food Bioscience, 3*, 59–74.

Santini, A., Tenore, G. C., & Novellino, E., (2017). Nutraceuticals: A paradigm of proactive medicine. *Eur. J. Pharm. Sci., 96*, 53–61.

Sanz, Y., & Palma, G. D., (2009). Gut microbiota and probiotics in modulation of epithelium and gut-associated lymphoid tissue function. *Int. Rev. Immunol., 28*(6), 397–413.

Sharma, R., Young, C., & Neu, J., (2010). Molecular modulation of intestinal epithelial barrier: Contribution of microbiota. *Journal of Biomedicine and Biotechnology*, 305879.

Shashkova, T., Popenko, A., Tyakht, A., Peskov, K., Kosinsky, Y., Bogolubsky, L., Raigorodskii, A., et al., (2016). Agent based modelling of human gut microbiome interactions and perturbations. *PLoS One, 11*(2), e0148386.

Shimizu, M., (2010). Interaction between food substances and the intestinal epithelium. *Biosci. Biotechnol. Biochem., 74*(2), 232–241.

Singh, P., Kesharwani, R. K., & Keservani, R. K., (2017a). Antioxidants and vitamins: Roles in cellular function and metabolism. In: *Sustained Energy for Enhanced Human Functions and Activity* (pp. 385–407). Academic Press. ISBN:9780128054130.

Singh, P., Kesharwani, R. K., & Keservani, R. K., (2017b). Protein, carbohydrates, and fats: Energy metabolism. In: *Sustained Energy for Enhanced Human Functions and Activity* (pp. 103–115). Academic Press. ISBN:9780128054130.

Sirotkin, A. V., & Harrath, A. H., (2014). Phytoestrogens and their effects. *Eur. J. Pharmacol., 741*, 230–236.

Steinmeyer, S., Lee, K., Jayaraman, A., & Alaniz, R. C., (2015). Microbiota metabolite regulation of host immune homeostasis: A mechanistic missing link. *Curr. Allergy. Asthma. Rep., 15*(5), 24.

Straniero, S., Cavallini, G., Donati, A., Metelli, M. R., Tamburini, I., Pietrini, P., & Bergamini, E., (2008). Deficiency and supplementation of PUFA in the diet have similar effects on the age-associated changes in rat-plasma cholesterol levels. *Mech. Ageing. Dev., 129*(12), 759–762.

Tang, W. H. W., & Hazen, S. L., (2014). The contributory role of gut microbiota in cardiovascular disease. *J. Clin. Invest., 124*(10), 4204–4211.

Thursby, E., & Juge, N., (2017). Introduction to the human gut microbiota. *Biochem. J., 474*(11), 1823–1836.

Tlaskalová-Hogenová, H., Stepánková, R., Hudcovic, T., Tucková, L., Cukrowska, B., Lodinová-Zádníková, R., Kozáková, H., et al., (2004). Commensal bacteria (normal microflora), mucosal immunity and chronic inflammatory and autoimmune diseases. *Immunol. Lett., 93*(2, 3), 97–108.

Tremaroli, V., & Backhed, F., (2012). Functional interactions between the gut microbiota and host metabolism. *Nature, 489*(7415), 242–249.

Vandenplas, Y., Zakharova, I., & Dmitrieva, Y., (2017). Oligosaccharides in infant formula: More evidence to validate the role of prebiotics. *Br. J. Nutr., 113*(9), 1339–1344.

Velasco, C., Lozano, M. A., Moreno, Y., Paron, L., Cuerda, C. D., & Bretón, I., (2012). Effect of a mixture of inulin and fructo-oligosaccharide on *lactobacillus* and *Bifidobacterium* intestinal microbiota of patients receiving radiotherapy; a randomized, double-blind, placebo-controlled trial. *Nutrición Hospitalaria., 27*(6), 1908–1915.

Vitale, D. C., Piazza, C., Melilli, B., Drago, F., & Salomone, S., (2013). Isoflavones: Estrogenic activity, biological effect and bioavailability. *Eur. J. Drug. Metab. Pharmacokinet., 38*(1), 15–25.

Wahab, A., Gao, K., Jia, C., Zhang, F., Tian, G., Murtaza, G., & Chen, J., (2017). Significance of resveratrol in clinical management of chronic diseases. *Molecules, 22*(8), 1–19.

Wan, L. Y. M., Chen, Z. J., Shah, N. P., & El-Nezami, H., (2016). Modulation of intestinal epithelial defense responses by probiotic bacteria. *Crit. Rev. Food. Sci. Nutr., 56*(16), 2628–2641.

Weitkunat, K., Stuhlmann, C., Postel, A., Rumberger, S., Fankhänel, M., Woting, A., Petzke, K. J., et al., (2017). Short-chain fatty acids and inulin, but not guar gum, prevent diet-induced obesity and insulin resistance through differential mechanisms in mice. *Sci. Rep., 7*(1), 6109.

Wilson, I. D., & Nicholson, J. K., (2009). The role of gut microbiota in drug response. *Curr. Pharm. Des., 15*(13), 1519–1523.

Woting, A., & Blaut, M., (2016). The intestinal microbiota in metabolic disease. *Nutrients, 8*(4), 202.

Wu, S., Rhee, K. J., Albesiano, E., Rabizadeh, S., Wu, X., Yen, H. R., Huso, D. L., et al., (2011). A human colonic commensal promotes colon tumorigenesis via activation of T helper type 17 T cell responses. *Nat. Med., 15*(9), 1016–1022.

Xu, J., Liu, X., Pan, W., & Zou, D., (2017). Berberine protects against diet-induced obesity by regulating metabolic endotoxemia and gut hormone levels. *Mol. Med. Rep., 15*(5), 2765–2787.

Yu, H. N., Zhu, J., Pan, W., Shen, S. R., Shan, W. G., & Das, U. N., (2014). Effects of fish oil with a high content of n-3 polyunsaturated fatty acids on mouse gut microbiota. *Arch. Med. Res., 45*(3), 195–202.

Zeng, H., Taussig, D. P., Cheng, W. H., Johnson, L. A. K., & Hakkak, R., (2017). Butyrate inhibits cancerous HCT116 colon cell proliferation but to a lesser extent in noncancerous NCM460 colon cells. *Nutrients, 9*(1), 25.

Zhang, X., Zhao, Y., Xu, J., Xue, Z., Zhang, M., Pang, X., Zhang, X., & Zhao, L., (2015). Modulation of gut microbiota by berberine and metformin during the treatment of high-fat diet induced obesity in rats. *Sci. Rep., 5*, 14405.

Zhou, K., (2017). Strategies to promote abundance of *Akkermansia muciniphila*, an emerging probiotic in the gut, evidence from dietary intervention studies. *J. Funct. Foods, 33*, 194–201.

CHAPTER 9

ANTIOXIDANT PROPERTIES OF NUTRACEUTICALS

SONIA SINGH,[1] ASHIMA AHUJA,[1] NITIN AGRAWAL,[2] SHIWANGI SHARMA,[1] and DEEP SHIKHAR VARSHNEY[2]

[1] *Institute of Pharmaceutical Research, GLA University, Mathura, Uttar Pradesh, India*

[2] *Faculty of Pharmacy, Raja Balwant Singh Engineering Technical Campuse, Bichpuri, Agra, Uttar Pradesh, India*

ABSTRACT

Food and drugs obtained from natural sources play a crucial role in worldwide health. The term "nutraceutical" represents foods and dietary supplements that give medical health benefits and even provide primary nutritional value present in the food products. Various significant bioactive components such as alkaloids, flavonoids, flavones, isoflavones, flavanols, anthocyanins, stilbenes, ellagic acid act as remarkable antioxidants, considered nutraceutical ingredients. Excess generation of reactive oxygen species (ROS) causes several chronic diseases like digestive disorders, cancer, neurodegenerative, and respiratory diseases. Therefore, antioxidant nutraceutical supplement has been shown to improve the endogenous defensive mechanism, which generally helps in reducing the oxidative damage observed in many clinical types of research. This chapter helps provide phytochemical classes of some significant nutraceuticals and antioxidants as nutraceuticals with the mode of action.

Immune-Boosting Nutraceuticals for Better Human Health: Novel Applications. Urmila Jarouliya, PhD, Raj K. Keservani, MPharm, Rajesh K. Kesharwani, PhD, Virendra K. Patel, PhD, and Adi D. Bharti, MPharm (Eds.)

9.1 INTRODUCTION

About 5% of the inhaled oxygen is ROS, like hydrogen peroxide (H_2O_2), O_2, and OH. Furthermore, exposure to sunlight X-rays, smoking, polluted environment are other factors responsible for producing such species that cause damage to proteins, nucleic acid, carbohydrates (Singh et al., 2017b), and lipids, affect the immune system. However, superoxide radicals are generated by cytochrome P-450, NADPH oxidase (NOX), lipoxygenase during mitochondrial respiration. Therefore, when the production of a radical level increases within the antioxidant defense system in the body, this may cause oxidative stress (OS). OS usually decreases cellular functions cause gastro-related disease, neuro-generative disorder, inflammation, cancer, metabolic dysfunction, early aging, and cataracts (Arts et al., 2005; Choe et al., 2006; Marnett et al., 2000; Bandyopadhyay et al., 1999; Willcox et al., 2004).

Antiradicals are used to scavenge free radicals that minimize the risk of OS-related diseases. They generally scavenge the reactive oxygen species (ROS), and at low concentrations, they inhibit the oxidative reactions, interrupting lipid peroxidation due to free radical chain reaction. Bioactive components are naturally present in food and show antioxidant activity and potent effects on human health. Some epidemiological studies reveal that regular whole grains, fruits, and vegetables can reduce disease risk associated with oxidative damage. Many antioxidants, such as tocopherols, lipoic acid, carotenoids, and polyphenolic components, are considered potent natural antioxidants are possessed radical scavenging properties. Some of the endogenous antioxidant enzymes, like catalase (CAT), glutathione reductase (GR) and peroxidase, vitamins (A, C, and E), minerals (Mn, Cu, Se, and Zn), superoxide dismutase (SOD), limonoids exhibit synergistic effects in defusing free radicals. The role of nutraceuticals for prevention of illness is shown in Figure 9.1 (Marnett et al., 2000; Bandyopadhyay et al., 1999; Willcox et al., 2004; Triantis et al., 2005; Olinski et al., 2002; Tiwari et al., 2001; Kris-Etherton et al., 2002; Halliwell et al., 2004; Scalbert et al., 2000, 2005; Cieslik et al., 2006). Some of the potential nutraceuticals (Keservani et al., 2010a, b, 2017, 2020) utilized to manage chronic diseases are discussed in subsections (Figures 9.3–9.5).

9.2 CAROTENOIDS

Around 700 naturally occurring carotenoids act as antioxidants, protect cells and tissues from the free radicals' harmful effects. Generally, they are classified into two groups based on the chemical structure: carotenes such as lycopene, beta-carotene, and oxycarotenoids such as lutein and xanthophylls. Tomato, carrot, orange, lemon, fenugreek, mustard, spinach, cabbage, turnip, and radish are the richest source of carotenoids (Kaur et al., 2004; Rietjens et al., 2002; George et al., 2004), shown in Figure 9.2((9), (10), (13)).

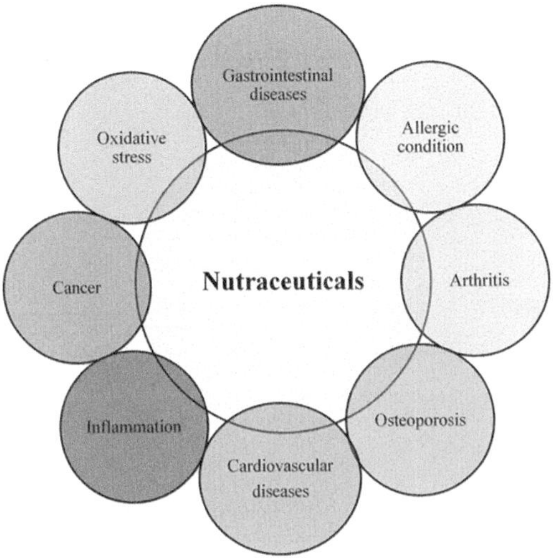

FIGURE 9.1 Nutraceutical role in illness prevention.

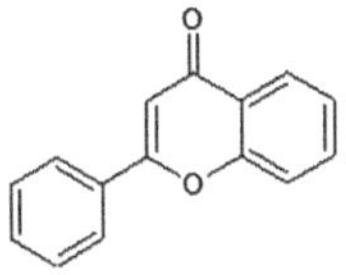

(1)

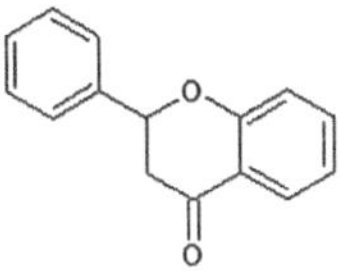

(2)

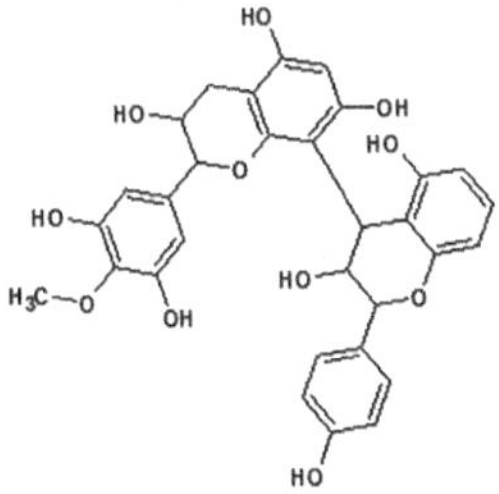

(3)

FIGURE 9.2 *(Continued)*

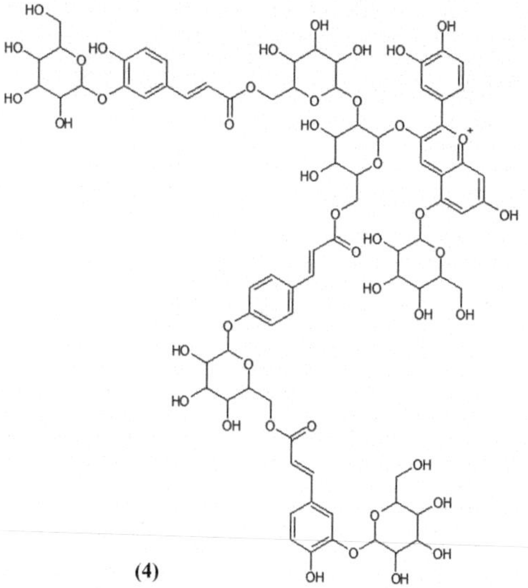

(4)

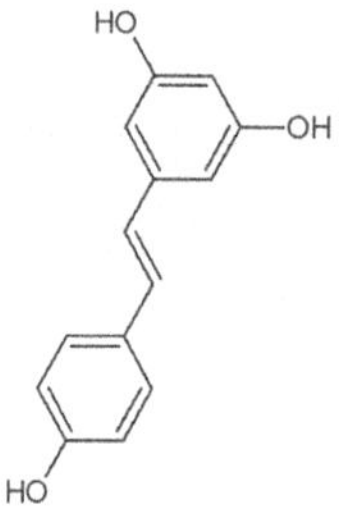

(5)

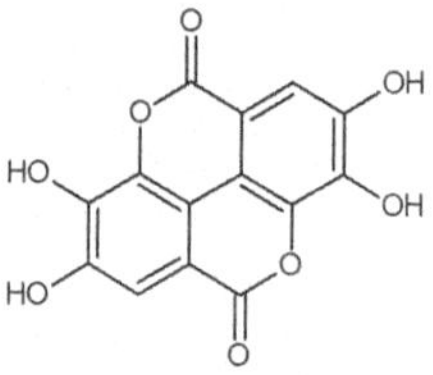

(6)

FIGURE 9.2 *(Continued)*

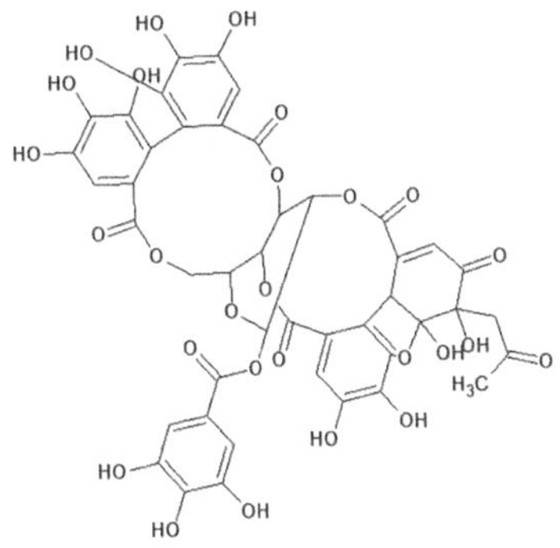

(7)

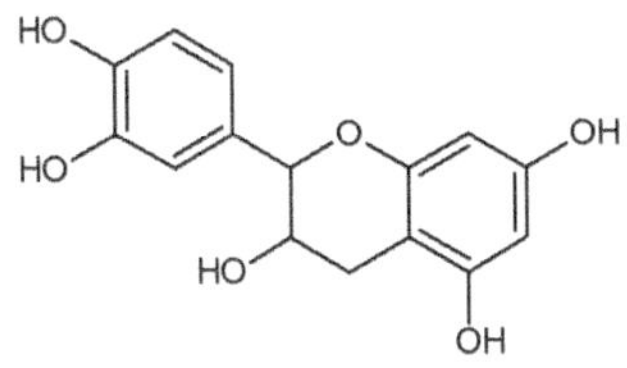

(8)

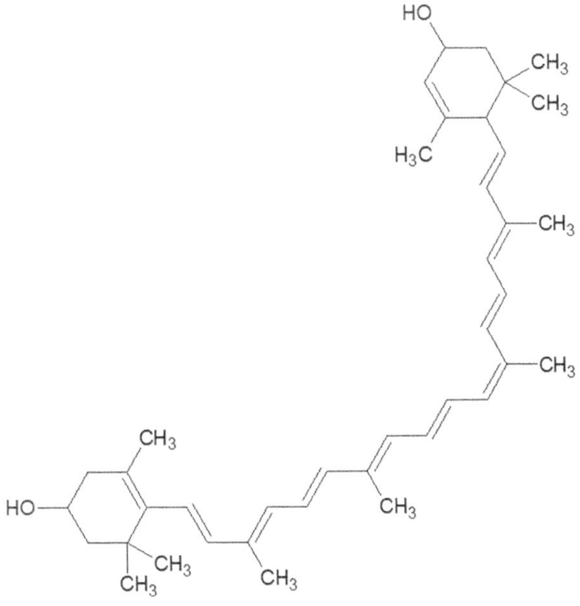

(9)

FIGURE 9.2 *(Continued)*

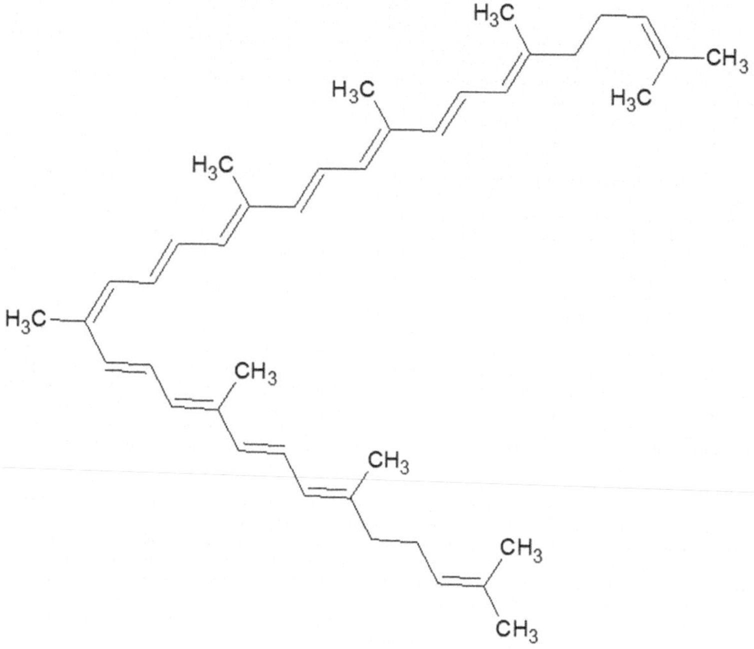

(10)

FIGURE 9.2 *(Continued)*

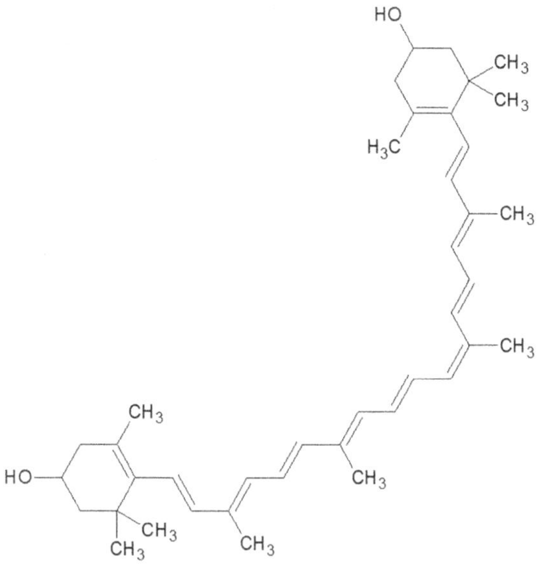

(11)

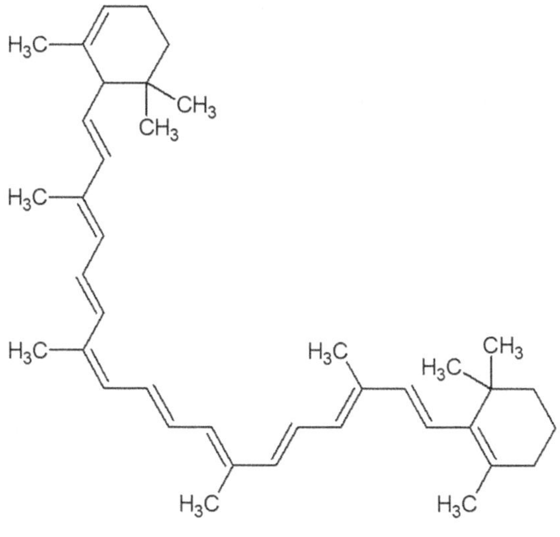

(12)

FIGURE 9.2 *(Continued)*

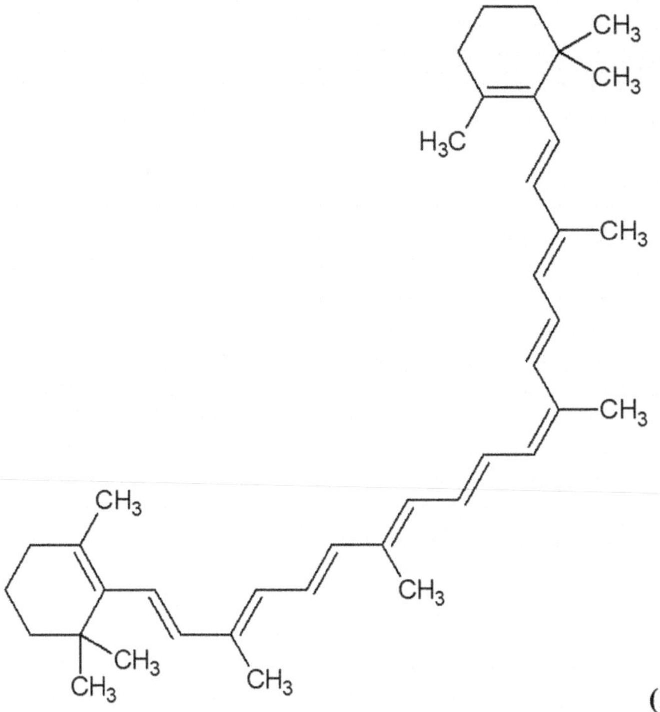

(13)

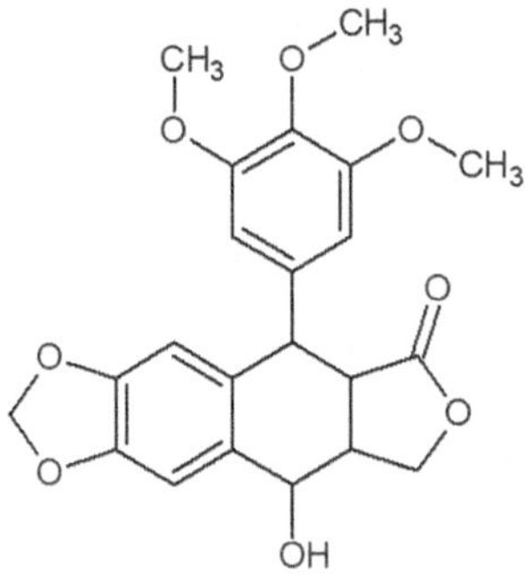

(14)

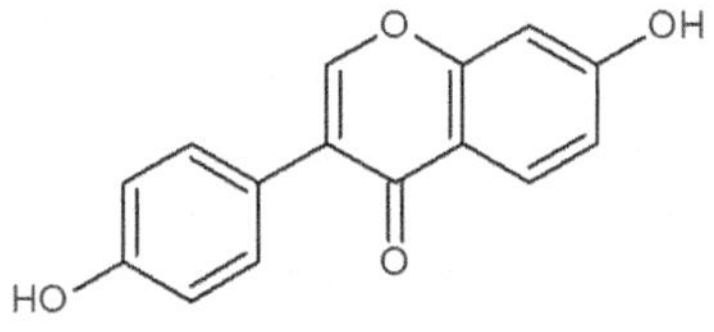

(15)

FIGURE 9.2 *(Continued)*

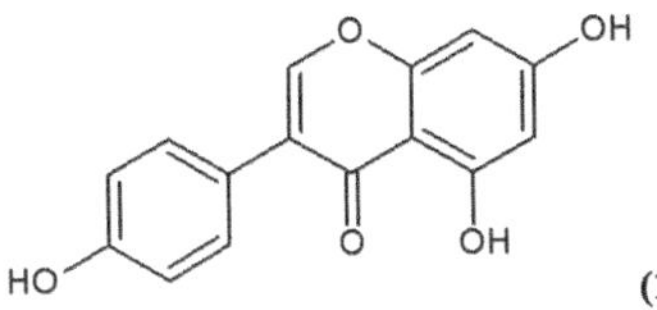

(16)

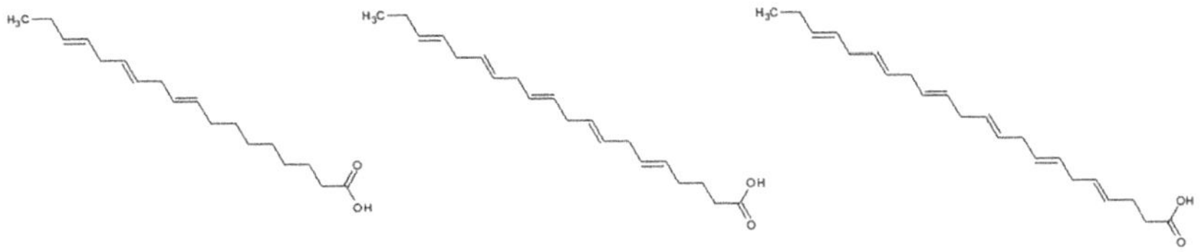

(17)

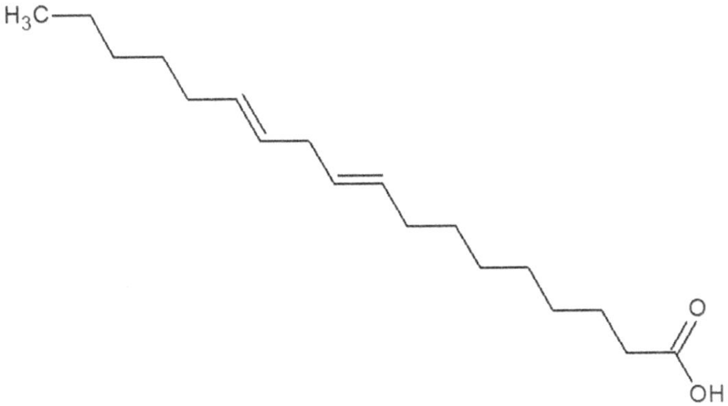

(18)

FIGURE 9.2 *(Continued)*

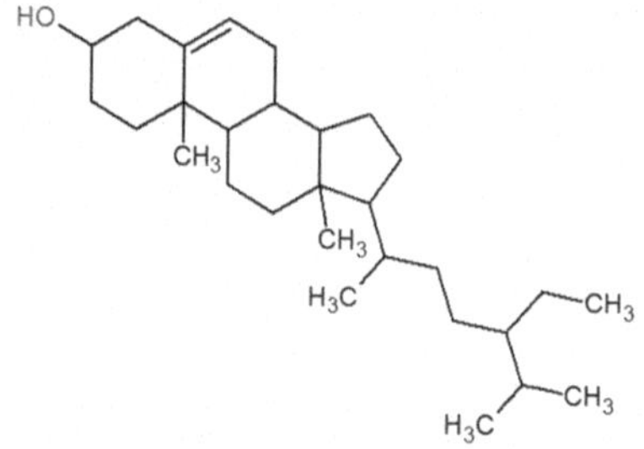

(19)

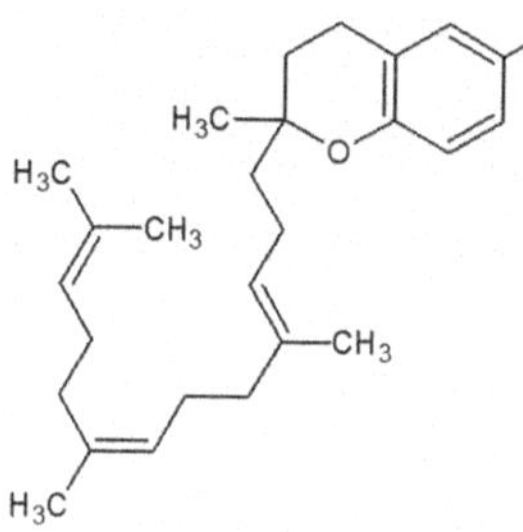

(20)

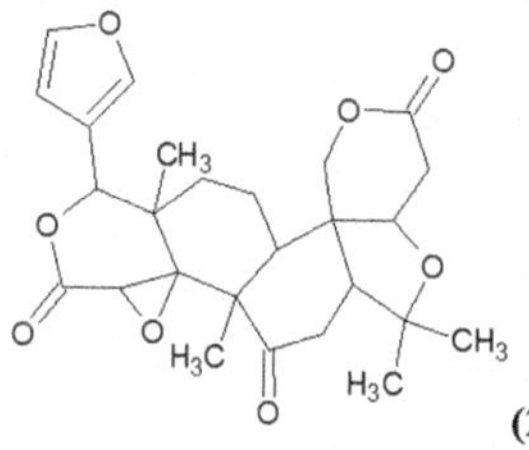

(21)

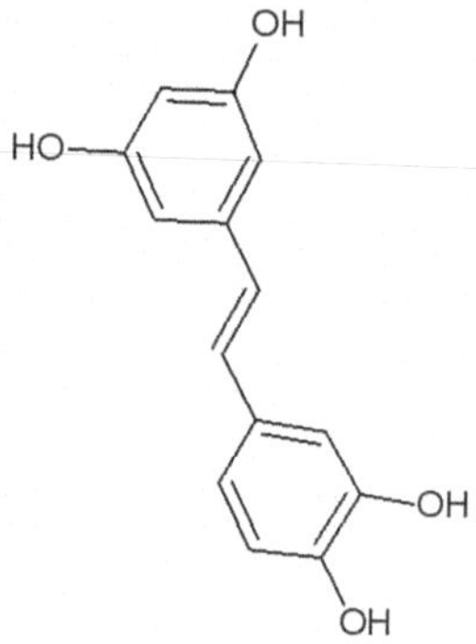

(22)

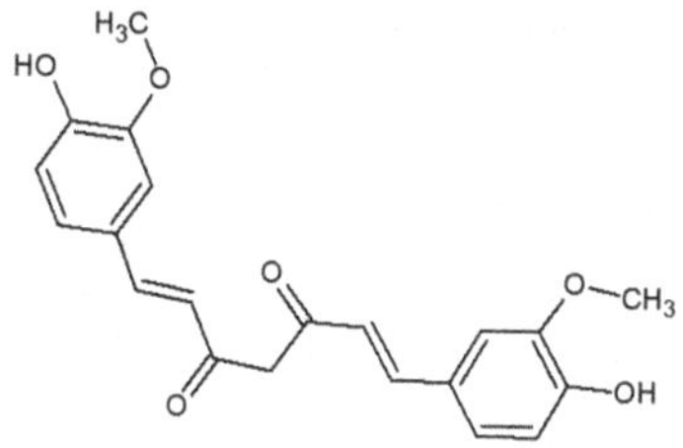

(23)

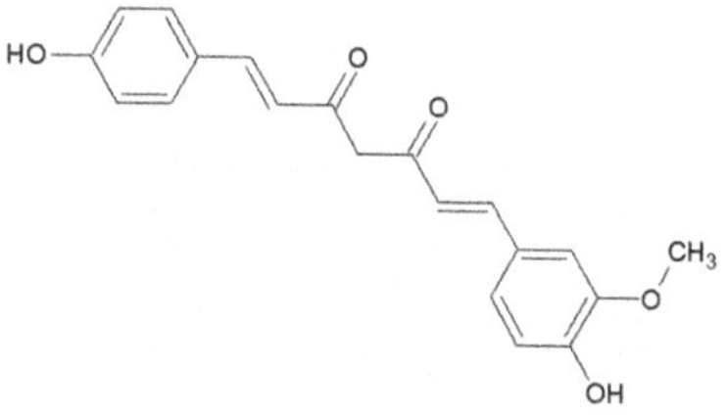

(24)

FIGURE 9.2 *(Continued)*

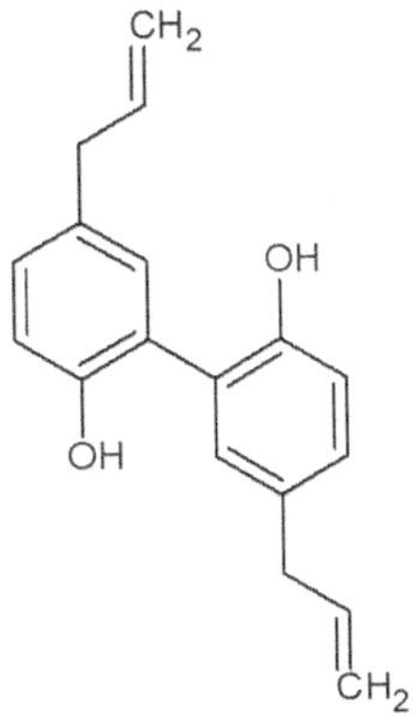

(25)

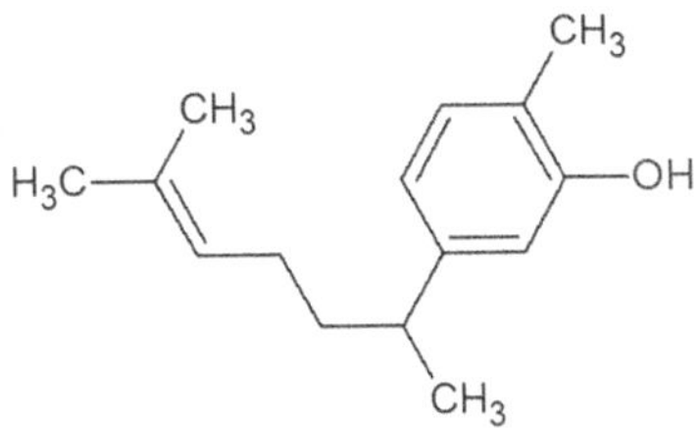

(26)

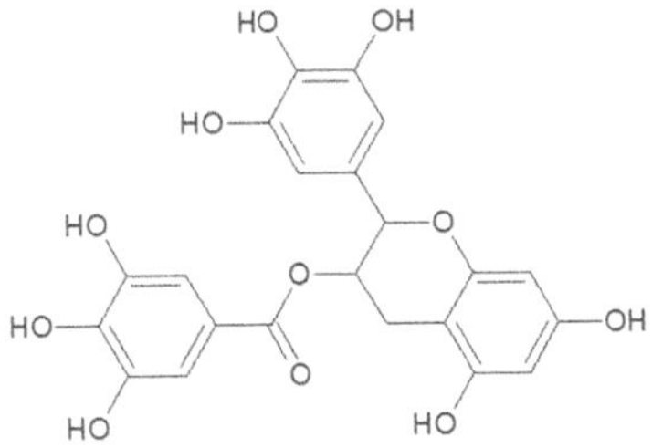

(27)

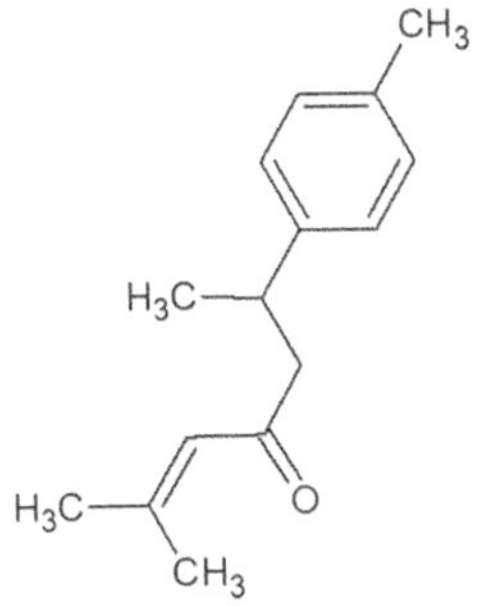

(28)

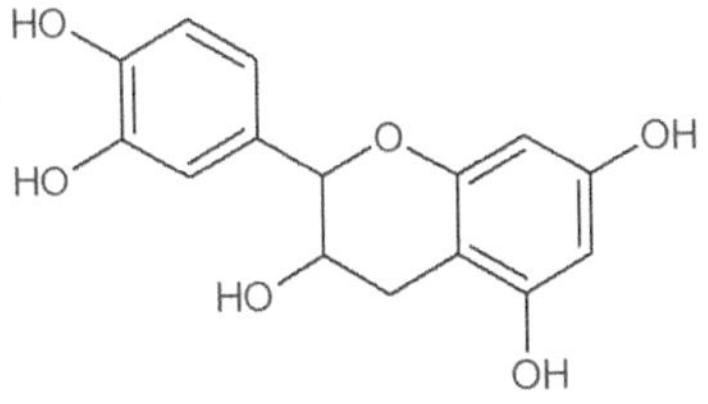

(29)

FIGURE 9.2 *(Continued)*

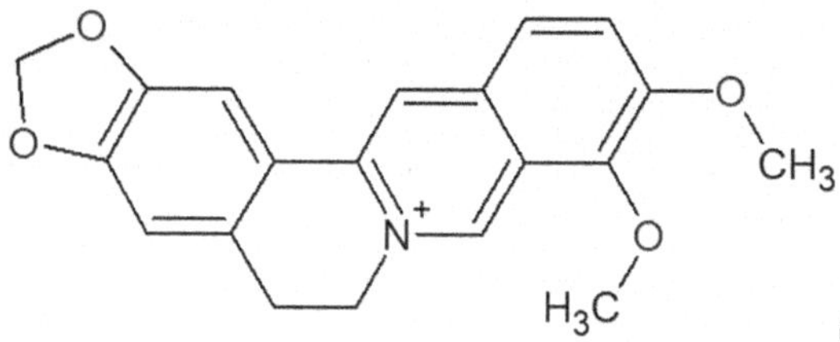

(30)

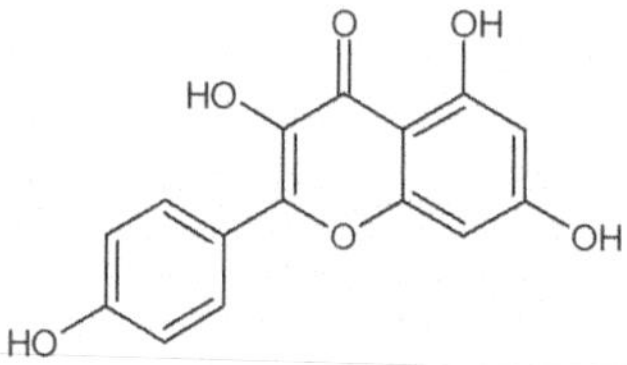

(31)

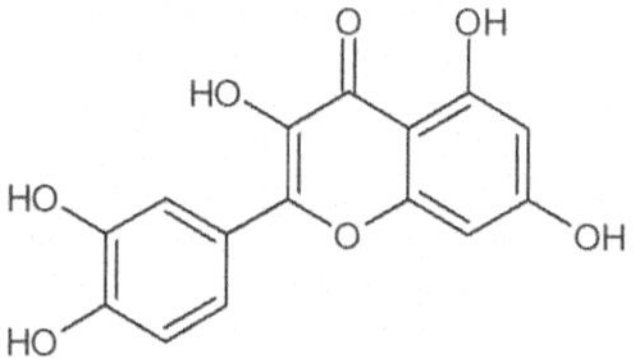

(32)

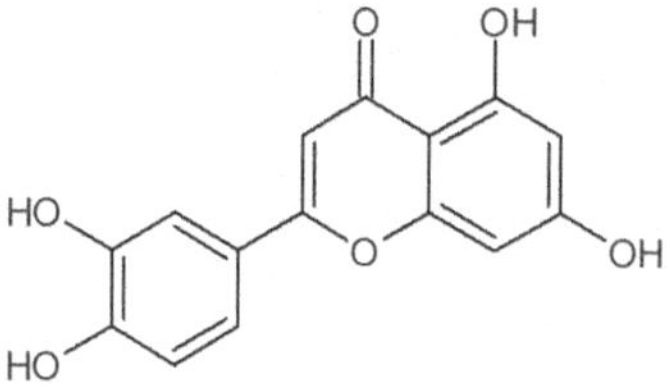

(33)

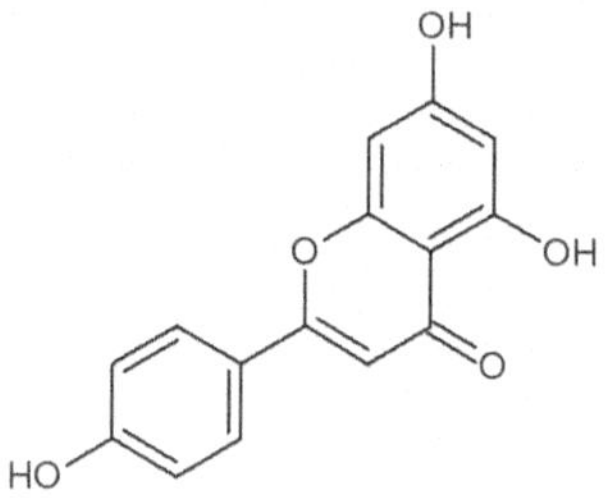

(34)

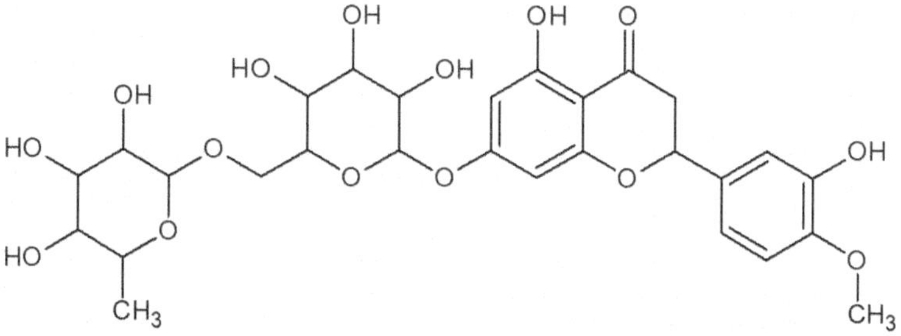

(35)

FIGURE 9.2 *(Continued)*

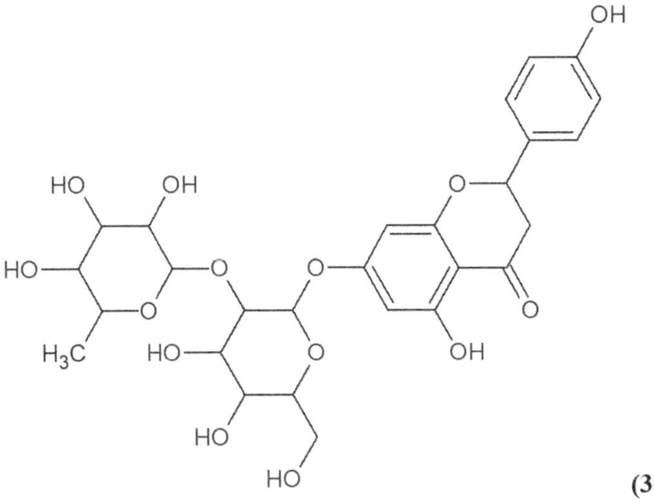

(36)

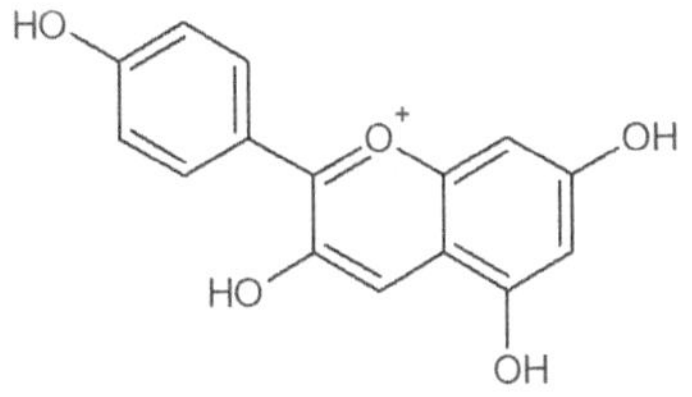

(37)

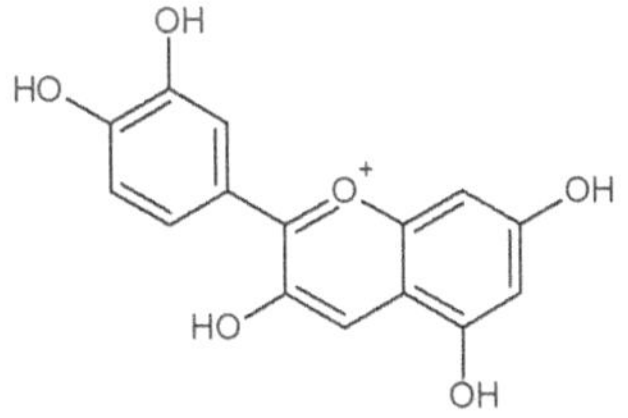

(38)

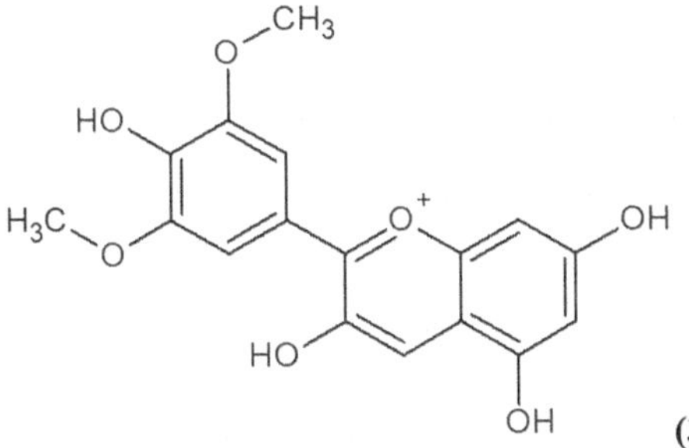

(39)

FIGURE 9.2 *(Continued)*

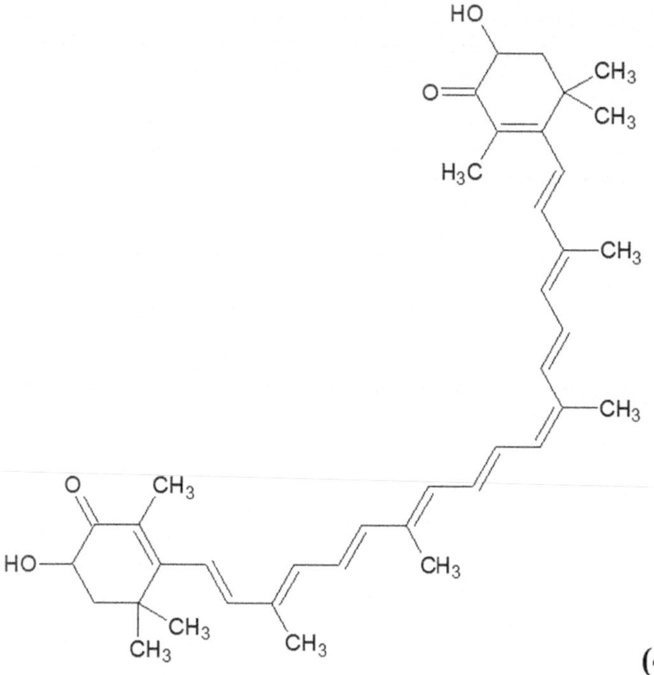

(40)

FIGURE 9.2 *(Continued)*

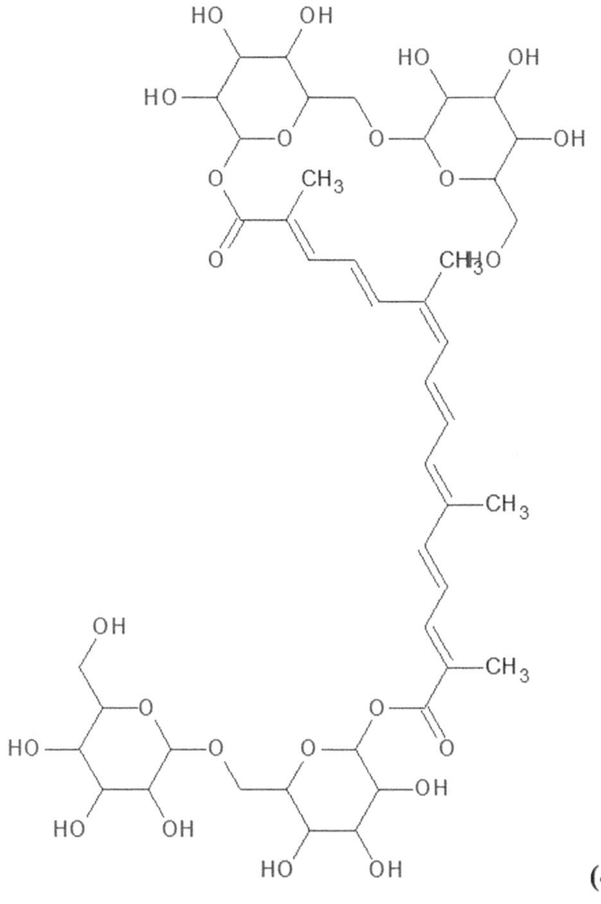

(41)

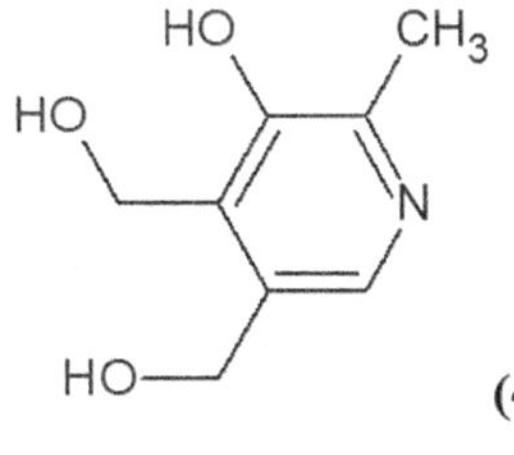

(42)

FIGURE 9.2 *(Continued)*

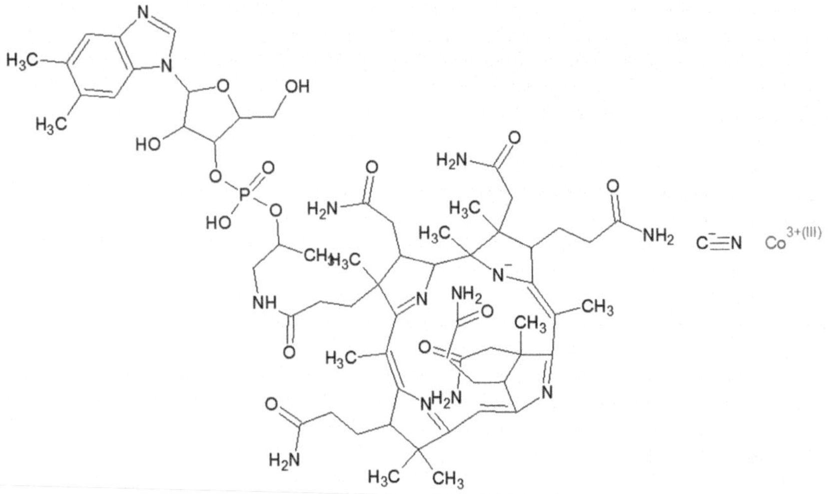

(43)

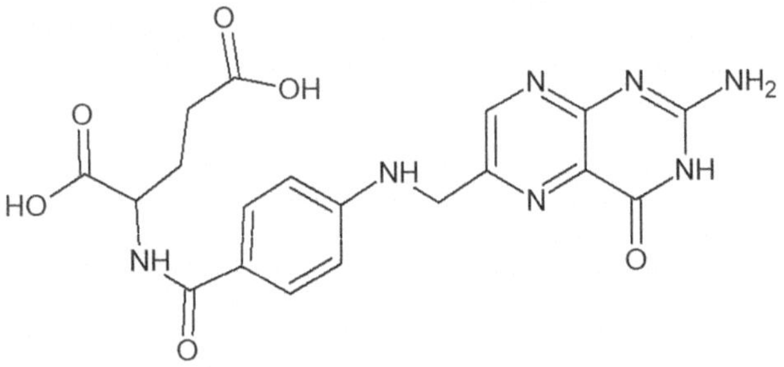

(44)

FIGURE 9.2 Structures of bioactive components containing nutraceutical benefits. (1) Flavones; (2) Flavanone; (3) Proanthocyanin; (4) Anthocyanin; (5) Resveratrol; (6) Ellagic acid; (7) Ellagitannin; (8) Catechin; (9) Lutein; (10) Lycopene; (11) Zeaxanthine; (12) Alpha-carotene; (13) Beta-carotene; (14) Lignan; (15) Daidzein; (16) Genistein; (17) Omega-3-fatty acid; (18) Linoleic acid; (19) Tocotrienol; (20) Phytosterol; (21) Limonoid; (22) Piceatannol; (23) Curcumin; (24) Demethoxycurcumin; (25) Magnolol; (26) Xanthorrhizol; (27) Epigallocatechin gallate; (28) Turmerone; (29) Epicatechin; (30) Berberine; (31) Kaempferol; (32) Quercetin; (33) Luteolin; (34) Apigenin; (35) Hesperidin; (36) Naringin; (37) Pelargonidin; (38) Cyanidin; (39) Malvidin; (40) Astaxanthin; (41) Crocin; (42) Vitamin B6; (43) Vitamin B12; and (44) Folic acid.

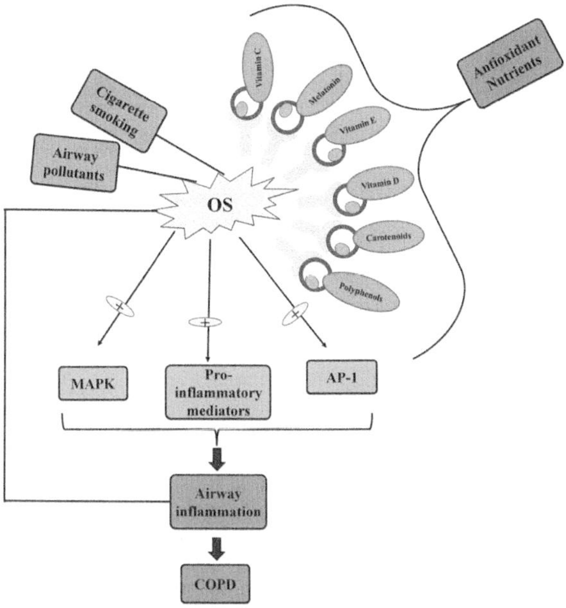

FIGURE 9.3 Significant role of nutraceutical antioxidants in the management of chronic obstructive pulmonary diseases (COPD).

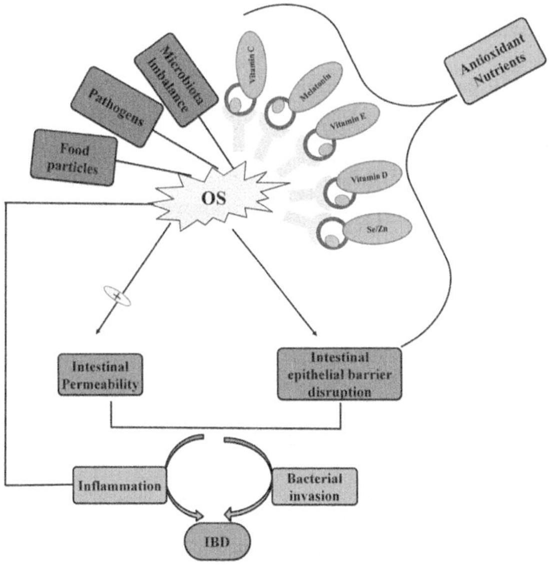

FIGURE 9.4 Significant role of nutraceutical antioxidant in the management of inflammatory bowel diseases (IBD).

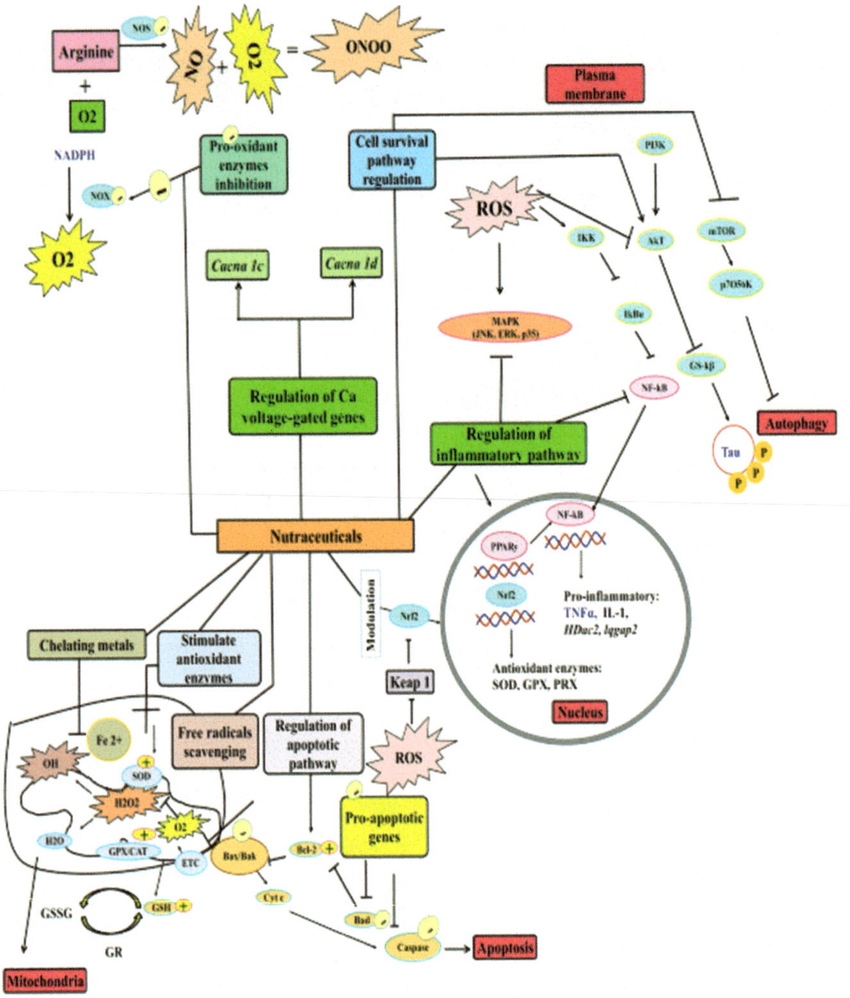

FIGURE 9.5 Significant role of nutraceutical antioxidants in the management of neurodegenerative diseases.

Beta-carotene is the precursor of vitamin A has a protective effect against eye and cancer diseases. In addition, Carotene may enhance the immune system and even protect the skin from the damaging effects of UV rays. Even they used to decrease the greater risk of CVD, inflammation, and asthma. Lycopene act as singlet oxygen quenchers present in a higher percentage in tomatoes. However, lutein and Carotene protect against breast, lungs, prostate, and uterus cancer. Xanthophylls act as a biological antioxidant and are present in leafy vegetables. Another xanthophyll, astaxanthin, is commonly present in shrimp, seafood, and salmon with potent antioxidants. Tocopherols and Tocotrienols, as well as their derivatives, has antiproliferative property. Even they have apoptotic effects, which help reduce the higher risk of breast cancer. Carotenes contain a diverse group of yellow and red pigments generally found in plants. Carotenoids are commonly called carotenes without oxygen and are widely distributed in animals and plants. These are fat-soluble pigments help decrease the risk of disorders including heart ailments, eye disorders, heart stroke, cancer, and enhance the immune responses, shown in Figure 9.2(19). The second primary class of terpenoids is limonoids which are actively present in citrus fruits as phytoconstituents and can protect lungs from free radicals. In addition, some research studies revealed that nomiline, limonoids, and limonin could inhibit cancer cell proliferation in breast cancer *in-vivo* (George et al., 2004; Stahl et al., 2005; Ortuno et al., 2006; Sun et al., 2005; Snodderly et al., 1995), shown in Figure 9.2(21).

9.3 VITAMIN E (TOCOPHEROL OR ALPHA-TOCOPHEROL)

Alpha-tocopherol is abundant in native plants, rich in vitamin E and acts as a singlet oxygen quencher. Tocopherols, also called tocols, have a saturated form. In contrast, tocotrienols have an unsaturated form (alpha, beta, gamma, and delta types), mainly in cereal, palm oil, rice bran oil, and barley. They can reduce the risk of CVD, Alzheimer, and cancer diseases. Additionally, they can lower cholesterol levels and inhibit LDL oxidation. Furthermore, gamma-tocopherol attained a significant effect in preventing breast cancer. Vitamin E is used to avoid coronary heart disease and atherosclerosis to prevent LDL-C oxidation (Tiwari et al., 2001; Albanes et al., 1996; Papas et al., 1999), shown in Figure 9.2(19).

9.4 ASCORBIC ACID

Ascorbic acid, commonly called Vitamin C, acts as a potent antioxidant whose mechanism of action is based on quenching the singlet oxygen and donating a hydrogen atom to lipid radicals. Vitamin C is used to prevent various forms of heart diseases and cancers. Both tocopherol and ascorbic acid supplements can effectively reduce oxidative damage (Kim et al., 2004; Lee et al., 2001; Larson et al., 1997).

9.5 LIPOIC ACID

Sulfur-containing components such as lipoic acid, dihydrolipoic acid, and GSH are usually found in the liver, heart, and meat that possess antioxidant properties. They can prevent oxidative protein damage and helps to regenerate GSH in kidney, liver, and lung tissues. Lipoic acid improves aging disorders, repairs mitochondrial membrane, and manages Alzheimer's and Parkinson's disorders. Further, the racemic form of lipoic acid treats metal intoxication, cirrhosis, and mushroom poisoning. Both the oxidized and reduced forms of lipoic acid act as powerful antioxidants (Larson et al., 1997; Kramer et al., 2001; Navari-Izzo et al., 2002).

9.6 TERPENES

This class of phytoconstituents is present in cereals, soy plants, and green vegetables. Chemically, terpenes undergo oxidation or rearrangement of the carbon skeleton to give terpenoids, also called isoprenoids (IPs). Terpenoids are essential for plants. Further, terpenoids can fix carbon via photosynthetic reactions using Carotene and chlorophyll using photosensitizing pigments. Additionally, these phytochemicals help to protect against certain diseases, growth dysregulation, and cell/tissue damage (Prakash et al., 2009; Dillard et al., 2000; Mcintire et al., 2000; Watzl et al., 2003; Bendich et al., 1989; Jian et al., 2005; Hertog et al., 1993).

9.7 PHYTOSTEROLS

Beta-sitosterols are an example of phytosterol as plants synthesize it. They are other subclasses of a terpene. Research studies revealed that these constituents are reported to inhibit the inflammatory enzymes (Dorow et al., 1987; Quilez et al., 2003), shown in Figure 9.2(20).

9.8 FLAVONOIDS

Generally, this class of phytochemicals is abundantly found in fruits and vegetables. Flavonoids such as luteolin, kaempferol, myricetin, and apigenin have antibacterial and anti-inflammatory activities. The pharmacological properties of flavonoids (Keservani & Sharma, 2014; Keservani et al.,

2016a, b) showed potency against inflammation, free radicals, and allergies. In addition, microbes platelet aggregation, tumors, hepatotoxins, and viral infections were reported. A dietary polyphenolic compound, curcumin, can suppress the tumor proliferation in rodent skin via inhibiting the protein kinase C and the expression of nuclear oncogenes (Prakash et al., 2009; Bouic et al., 2001; Donovan et al., 2006; Aggarwal et al., 2006; Patel et al., 2008; Kinsella et al., 1993; Lin et al., 2007), shown in Figure 9.2((23), (31), (33)).

9.9 FLAVANONES

Flavanones, including hesperetin and naringenin, are found in oranges and grapefruits. In addition, these bioactive components have hypoglycemic, antioxidant (Singh et al., 2017a), antitumor, and hypocholesterolemic properties. These compounds exhibit beneficial effects in particular diseases, including osteoporosis, cancer, diabetes, and CVD (Lin et al., 2007), shown in Figure 9.2((2), (35)).

9.10 CATECHINS AND STILBENES

Tea contains a high percentage of gallic acid esters such as epigallocatechin gallate (EGCG), epicatechin, and epicatechin. In addition, significant phytonutrients are present in berries, grapes, green tea, and cocoa. For example, the presence of EGCG in tea showed beneficial effects in human disease. It further reduces the harmful effects of memory loss, neurodegenerative, anxiety, sclerosis. Furthermore, it helps in insulin dysregulation, alters the intracellular signal transduction pathway, scavenges ROS and nitrogen species, and modifies cytokines and neurotransmitter expression production. Catechin exhibits protective effects against cardiovascular disease (CVD) and atherosclerosis, which are abundant in red wines and grapes. It possesses substantial biological properties against cancer and CVDs. In addition, grapes red wines also contain other polyphenolic components such as stilbenes. Stilbenes are considered an influential group of monomers and oligomers obtained from the oxidative condensation of resveratrol monomers (Espin et al., 2007; Jain et al., 2013; Ramli et al., 2020), shown in Figure 9.2((6)–(8), (29)).

9.11 ANTHOCYANINS

This bioactive chemical constituent imparts colors to several fruits and flowers. It contains various aglycone moieties such as cyanidin, pelargonidin, peonidin, petunidin, delphinidin, and malvidin. This component is present in the wealthiest form in berries and protects against CVD, oxidative cell damage, ROS, reduced blood pressure, lung cancer, and enhances the peripheral blood supply (Espin et al., 2007; Jain et al., 2013; Ramli et al., 2020) shown in Figure 9.2((3), (4), (37)–(39)).

9.12 PHENOLS

The phenolic contents are extensively found and distributed in various fruits and vegetables. They are classified as flavones, anthocyanidin, flavanones, isoflavones, phenolic acids, and some non-flavonoid contents. Polyphenolic components have a well-established correlation between antioxidant properties and health benefits; thereby, this leads to beneficiary effects of commercial containing free radical scavenging activities (Espin et al., 2007; Jain et al., 2013; Ramli et al., 2020) shown in Figure 9.2(2).

9.13 ISOFLAVONES

Isoflavones are the most extensively investigated polyphenolic compound, commonly called Phytoestrogens. They are widely present in pulses soybeans. Further, in Asian countries, isoflavones are consumed widely. Isoflavones have attracted significant attention and play a crucial role in facilitating post-menopausal symptoms of osteoporosis and hot flushes. Besides this, they exhibit protective effects against CVDs, prostate cancer, breast cancer, and cognitive functions. Furthermore, several studies showed significant results in improving bone mineral density, reducing the total cholesterol, colon cancer, and enhancing immune response. Resveratrol is one of the potent hermetic components, widely present in broccoli, berries, green tea, and cocoa. It enhances the stress resistance capacity and stimulates the sirtuins (proteins class containing either mono-ribosyltransferase or Hdac2 activity). In addition, the compound may cause improvement in mitochondrial function and possess protective effects against lipid peroxidation, atherosclerosis, apoptosis, and OS, shown in Table 9.1 (Espin et al., 2007; Jain et al., 2013; Ramli et al., 2020); Table 9.2 (Kim et al., 2007; Sharma et al., 2002; Lin et al., 2018; Arab et al., 2016; Frautschy et al., 2001; Seo et al., 2010; Ramkumar et al., 2017; Balu et al., 2006; Balu et al., 2005; Muroyama et al., 2012; Dong et al., 2013; Lim et al., 2005; Bittencourt et al., 2014; Boasquívis et al., 2018; Goon et al., 2017; Damanhuri et al., 2016; Khan et al., 2019; Liu et al., 2016; Ikram et al., 2019; Levites et al., 2002; Ortiz-Lopez et al., 2016; Park et al., 2012; Cong et al., 2016; Jia et al., 2012; Xie et al., 2020; Nasri et al., 2019; Gupta et al., 2020; Mokni et al., 2007; Cosin-Tomas et al., 2019; Chen et al., 2005, 2006; Jiang et al., 2011, 2013; Cui et al., 2016; Reyes-Fermín et al., 2012; Panchal et al., 2008; Peng et al., 2015; Singh et al., 2011; Nor et al., 2018; Li et al., 2004; Biasibetti et al., 2013; Esmaeelpanah et al., 2018); Table 9.3 (Kim et al., 2010; Ramli et al., 2020; Mecocci et al., 2014; Liu et al., 2018) (Figure 9.2((5), (11), (12), (14)–(18), (21)–(44)).

TABLE 9.1 Bioactive Compounds: Source and Health Benefits

Bioactive Compounds	Source	Health Benefits
Flavones	Soybean, fruits, and vegetables	Decrease the cancer risk, antioxidant.
Flavanones	Citrus fruits	Anticancer, antioxidant.
Proanthocyanidins	Beans, nuts, grapes, strawberries, cocoa	Protective effect against cardiovascular diseases and urinary tract diseases.
Anthocyanins	Berries, raspberries, blueberries	Improve visual disorders, anticancer, antioxidant.
Resveratrol	Berries, grapes, raisins, peanuts	Decrease the risk of cancer, antioxidant.
Ellagic acid, ellagitannins	Walnuts, pomegranates, strawberries, raspberries, blackberries	Antioxidant, antiproliferation.
Catechins	Mustard cake, tea, wine, fruits, and vegetables	Antioxidant, reduce the risk of cancer and provide skin protection from UV radiation.
Lutein	Corn, egg yolk, green vegetables	Decrease the risk of macular degeneration and cancer.
Lycopene	Papaya, guava, tomatoes, watermelons	Effective against prostate cancer, antioxidant.

TABLE 9.1 *(Continued)*

Bioactive Compounds	Source	Health Benefits
Zeaxanthin	Corn, citrus fruits, eggs	Improvement in the visual disorders.
Alpha-carotene	Lettuce, turnips, peas, spinach, broccoli, carrots, sweet potatoes, pumpkin, squash	Antioxidant
Beta-carotene	Pumpkin, sweet potatoes, carrots, oat	Antioxidant
Lignans	Vegetables, rye, flaxseeds	Decrease the risk of cancer and cardiovascular disease.
Isoflavones: Daidzein, Genistein	Legumes, soybeans	Reduce the LDL-C, enhance the bone strength, antioxidant.
Omega-3 fatty acid	Olive oil, flaxseed, salmon	Decrease the cholesterol level, improve the brain functions, antioxidant.
Linoleic acid	Meat, milk, cheese	Decrease the risk of cancers.
Tocotrienols	Palm oil, grains	Reduce the cholesterol level, antiproliferative.
Phytosterols	Yellow and green vegetables	Inhibit the uptake of cholesterol, anti-inflammatory.
Limonoids	Citrus fruits and vegetables	Protective effect against lung cancer.

Source: Esprin et al. (2007); Jain et al. (2013); and Ramli et al. (2020).

TABLE 9.2 Nutraceuticals as Bioactive Components with Their Mode of Actions

SL. No.	Nutraceuticals as Bioactive Components	Experimental Models	Mode of Actions	References
Nutraceuticals Acted via Direct Antioxidative Mechanisms				
1.	Piceatannol	Amyloid beta-induced PC12 neuronal cell death.	Decreased ROS generation.	Kim et al. (2007)
2.	Resveratrol	STZ induced cognitive impairment, oxidative stress in Wistar albino rats; angiotensin II-induced early AD in rat model.	Decreased the MDA and superoxide levels in the rat brains.	Sharma et al. (2002); Lin et al. (2018)

TABLE 9.2 *(Continued)*

SL. No.	Nutraceuticals as Bioactive Components	Experimental Models	Mode of Actions	References
3.	Green tea	Patients with severe symptoms of Alzheimer's disease.	Antioxidant activity; decreased the MDA, protein carbonyls and 8-hydroxy-2′-deoxyguanosine levels.	Arab et al. (2016); Kesharwani et al. (2016)
4.	Curcumin	Amyloid beta-protein-induced damage in Sprague-Dawley rats; transgenic mouse Tg2576 model of AD.	Reduced the LPO in the spinal cord of rats; decreased the isoprostane levels in the mouse brain.	Frautschy et al. (2001); Seo et al. (2010)
5.	Demethoxycurcumin	Rotenone induced cell death in SH-SY5Y cells; rotenone induced neurotoxicity.	Decreased the ROS generation significantly, neuroprotective effects.	Ramkumar et al. (2017)
6.	Grape seed (Proathocyanidine)	Young and aged groups of Wistar rats.	Reduced the ROS and protein carbonyls levels; decreased the DNA protein cross-links and 8-hydroxy-2′-deoxyguanosine levels in the aged rat's brain.	Balu et al. (2005)
7.	Magnolol	1-Methyl-4-phenyl-1,2,3,6-tetrahydropyridine MPTP-induced neurodegeneration in mice and 1-methyl-4-phenylpyridinium ion MPP^+-induced cytotoxicity to human neuroblastoma SH-SY5Y cells; apoptosis induced in human neuroblastoma SH-SY5Y cells via neurotoxin acrolein.	Reduced the LPO level in the mice brain; decreased the 4-hydroxy-2-nonrenal, protein carbonyls and superoxide levels.	Muroyama et al. (2012); Dong et al. (2013)

TABLE 9.2 *(Continued)*

SL. No.	Nutraceuticals as Bioactive Components	Experimental Models	Mode of Actions	References
8.	Xanthorrhizol	Glutamate-induced neurotoxicity; murine hippocampal HT22 cell line; hydrogen peroxide-induced lipid peroxidation in rat brain.	Reduced the ROS production in the hippocampal neurons; decreased the LPO levels in rats.	Lim et al. (2005)
9.	Guarana seed extract	Acrolein-induced toxicity on neuronal-like cells; *Caenorhabditis elegans* models of PD, AD, and HD.	Reduced the ROS generation.	Bittencourt et al. (2014); Boasquivis et al. (2018)
10.	Tocotrienol	Aged male and female adults between 50 and 55 years old; hippocampus of an AD APPswe/PS1dE9 double transgenic mice.	Decreased the DNA damage and MDA level in plasma; reduced the DNA strand breaks in the mouse blood.	Goon et al. (2017); Damanhuri et al. (2016)
Nutraceuticals Acted via Reducing Inflammation and Regulation of Apoptosis				
1.	Curcumin	LPS-induced ROS-mediated, neurodegeneration, neuroinflammation, and memory deficits in the rat hippocampus.	Reduced MDA and 2,7-dichlorofluorescein levels; ameliorated neuronal cell death and neuroinflammation through regulation of the JNK/NF-κB/Akt pathway; improved LPS-induced memory deficits.	Khan et al. (2019)
		APP/PS1 transgenic mice; amyloid-beta-induced neuroinflammation in neuronal or glial cultures.	Reduced the microglia and astrocytes activation, enhanced the cytokine production, inhibited nuclear factor kappa B-pathway, enhanced the spatial memory impairment in mice.	Liu et al. (2016)

TABLE 9.2 *(Continued)*

SL. No.	Nutraceuticals as Bioactive Components	Experimental Models	Mode of Actions	References
		Ethanol induced oxidative stress; inflammation-mediated cognitive dysfunction in mice.	Regulated the expression of the glial cell markers; downregulated the expression of p-JNK, p-NF-kB.	Ikram et al. (2019)
		A53T α-synuclein protein-linked Parkinsonism in SH-SY5Y cells.	Reduced the A53T α-synuclein accumulation via downregulation of the mTOR/p70S6K signaling.	Jiang et al. (2013)
2.	Green tea polyphenol, epigallo-catechin 3-gallate.	Parkinsonism-inducing neurotoxin, 6-hydroxy-dopamine induced neurotoxicity.	Stimulated modulation of cell cycle genes; prevented 6-OHDA-induced expression of Bax, Bad, Mdm2; decreased the Bcl-w, Bcl-2, and Bcl-x (L).	Levites et al. (2002)
		Hippocampal neurogenesis in Balb/C male mice model.	Enhanced the neurogenesis in the hippocampus.	Ortiz-Lopez et al. (2016)
3.	Turmerone	Aβ-stimulated microglial cells.	Inhibited Aβ-induced expression and activation of iNOS, COX-2, MMP-9; decreased TNF-α, IL-1β, IL-6, and MCP-1; inhibited ROS; inhibited JNK, p38 MAPK and NF-κB.	Park et al. (2012)
4.	Green tea polyphenols.	Glutamate induced neurotoxicity in cortical neurons.	Inhibited ROS generation as well as SOD level in neuron.	Cong et al. (2016)

TABLE 9.2 *(Continued)*

SL. No.	Nutraceuticals as Bioactive Components	Experimental Models	Mode of Actions	References
5.	Berberine	Aβ-induced neuroinflammation in primary microglia cells of murine as well as BV2 microglia cells.	Inhibited Aβ-stimulated production of interleukin-6 and monocyte chemotactic protein-1; down-regulated the COX-2 expression, inhibited the activation of nuclear factor-kappa-B.	Jia et al. (2012)
6.	Magnolol	Aβ-induced AD	Enhanced PPAR-γ luciferase; inhibited the activation of NF-κB.	Xie et al. (2020)
7.	Tocotrienol fraction	AD mouse model	Modulate hippocampal gene expression as well as biological processes that can potentially delay the progression of AD.	Wan et al. (2019)
Nutraceuticals Acted via Indirect Antioxidative Mechanisms				
1.	Resveratrol	LPO as well as antioxidant enzyme level in rats' brain.	Possessed neuroprotective properties through up-regulation of detoxifying enzymes.	Gupta et al. (2020)
		Lymphoblastoid cell lines LCLs from AD patients.	Inhibited ROS; anti-aging factors in AD patients.	Mokni et al. (2007)
		Cultured PC12 cells	Induced HO-1 expression; augmented cellular antioxidant ability via Nrf2-ARE signaling.	Cosin-Tomas et al. (2019)

TABLE 9.2 *(Continued)*

SL. No.	Nutraceuticals as Bioactive Components	Experimental Models	Mode of Actions	References
2.	Curcumin	MPP(+)-induced apoptosis	Reduced the loss of MMP.	Chen et al. (2005)
		OS toxicity	Activated Nrf2 target genes, decreased ROS, attenuated oxidative stress damage as well as mitochondrial dysfunction.	Chen et al. (2006)
		Behavioral deficits, oxidative damage in rats.	Ameliorated neuronal oxidative damage in the SNpc of rats by activation of Akt/Nrf2 pathway.	Jiang et al. (2011)
3.	α-Mangostin, curcumin	IAA-induced cell death.	Improved the neuronal death induced IAA.	Cui et al. (2016)
4.	Curcumin	C6 rat glioma cell cultures.	Antioxidant; anti-proliferative properties.	Reyes-Fermin et al. (2012)
5.	6-shogaol	OS-induced cell damage in PC12 cells of rats.	Enhanced the Nrf2 translocation.	Panchal et al. (2008)
6.	*Alpinia galanga* (L.) Wild (EAG)	OS induced amnesia in mice.	Antiamnesic activity.	Peng et al. (2015)
7.	Tocotrienol fraction containing α-tocopherol.	Healthy individuals between 50 and 55 years of aged group.	Antioxidant property.	Hanish et al. (2011)
8.	Epigallocatechin gallate	Inflammation-mediated dopaminergic neuronal injury.	Reduced LPS-activated microglial secretion of NO and TNF-alpha.	Nor Azman et al. (2018)
		STZ-induced dementia in rats' model.	Reversed the oxidative stress; reduced ach.	Li et al. (2004)

TABLE 9.2 *(Continued)*

SL. No.	Nutraceuticals as Bioactive Components	Experimental Models	Mode of Actions	References
9.	Epigallocatechin gallate, epicatechin gallate.	Acrylamide neurotoxicity *in-vitro* as well as *in-vivo*.	Suppressed the ACR induced cytotoxicity PC12 cells; enhanced the GSH level; reduced the LPO in cerebral cortex of rats.	Biasibetti et al. (2013)
10.	Epigallocatechin gallate	1-methyl-4-phenyl-1,2,3,6-tetrahydropyridine induced Parkinson's disease in mouse.	Suppressed iNOS	Esmaeelpanah et al. (2018)

TABLE 9.3 Health Benefits of Flavonoids and Their Derivatives as Nutraceuticals

Chemical Class of Phytoconstituents	Bioactive Constituents	Source	Health Benefits
Flavanols	Catechin, epicatechin	Cocoa, green tea, grapes. Grapes seeds, grape juice.	Improved the vascular functions; Influenced the cerebral hemodynamics; improved the performance on cognitive and visual tasks.
	Vitis vinifera	–	Inhibited the glutamate excitotoxicity; increased the serum antioxidant level, reduced LDL oxidation, improved in the endothelial function and reduced in platelet aggregation in patients with cerebro-vascular diseases.
	Epigallocatechin gallate	–	Decreased the increased level of OS induced by hypertension; decreased the ROS on hippocampus of rats' brain; effective in vascular learning and memory impairment.
	Catechin	Black and green tea	Protective effect in AD.

TABLE 9.3 *(Continued)*

Chemical Class of Phytoconstituents	Bioactive Constituents	Source	Health Benefits
Polyphenols	Polyphenols	Tea	Effective in neurodegenerative diseases; decreased the risk of PD and cognitive impairment.
Flavonols	Quercetin	Onion, green tea, apples, capers	Inhibited the endothelial apoptosis via antioxidant and cytoprotective properties; increased the cell survival in neurotoxic conditions; decreased the size of ischemic lesions in vascular dementia; improved the hippocampal synaptic plasticity and memory in impairment induced model; neuronal repairment in spinal cord injury models.
	Kaempferol	Fruits, tea, vegetables, beverages.	Protective effect on PC12 cells against OS induced by hydrogen peroxide; improved the memory and cognitive learning in mice.
–	Myricetin, kaempferol, quercetin	–	Improved the cognitive performance.
–	Quercetin, kaempferol, terpenoid lactones.	Ginkgo biloba extract.	Improved the memory and cognitive functions; neuroprotective functions, neurotransmitter, free radical scavenging, hemodynamic effects.
Flavones	Luteolin, apigenin	Celery, parsley, rosemary.	Protective role of DNA from hydrogen peroxide induced oxidative damage; anti-inflammatory effect.
	Apigenin	–	Protective effect of neurons against Aβ-mediated toxicity produced by copper; increased neuronal viability; improved mitochondrial membrane dissipation; modulated glutamatergic and GABAergic transmission in cultural cortical neurons.

TABLE 9.3 *(Continued)*

Chemical Class of Phytoconstituents	Bioactive Constituents	Source	Health Benefits
Isoflavones	Genistein, daidzein, glycitin	Soybean	Richest source of phytoestrogens; agonists of estrogen receptors; improved the memory and cognitive functions; improved the cholinergic function by increasing the potassium-stimulated acetyl-choline release and choline uptake; improved the learning and visual spatial memory ability; improved the cognitive ability; decreased the TBARs; increased the glutathione peroxidase in plasma levels via antioxidant effects.
–	Crocin	Saffron	Improved the learning and memory ability; improvement in the neurodegenerative diseases such as AD; antagonist effects in cognitive deficits produced by STZ in animal models; modulated the oxidative maker levels in hippocampus; decreased the OS on learning and memory.
Diterpenes	Carnosic acid, rosmarinic acid	–	Free reactive nitrogen species scavenger; protective effect of neuroblastoma cells against hydrogen peroxide-induced OS; delayed the onset of AD; potentially increased the memory impairment seen in Aβ-neurotoxicity.
–	Carnosic acid	–	Antiangiogenic effects via antioxidant properties.
Non-flavonoid polyphenols	Resveratrol	Grapes	Cardiovascular effects; antidepressant effect by increasing 5-HT activity; inhibited 5-HT and noradrenaline reuptake in rat models; increased the serotonin levels in hippocampus; decreased the inflammation and protect the neuron death in OS-induced neuronal toxicity in animal model; modulated Aβ and upregulation of sirtuin 1.

TABLE 9.3 *(Continued)*

Chemical Class of Phytoconstituents	Bioactive Constituents	Source	Health Benefits
	Curcumin	*Curcuma longa*	Increased the cell viability; decreased the apoptotic signals in encephalitis in mice models; decreased the Aβ processing in cerebral and inflammation in transgenic mice induced with AD; inhibited the MAO levels; decreased the depression by modulating dopaminergic, noradrenergic, and serotonergic transmission.
Carotenoids	Lutein, lycopene, alpha-carotene, beta-cryptoxanthine, zeaxanthine, beta-carotene.	Fruits and vegetables.	Antioxidant status; lutein and zeaxanthin act as photoprotectants in preventing age-related degeneration of macula lutea; lycopene act as a sun protective effect.
	Astaxanthin	Seafood	Protective effects in mitochondrial and microcirculation functions; antioxidant; anti-inflammatory; effective against neurodegenerative disorders; significant effect in visual problems like decreased visual performance and blurred vision; improved the muscle strength.
Anthocyanidins	Pelargonidin, cyanidin, malvidin.	Blueberry, cranberry, elderberry, bilberry, raspberry, strawberry.	Improved the neuronal and cognitive brain function; protection in genomic DNA integrity; improved the memory and learning in aged animals; anti-inflammatory, reduced the depression symptoms.
Vitamin-B	Folate	–	Improved the mental functions in depression, tension, fatigue; in the management of neurological disorders.

'–': not known.

Source: Kim et al. (2010); Ramli et al. (2020); Mecocci et al. (2014); and Liu et al. (2018).

9.14 CONCLUSION

Nutraceutical supplements play a remarkable role in the healthcare system of medicine. Furthermore, these supplements are highly active and significantly affect cellular metabolism with minimized side effects. Several regulatory bodies are required to play a significant role in the safety aspects of nutraceutical products. Nutraceuticals may be found helpful in enhancing health problems, delaying the aging process, and preventing the chronic many biological pathways such as antioxidant defense, cell proliferation, cell differentiation, and soon. Though the mechanisms of action of nutraceuticals are not clearly stated; however, they might be involved in treating diseases. The present manuscript has provided the possible mechanism of action of nutraceuticals as antioxidants in the prevention and management of chronic diseases.

CONFLICT OF INTEREST

The authors have no conflict of interest.

KEYWORDS

- **antioxidants**
- **flavonoids**
- **nutraceuticals**
- **supplements**

REFERENCES

Aggarwal, B. B., Ichikawa, H., Garodia, P., Weerasinghe, P., Sethi, G., Bhatt, I. D., Pandey, M. K., et al., (2006). From traditional Ayurvedic medicine to modern medicine: Identification of therapeutic targets for suppression of inflammation and cancer. *Expert Opin. Ther. Targets, 10*, 87–118.

Albanes, D., Heinonen, O. P., Taylor, P. R., Virtamo, J., Edwards, B. K., Rautalahti, M., Hartman, A. M., et al., (1996). Alpha – tocopherol and beta-carotene supplements and lung cancer incidence in the alpha-tocopherol, beta-carotene cancer prevention study: Effects of base-line characteristics and study compliance. *J. Natl. Cancer Inst., 88*, 1560–1570.

Arab, H., Mahjoub, S., Hajian-Tilaki, K., & Moghadasi, M., (2016). The effect of green tea consumption on oxidative stress markers and cognitive function in patients with Alzheimer's disease: A prospective intervention study. *Caspian J. Int. Med., 7*, 188–194.

Arts, I. C. W., & Hollman, P. C. H., (2005). Polyphenols and disease risk in epidemiologic studies. *Am. J. Clin. Nutr., 81,* 317S–325S.

Balu, M., Sangeetha, P., Murali, G., & Panneerselvam, C., (2005). Age-related oxidative protein damages in central nervous system of rats: Modulatory role of grape seed extract. *Int. J. Dev. Neurosci., 23*, 501–507.

Balu, M., Sangeetha, P., Murali, G., & Panneerselvam, C., (2006). Modulatory role of grape seed extract on age-related oxidative DNA damage in central nervous system of rats. *Brain Res. Bull., 68*, 469–473.

Bandyopadhyay, U., Das, D., & Banerjee, R. K., (1999). Reactive oxygen species: Oxidative damage and pathogenesis. *Curr. Sci., 77*, 658–666.

Bendich, A., (1989). Carotenoids and the immune response. *J. Nutr., 119*, 112–115.

Biasibetti, R., Tramontina, A. C., Costa, A. P., Dutra, M. F., Quincozes-Santos, A., Nardin, P., Bernardi, C. L., et al., (2013). Green tea (−) epigallocatechin-3-gallate reverses oxidative stress and reduces acetylcholinesterase activity in a streptozotocin-induced model of dementia. *Behav. Brain Res., 236*, 186–193.

Bittencourt, L. D., Zeidan-Chulia, F., Yatsu, F. K., Schnorr, C. E., Moresco, K. S., Kolling, E. A., Gelain, D. P., et al., (2014). Guarana (*Paullinia cupana* Mart.) prevents β-amyloid aggregation, generation of advanced glycation-end products (AGEs), and acrolein-induced cytotoxicity on human neuronal-like cells. *Phytother. Res., 28*, 1615–1624.

Boasquívis, P. F., Silva, G. M., Paiva, F. A., Cavalcanti, R. M., Nunez, C. V., & De Paula, O. R., (2018). Guarana (*Paullinia cupana*) extract protects *Caenorhabditis elegans* models for Alzheimer's disease and Huntington's disease through activation of antioxidant and protein degradation pathways. *Oxid. Med. Cell. Longev.*

Bouic, P. J., (2001). The role of phytosterols and phytosterolins in immune modulation: A review of the past 10 years. *Curr. Opin. Clin. Nutr. Metab. Care, 4*, 471–475.

Chen, C. Y., Jang, J. H., Li, M. H., & Surh, Y. J., (2005). Resveratrol upregulates heme oxygenase-1 expression via activation of NF-E2-related factor 2 in PC12 cells. *Biochem. Biophys. Res. Commun., 331*, 993–1000.

Chen, J., Tang, X. Q., Zhi, J. L., Cui, Y., Yu, H. M., Tang, E. H., Sun, S. N., et al., (2006). Curcumin protects PC12 cells against 1-methyl-4-phenylpyridinium ion-induced apoptosis by bcl-2-mitochondria-ROS-iNOS pathway. *Apoptosis., 11*, 943–953.

Choe, E., & Min, D. B., (2006). Chemistry and reactions of reactive oxygen species in foods. *Crit. Rev. Food Sci. Nutr., 46*, 1–22.

Cieslik, E., Gręda, A., & Adamus, W., (2006). Contents of polyphenols in fruit and vegetables. *Food Chem., 94*, 135–142.

Cong, L., Cao, C., Cheng, Y., & Qin, X. Y., (2016). Green tea polyphenols attenuated glutamate excitotoxicity via antioxidative and antiapoptotic pathway in the primary cultured cortical neurons. *Oxid. Med. Cell. Longe.*, 2050435.

Cosin-Tomas, M., Senserrich, J., Arumi-Planas, M., Alquezar, C., Pallas, M., Martin-Requero, A., Sunol, C., et al., (2019). Role of resveratrol and selenium on oxidative stress and expression of antioxidant and anti-aging genes in immortalized lymphocytes from Alzheimer's disease patients. *Nutrients, 11,* 1764.

Cui, Q., Li, X., & Zhu, H., (2016). Curcumin ameliorates dopaminergic neuronal oxidative damage via activation of the Akt/Nrf2 pathway. *Mol. Med. Rep., 13*, 1381–1388.

Damanhuri, H. A., Rahim, N. A., Nasri, W. N., Tan, J. K., Makpol, S., Mazlan, M., Tooyama, I., & Ngah, W. Z., (2016). Tocotrienol-rich fraction supplementation modulates antioxidant enzymes activity and reduces DNA damage in APPswe/PS1dE9 Alzheimer's disease mouse model. *Sains Malays., 45*, 1363–1370.

Dillard, C. J., & German, J. B., (2000). Phytochemicals: Nutraceuticals and human health. *J. Sci. Food Agri., 80*, 1744–1756.

Dong, L., Zhou, S., Yang, X., Chen, Q., He, Y., & Huang, W., (2013). Magnolol protects against oxidative stress-mediated neural cell damage by modulating mitochondrial dysfunction and PI3K/Akt signaling. *J. Mol. Neurosci., 50*, 469–481.

Donovan, J. L., Manach, C., Faulks, R. M., & Kroon, P. A., (2006). Absorption and metabolism of dietary plant secondary metabolites. In: Crozier, A., Clifford, M. K., & Ashihara, H., (eds.), *Plant Secondary Metabolites: Occurrence, Structure and Role in the Human Diet* (Vol. 12, pp. 303–351). Wiley: New York.

Dorow, P., Weiss, T., Felix, R., & Schmutzler, H., (1987). Effect of a secretolytic and a combination of pinene, limonene and cineole on mucociliary clearance in patients with chronic obstructive pulmonary disease. *Arzneimittelforschung, 37*, 1378–1381.

Esmaeelpanah, E., Razavi, B. M., Vahdati, H. F., & Hosseinzadeh, H., (2018). Evaluation of epigallocatechin gallate and epicatechin gallate effects on acrylamide-induced neurotoxicity in rats and cytotoxicity in PC 12 cells. *Drug Chem. Toxicol., 41*, 441–448.

Espin, J. C., Garcia-Conesa, M. T., & Tomas-Barberan, F. A., (2007). Nutraceuticals: Facts and fiction. *Phytochemistry, 68*, 2986–3008.

Frautschy, S. A., Hu, W., Kim, P., Miller, S. A., Chu, T., Harris-White, M. E., & Cole, G. M., (2001). Phenolic anti-inflammatory antioxidant reversal of Aβ-induced cognitive deficits and neuropathology. *Neurobiol. Aging., 22*, 993–1005.

George, B., Kaur, C., Khurdiya, D. S., & Kapoor, H. C., (2004). Antioxidants in tomato (*Lycopersium esculentum*) as a function of genotype. *Food Chem., 84*, 45–51.

Goon, J. A., Azman, N. H., Ghani, S. M., Hamid, Z., & Ngah, W. Z., (2017). Comparing palm oil tocotrienol rich fraction with α-tocopherol supplementation on oxidative stress in healthy older adults. *Clin. Nutr. ESPEN., 21*, 1–12.

Gupta, J., & Gupta, R., (2020). Nutraceutical status and scientific strategies for enhancing production of omega-3 fatty acids from microalgae and their role in healthcare. *Curr. Pharm. Biotechnol., 21*, 1616–1631.

Halliwell, B., & Whiteman, M., (2004). Measuring reactive species and oxidative damage *in vivo* and in cell culture: How should you do it and what do the results mean? *Br J. Pharmacol., 142*, 231–255.

Hanish, S. J. C., Alagarsamy, V., Sathesh, K. S., & Narsimha, R. Y., (2011). Neurotransmitter metabolic enzymes and antioxidant status on Alzheimer's disease induced mice treated with *Alpinia galanga* (L.) Willd. *Phytother. Res., 25*, 1061–1067.

Hertog, M. G., Feskens, E. J., Kromhout, D., Hollman, P. C., & Katan, M. B., (1993). Dietary antioxidant flavonoids and risk of coronary heart disease: The zutphen elderly study. *Lancet., 342*, 1007–1011.

Ikram, M., Saeed, K., Khan, A., Muhammad, T., Khan, M. S., Jo, M. G., Rehman, S. U., & Kim, M. O., (2019). Natural dietary supplementation of curcumin protects mice brains against ethanol-induced oxidative stress-mediated neurodegeneration and memory impairment via Nrf2/TLR4/RAGE signaling. *Nutrients, 11*, 1082.

Jain, N., & Ramawat, K. G., (2013). Nutraceuticals and antioxidants in prevention of diseases. *Natural Products* (pp. 2559–2580). Springer, Berlin.

Jia, L., Liu, J., Song, Z., Pan, X., Chen, L., Cui, X., & Wang, M., (2012). Berberine suppresses amyloid-beta-induced inflammatory response in microglia by inhibiting nuclear factor-kappa-B and mitogen-activated protein kinase signaling pathways. *J. Phar. Pharmacol., 64*, 1510–1521.

Jian, L., Du, C. J., Lee, A. H., & Binns, C. W., (2005). Do dietary lycopene and other carotenoids protect against prostate cancer? *Int. J. Cancer., 113*, 1010–1014.

Jiang, H., Tian, X., Guo, Y., Duan, W., Bu, H., & Li, C., (2011). Activation of nuclear factor erythroid 2-related factor 2 cytoprotective signaling by curcumin protect primary spinal cord astrocytes against oxidative toxicity. *Biol. Pharm. Bull., 34*, 1194–1197.

Jiang, T. F., Zhang, Y. J., Zhou, H. Y., Wang, H. M., Tian, L. P., Liu, J., Ding, J. Q., & Chen, S. D., (2013). Curcumin ameliorates the neurodegenerative pathology in A53T α-synuclein cell model of Parkinson's disease through the downregulation of mTOR/p70S6K signaling and the recovery of macroautophagy. *J. Neuroimmune Pharmacol., 8*, 356–369.

Kaur, C., George, B., Deepa, N., Singh, B., & Kapoor, H. C., (2004). Antioxidant status of fresh and processed tomato: A review. *J. Food Sci. Technol.-Mysore., 41*, 479–486.

Keservani, R. K., & Sharma, A. K., (2014). Flavonoids: Emerging trends and potential health benefits. *Journal of Chinese Pharmaceutical Sciences, 23*(12), 815.

Keservani, R. K., Kesharwani, R. K., Sharma, A. K., Vyas, N., & Chadoker, A., (2010b). Nutritional supplements: An overview. *International Journal of Current Pharmaceutical Review and Research, 1*(1), 59–75.

Keservani, R. K., Kesharwani, R. K., Vyas, N., Jain, S., Raghuvanshi, R., & Sharma, A. K., (2010a). Nutraceutical and functional food as future food: A review. *Der Pharmacia Lettre, 2*(1), 106–116.

Keservani, R. K., Sharma, A. K., & Kesharwani, R. K., (2016a). Nutraceutical and functional foods for cardiovascular health. In: *Food Process Engineering* (pp. 257–278, 291–312). Apple Academic Press, CRC Press. ISBN: 9781771884020.

Keservani, R. K., Sharma, A. K., & Kesharwani, R. K., (2016b). Medicinal effect of nutraceutical fruits for the cognition and brain health. *Scientifica*, 10. Article ID 3109254. doi: 10.1155/2016/3109254.

Keservani, R. K., Sharma, A. K., & Kesharwani, R. K., (2017). An overview and therapeutic applications of nutraceutical and functional foods. *Recent Advances in Drug Delivery Technology*, 160–201. ISBN:9781522507543.

Keservani, R. K., Sharma, A. K., & Kesharwani, R. K., (2020). *Nutraceuticals and Dietary Supplements: Applications in Health Improvement and Disease Management*. CRC Press. ISBN: 9781771888738.

Kesharwani, R. K., Singh, P., & Keservani, R. K., (2016). Green coffee bean extract and chlorogenic acids: Chemistry and novel antioxidant benefits. In: Bagchi, D., Moriyama, H., & Swaroop, A., (eds.), *Green Coffee Bean Extract in Human Health* (pp. 1–17). Boca Raton, Taylor and Francis, USA. Chapter 1. ISBN: 9781498716376.

Khan, M. S., Muhammad, T., Ikram, M., & Kim, M. O., (2019). Dietary supplementation of the antioxidant curcumin halts systemic LPS-induced neuroinflammation-associated neurodegeneration and memory/synaptic impairment via the JNK/NF-κB/Akt signaling pathway in adult rats. *Oxid. Med. Cell. Longev.*, 7860650.

Kim, D. O., & Lee, C. Y., (2004). Comprehensive study on vitamin C equivalent antioxidant capacity (VCEAC) of various polyphenolics in scavenging a free radical and its structural relationship. *Crit. Rev. Food Sci. Nutr., 44*, 253–273.

Kim, H. J., Lee, K. W., & Lee, H. J., (2007). Protective effects of piceatannol against beta-amyloid–induced neuronal cell death. *Ann. N. Y. Acad. Sci.*, 473–482.

Kim, J. S., Kim, J. M., Jeong-Ja, O., & Jeon, B. S., (2010). Inhibition of inducible nitric oxide synthase expression and cell death by (−)-epigallocatechin-3-gallate, a green tea catechin, in the 1-methyl-4-phenyl-1, 2, 3, 6-tetrahydropyridine mouse model of Parkinson's disease. *J Clini. Neurosci., 17*, 1165–1168.

Kinsella, J. E., Frankel, E., German, B., & Kanner, J., (1993). Possible mechanisms for the protective role of antioxidants in wine and plant foods. *Food Technol., 47*, 85–89.

Kramer, K., Packer, L., & Hoppe, P. (2001). R-alpha-lipoic acid. *Oxidative Stress and Disease, 6,* 129–164.

Kris-Etherton, P. M., Hecker, K. D., Bonanome, A., Coval, S. M., Binkoski, A. E., Hilpert, K. F., Griel, A. E., & Etherton, T. D., (2002). Bioactive compounds in foods: Their role in the prevention of cardiovascular disease and cancer. *Am. J. Med., 113*, 71–88.

Larson, R. A., (1997). *Naturally Occurring Antioxidants* (pp. 83–186). Lewis Publishers, New York.

Lee, S. H., Oe, T., & Blair, I. A., (2001). Vitamin C-induced decomposition of lipid hydroperoxides to endogenous genotoxins. *Science, 292*, 2083–2086.

Levites, Y., Amit, T., Youdim, M. B., & Mandel, S., (2002). Involvement of protein kinase C activation and cell survival/cell cycle genes in green tea polyphenol (−)-epigallocatechin 3-gallate neuroprotective action. *J. Bio. Chem., 277,* 30574–30580.

Li, R., Huang, Y. G., Fang, D., & Le, W. D., (2004). (−)-epigallocatechin gallate inhibits lipopolysaccharide-induced microglial activation and protects against inflammation-mediated dopaminergic neuronal injury. *J. Neurosci. Res., 78*, 723–731.

Lim, C. S., Jin, D. Q., Mok, H., Oh, S. J., Lee, J. U., Hwang, J. K., Ha, I., & Han, J. S., (2005). Antioxidant and anti-inflammatory activities of xanthorrhizol in hippocampal neurons and primary cultured microglia. *J. Neurosci. Res., 82*, 831–838.

Lin, Y. G., Kunnumakkara, A. B., Nair, A., Merritt, W. M., Han, L. Y., Armaiz-Pena, G. N., Kamat, A. A., et al., (2007). Curcumin inhibits tumor growth and angiogenesis in ovarian carcinoma by targeting the nuclear factor kB pathway. *Clin. Cancer Res., 13,* 3423–3430.

Lin, Y. T., Wu, Y. C., Sun, G. C., Ho, C. Y., Wong, T. Y., Lin, C. H., Chen, H. H., et al., (2018). Effect of resveratrol on reactive oxygen species-induced cognitive impairment in rats with angiotensin II-induced early Alzheimer's disease. *J. Clin. Med., 7,* 329.

Liu, Z. J., Li, Z. H., Liu, L., Tang, W. X., Wang, Y., Dong, M. R., & Xiao, C., (2016). Curcumin attenuates beta-amyloid-induced neuroinflammation via activation of peroxisome proliferator-activated receptor-gamma function in a rat model of Alzheimer's disease. *Front. Pharmacol., 7,* 261.

Liu, Z., Ren, Z., Zhang, J., Chuang, C. C., Kandaswamy, E., Zhou, T., & Zuo, L., (2018). Role of ROS and nutritional antioxidants in human diseases. *Front. Physiol., 9,* 477.

Marnett, L. J., (2000). Oxyradicals and DNA damage. *Carcinogenesis., 21*, 361–370.

Mcintyre, B. S., Briski, K. P., Gapor, A., & Sylvester, P. W., (2000). Antiproliferative and apoptotic effects of tocopherols and tocotrienols on preneoplastic and neoplastic mouse mammary epithelial cells. *Proc. Soc. Exp. Biol. Med., 224*, 292–301.

Mecocci, P., Tinarelli, C., Schulz, R. J., & Polidori, M. C., (2014). Nutraceuticals in cognitive impairment and Alzheimer's disease. *Front. Pharmacol., 5,* 147.

Mokni, M., Elkahoui, S., Limam, F., Amri, M., & Aouani, E., (2007). Effect of resveratrol on antioxidant enzyme activities in the brain of healthy rat. *Neurochem. Res., 32*, 981–987.

Muroyama, A., Fujita, A., Lv, C., Kobayashi, S., Fukuyama, Y., & Mitsumoto, Y., (2012). Magnolol protects against MPTP/MPP+-Induced toxicity via inhibition of oxidative stress in *in vivo* and *in vitro* models of Parkinson's disease. *Parkinson's Disease.,* 985157.

Navari-Izzo, F., Quartacci, M. F., & Sgherri, C., (2002). Lipoic acid: A unique antioxidant in the detoxification of activated oxygen species. *Plant Physiol. Biochem., 40,* 463–470.

Nor, A. N. H., Goon, J. A., Abdul, G. S. M., Hamid, Z., & Wan, N. W. Z., (2018). Comparing palm oil, tocotrienol-rich fraction and α-tocopherol supplementation on the antioxidant levels of older adults. *Antioxidants (Basel), 7,* 74.

Olinski, R., Gackowski, D., Foksinski, M., Rozalski, R., Roszkowski, K., & Jaruga, P., (2002). Oxidative DNA damage: Assessment of the role in carcinogenesis, atherosclerosis, and acquired immunodeficiency syndrome. *Free Radic. Biol. Med., 33,* 192–200.

Ortiz-Lopez, L., Marquez-Valadez, B., Gomez-Sanchez, A., Silva-Lucero, M. D., Torres-Perez, M., Tellez-Ballesteros, R. I., Ichwan, M., et al., (2016). Green tea compound epigallo-catechin-3-gallate (EGCG) increases neuronal survival in adult hippocampal neurogenesis *in vivo* and *in vitro. Neuroscience., 322,* 208–220.

Ortuno, A., Baidez, A., Gomez, P., Arcas, M. C., Porras, I., Garcia-Lidon, A., & Del, R. J. A., (2006). *Citrus paradisi* and *Citrus sinensis* flavonoids: Their influence in the defense mechanism against *Penicillium digitatum. Food Chem., 98,* 351–358.

Panchal, H. D., Vranizan, K., Lee, C. Y., Ho, J., Ngai, J., & Timiras, P. S., (2008). Early anti-oxidative and anti-proliferative curcumin effects on neuroglioma cells suggest therapeutic targets. *Neurochem. Res., 33,* 1701–1710.

Papas, A. M., (1999). *Antioxidants Status, Diet, Nutrition and Health* (pp. 189–210). CRC Press, Boca Raton, London, New York, Washington, D. C.

Park, S. Y., Jin, M. L., Kim, Y. H., Kim, Y., & Lee, S. J., (2012). Anti-inflammatory effects of aromatic-turmerone through blocking of NF-κB, JNK, and p38 MAPK signaling pathways in amyloid β-stimulated microglia. *Int. Immunopharmacol., 14,* 13–20.

Patel, J. M., (2008). A review of potential health benefits of flavonoids. *Lethbridge Undergrad Research Journal, 3,* 1–5.

Peng, S., Yao, J., Liu, Y., Duan, D., Zhang, X., & Fang, J., (2015). Activation of Nrf2 target enzymes conferring protection against oxidative stress in PC12 cells by ginger principal constituent 6-shogaol. *Food Funct., 6,* 2813–2823.

Prakash, D., & Gupta, K. R., (2009). The antioxidant phytochemicals of nutraceutical importance. *Open Nutraceuticals J., 2,* 20–35.

Quilez, J., Garcia-Lorda, P., & Salas-Salvado, J., (2003). Potential uses and benefits of phytosterols in diet: Present situation and future directions. *Clin. Nutr., 22,* 343–351.

Ramkumar, M., Rajasankar, S., Gobi, V. V., Dhanalakshmi, C., Manivasagam, T., Justin, T. A., Essa, M. M., et al., (2017). Neuroprotective effect of demethoxycurcumin, a natural derivative of curcumin on rotenone induced neurotoxicity in SH-SY 5Y neuroblastoma cells. *BMC Complement. Altern. Med., 17,* 217.

Ramli, N. Z., Yahaya, M. F., Tooyama, I., & Damanhuri, H. A., (2020). A mechanistic evaluation of antioxidant nutraceuticals on their potential against age-associated neurodegenerative diseases. *Antioxidants (Basel), 9,* 1019.

Reyes-Fermín, L. M., Gonzalez-Reyes, S., Tarco-Alvarez, N. G., Hernandez-Nava, M., Orozco-Ibarra, M., & Pedraza-Chaverri, J., (2012). Neuroprotective effect of α-mangostin and curcumin against iodoacetate-induced cell death. *Nutr. Neurosci., 15,* 34–41.

Rietjens, I. M. C. M., Boersma, M. G., Haan, L. De., Spenkelink, B., Awad, H. M., Cnubben, N. H. P., Zanden, J. J. V., et al., (2002). The pro-oxidant chemistry of the natural antioxidants vitamin C, vitamin E, carotenoids and flavonoids. *Environ. Toxicol. Pharmacol., 11*, 321–333.

Scalbert, A., & Williamson, G., (2000). Dietary intake and bioavailability of polyphenols. *J. Nutr., 130,* 2073S–2085S.

Scalbert, A., Manach, C., Morand, C., Remesy, C., & Jimenez, L., (2005). Dietary polyphenols and the prevention of diseases. *Crit. Rev. Food Sci. Nut., 45*, 287–306.

Seo, J. S., Leem, Y. H., Lee, K. W., Kim, S. W., Lee, J. K., & Han, P. L., (2010). Severe motor neuron degeneration in the spinal cord of the Tg2576 mouse model of Alzheimer disease. *J. Alzheimers Dis., 21*, 263–276.

Sharma, M., & Gupta, Y. K., (2002). Chronic treatment with trans resveratrol prevents intracerebroventricular streptozotocin induced cognitive impairment and oxidative stress in rats. *Life Sci., 71*, 2489–2498.

Singh, P., Kesharwani, R. K., & Keservani, R. K., (2017a). Antioxidants and vitamins: Roles in cellular function and metabolism. In: *Sustained Energy for Enhanced Human Functions and Activity* (pp. 385–407). Academic Press. ISBN:9780128054130.

Singh, P., Kesharwani, R. K., & Keservani, R. K., (2017b). Protein, carbohydrates, and fats: Energy metabolism. In: *Sustained Energy for Enhanced Human Functions and Activity* (pp. 103–115). Academic Press. ISBN:9780128054130.

Snodderly, D. M., (1995). Evidence for protection against age-related macular degeneration by carotenoids and antioxidant vitamins. *Am. J. Clin. Nutr., 62,* 1448S-1461S.

Stahl, W., & Sies, H., (2005). Bioactivity and protective effects of natural carotenoids. *Biochim. Biophys. Acta., 1740*, 101–107.

Sun, C., Chen, K., Chen, Y., & Chen, Q. J., (2005). Contents and antioxidant capacity of limonin and nomilin in different tissues of citrus fruit of four cultivars during fruit growth and maturation. *Food Chem., 93*, 599–605.

Tiwari, A., (2001). Imbalance in antioxidant defense and human diseases: Multiple approach of natural antioxidants therapy. *Curr. Sci., 81*, 1179–1187.

Triantis, T., Stelakis, A., Dimotikali, D., & Papadopoulos, K., (2005). Investigations on the antioxidant activity of fruit and vegetable aqueous extracts on superoxide radical anion using chemiluminescence techniques. *Anal. Chim. Acta., 536*, 101–105.

Wan, N. W. N., Makpol, S., Mazlan, M., Tooyama, I., Wan, N. W. Z., & Damanhuri, H. A., (2019). Tocotrienol rich fraction supplementation modulates brain hippocampal gene expression in APPswe/PS1dE9 Alzheimer's disease mouse model. *J. Alzheimers Dis., 70,* S239-S254.

Watzl, B., Bub, A., Briviba, K., & Rechkemmer, G., (2003). Supplementation of a low-carotenoid diet with tomato or carrot juice modulates immune functions in healthy men. *Ann. Nutr. Metab., 47*, 255–261.

Willcox, J. K., Ash, S. L., & Catignani, G. L., (2004). Antioxidants and prevention of chronic disease. *Crit. Rev. Food Sci. Nutr., 44*, 275–295.

Xie, Z., Zhao, J., Wang, H., Jiang, Y., Yang, Q., Fu, Y., Zeng, H., et al., (2020). Magnolol alleviates Alzheimer's disease-like pathology in transgenic C. elegans by promoting microglia phagocytosis and the degradation of beta-amyloid through activation of PPAR-γ. *Biomed. Pharmacother., 124,* 109886.

CHAPTER 10

UNDERSTANDING THE ROLE OF ω-3 POLYUNSATURATED FATTY ACIDS ON AUTOIMMUNITY

AARTI TIWARI, DEEPTI DWIVEDI, AJAY KUMAR SHUKLA, SATYA PRAKASH SINGH, and SANDEEP KUMAR SINGH

Institute of Pharmacy, Dr. Ram Manohar Lohia Avadh University, Ayodhya, Uttar Pradesh, India

ABSTRACT

More than 80 years ago, ω-3 polyunsaturated acids (PUFAs) were recognized as important fatty acids for appropriate growth and well-being. However, since 1980, people have increased the use of polyunsaturated acids in regular diet, decreasing the risk of very long and complicated diseases. The importance of reducing the risk of the number of chronic human illnesses has expanded in recent years. Despite the enormous epidemiological data, several attempts to use fish-oil supplements to intervene in human disorders have resulted in inconsistent and frequently unclear results; null or weak supportive conclusions were occasionally drawn in the META analysis. Various doses and fish-oil sources may have contributed to the disparate effects of intervention performed at different clinics. A number of researchers have been reported in their genetic and clinical selected models' study under the ω-3 PUFAs, and were found that it is able to reduce the pathological complications of various diseases. These various long-term diseases are diabetic-1, rheumatoid arthritis, systemic lupus erythematosus (SLE) and multiple sclerosis. This type of valuable information is summarized and presented in chapters.

Immune-Boosting Nutraceuticals for Better Human Health: Novel Applications. Urmila Jarouliya, PhD, Raj K. Keservani, MPharm, Rajesh K. Kesharwani, PhD, Virendra K. Patel, PhD, and Adi D. Bharti, MPharm (Eds.)

All the information are concerning of ω-3 PUFAs roles and importance and which type of research going on ω-3 PUFAs presented. Some significant research finding are also presented in this chapter, especially on autoimmune diseases.

10.1 INTRODUCTION

Fatty acids are hydrocarbon chains having one end having a carboxyl group and the other a methyl group. The carboxyl group more easily forms ester linkages with alcohol groups; examples are glycerol or cholesterol, resulting in cholesteryl esters and acyl-glycerols (e.g., phospholipids and triacylglycerols (TAGs)) (Scrimgeour & Harwood, 2007).

Single double bond is present in the hydrocarbon chain of unsaturated fatty acids, whereas more than two double bonds are of polyunsaturated fatty acids (PUFAs). On the 3rd carbon, where first double bond located referred as the methyl carbon because of that 1st carbon, this is known as Omega-3 fatty acids. A simplest example of the Omega-3 fatty acid is α-linolenic acid (ALA) (18: 3ω-3) (Calder & Yaqoob, 2009).

Polyunsaturated fatty acids are classified into two major categories:

- ω-3 polyunsaturated fatty acids (PUFAS); and
- ω-6 PUFAS.

With some conformational differences as the double bond position at carbon chain and the structural differences in their skeleton. The linolenic acid arachidonic acid (ARA) is the best example of consumable ω-6 PUFAS are ALA, eicosapentaenoic acid (EPA), and docosahexaenoic acid (DHA) are the best example of consumable ω-3 PUFAS (Beare-Rogers et al., 2001). The mammals are not able to synthesize the a-linolenic acid so regarded as essential fatty acids, and with the help of desaturation followed by δ-15 desaturase (interestingly, the enzymes desaturase is named according to the newly inserted double bond at the first carbon, counting as the carbon number 1st as carboxyl carbon), from that ALA is synthesized from linoleic acid (18:2ω-6). All the animals (including humans) do not possess the enzyme: delta-15 desaturase thus never synthesize ALA, whereas plants bearing δ-15 desaturase and subsequently can synthesize ALA (Calder & Yaqoob, 2009). Even though, animals neither synthesize ALA nor attain, erstwhile can metabolize it by elongation and desaturation. The site of desaturation primarily in the liver, stoichiometrically carbon atoms ranges carbon number

9 (numbering from carboxyl carbon) (Nitbani et al., 2020; Sartorio et al., 2022; Bali et al., 2020; Bahreynian et al., 2022; Halken et al., 2021).

10.2 OMEGA-3 FATTY ACIDS AS DIETARY SOURCES

Although green leaves contain a substantial amount of ALA (usually 50%), they are not high in fat. Abundant amount of ALA is accessible in biodiversity of plants as in seeds, seed oils, and some nuts (Nitbani et al., 2020). An emerging amount of 45% to 55% of fatty acids as ALA is extracted from Flaxseeds (Linseeds) their oil, whereas 5% to 10% of fatty acids as ALA obtained from soybean oil. ALA is also rich in rapeseed oil and walnuts. Linoleic acid is abundant in corn oil, sunflower oil, and safflower oils, whereas ALA is limited. ALA consumption ranges from 0.5 to 2 g/d in Western people (Asif, 2011). The ω-6 fatty acid (18:2ω-6) is commonly taken in 5- to 20-fold larger levels than ALA and is the most common PUFA in most Western diets. Seafood contains a longer chain, more unsaturated ω-3 PUFAs. Fish can be categorized as "lean" (or "oily") if they store lipid as "fatty" (oily) or in "liver" (viz. cod) if they store lipid in their flesh (e.g., mackerel, herring, salmon, tuna, and sardines) (Micaleff et al., 2012). Fish and other seafood are good providers of the extremely long-chain omega-3 fatty acids (LCOFAs) EPA, DPA, and DHA when compared to other diets. In general, from different EPA/DHA ratios, the various varieties of fish are found keeping diverging levels of these fatty acids at, it depends on the fish's metabolic properties, as well as its nutrition, water temperature, season, and other factors.

Different fish species are having different level of big-chain omega-3 fatty acids level for the examples one species lean fish having is the good source of big-chain omega-3 fatty acids level, whereas salmon or mackerel species of fish having 1.5 to 3 g big-chain omega-3 fatty acids level. The major source of big-chain omega-3 fatty acids is cod liver oil which is obtained from cod fish, and oil obtained from liver of cod fish (Calder, 2014).

Fish oil is key sources of LCOFAs. The long chain-3 fatty acids EPA and DHA are mainly present in fish oil. It is commonly available in the market. Fish oil capsules having 30% of LCOFAs and is approximately 1-g fish oil of capsule contains 0.3 g of EPA plus DHA. The amount of LCOFA is found in different fish species in different amount. It is dependent on the percentage of fish oil quantity. Fish oil is the chief sources of all LCOFAs. These are omega-3 PUFAs (EPA, DPA, and DHA) respectively. The fish cod liver oil is the richest source of EPA than DHA. Whereas tuna fish oil is containing DHA higher amount as compare than EPA amount (Suseno et al., 2014). The fish liver oils having a major quantity of fat-soluble vitamins also, like vitamins A and D. The marketed encapsulated oil preparations key sources of omega-3 fatty acids in higher amounts. All the fatty acids are commonly found in the form of TAGs in fish oil capsules. Therefore, those people are not eating fish; he can get the all types of LCOFAs from the fish oil. It is having long-chain omega-3 fatty acid 5-times more than fish. The algal oils that are chiefly contain DHA (Jacobsen et al., 2013).

10.3 AUTOIMMUNITY

In general, the immune system separate out friend and enemy of body, and easily to detect body enemies, and after identification of foreign infected microorganism kill them but sometimes these all of body friends act against the body this process created very much complications in our body. This type of disease we are commonly called auto-immune diseases (Downing & Mujic, 2009).

The autoimmune diseases are effected the body cells, tissues in various parts of body such as brain, spinal card, joints, pancreatic cells, heart muscles and many more important parts of body. The world are suffering from various autoimmune which are challenging task for discovery and curing of these diseases like diabetes, arthritis, psoriasis, asthma, and many more diseases (Marrack et al., 2001).

The others autoimmune disease associated problems is myasthenia gravis. In this disease patient not properly walk, and perform normal functions of body. In myasthenia gravis disease mainly affected by neural cells, tissue, and membrane functions which can be improved by using long chain fatty acids. In Graves' disease autoimmune disease the nerve muscle are induced the thyroid cells which secreted the higher amount of thyroxin hormone, and it produced various effect in the body which increase the blood pressure, and many more complications in body. The higher amount of thyroxin hormone also causes the blister in the skin, destroys the kidneys and other complications in body organs. This type of complicated condition called systemic lupus erythematosus (SLE) (Mathur et al., 2020).

The autoimmune diseases have been found that, these chronic problems also induce much more expenses of money, time, and enjoyment of life. Many people are ignoring this type of diseases suffering patients. Modern research some give hope for the treatment that are cytokine antagonists therapy, under this therapy have been found good results under the clinical research work (Goodnow, 2007).

TNF-α is responsible for the cause of inflammation like complications in rheumatoid arthritis autoimmune diseases. The TNF-α comes formed in during cell signaling when any foreign particles comes into the body but in autoimmune diseases it is formed against the body which create the inflammation. Therefore, currently many autoimmunity diseases like rheumatoid arthritis are treated by using cytokine antagonists. The cytokine antagonists are playing a significant role in the treatment and management of rheumatoid arthritis disease. In addition, it also decreased the action of TNF-α. Earlier time researchers were doing research based on anti-inflammation action of drug only; they do not target on cause of inflammation which key factor of rheumatoid arthritis disease (Rosenblum et al., 2015).

Many autoimmune diseases were treated by doctors, and researchers doing work in clinical research. The available medicine are only reduced the symptoms only, and give relax to patient for some time. In the treatment of autoimmune diseases, patients are taking a lot of medicine for a long time, long taking medicine created more complications in patients. Therefore, we need to find out which is the main cause of this autoimmunity problems. These types of autoimmunity problems may come from cell response and it may be associated with cell signaling, so need to find out and remove this problem (Goris & Liston, 2012).

All the autoimmunity diseases research need to do research in systemically step by step like first need to find out initial cause of problem, circulation of diseases, then need to solve the problems. These all steps are associated with cell signaling, receptors, and immune response. Consequently need to do more clinical trial and find to solve these problems (Mackay et al., 2008).

The mechanism of diseases causing is the main goal of research of many researchers, but they need to find out early symptoms of autoimmune diseases. These types of auto-immune diseases suffering patients are carrying many complications and symptoms and not share the problems to doctors. This is one of the major problems which make it more complicated. In long time diseases are not easy to treat because body cells, receptors, and immune response create problems. So need to find out the primary symptoms of autoimmune disease symptoms in the early phase (Thewissen et al., 2007).

These types of autoimmune diseases are induced or affected by many factors such as genetic and environmental factors. These disorders factors are supported by some scientific hypothesis. Scientific hypothesis based on the changes come in the many genes due to improper regulation or decreased the threshold for lymphocyte activation, and surrounding environmental factors that bring or inducing activation of self-responsive lymphocytes that have remove control and are arise to react against their self-constituents (Doyle & Mamula, 2012).

The pathology of autoimmune diseases causes behind associated may factors that are intrinsic (it is associated with genetic problems) and extrinsic (it is associated with environmental). These factors are inducing autoimmune diseases and contribute to disease spreading and pathogenesis complications. These intrinsic or genetic abnormalities are very complicated, due to changes of genetic characters of genes. It is vary to genes to genes, and it has been seen in different ethnic groups in during clinical research. It is strongly recommended that the same autoimmunity symptoms problems are not be the same for all patient associated problems, so need to collect clinical data systemically (Godessart & Kunkel, 2001).

The clinical research revealed and suggested that the autoimmunity is associated with sex factors also. It is varying with gender to gender. The activation and response of immune is more easily in females than in males. It means females of cells are more sensitive as compare than male. The immune response is related to lymphocytes and monocytes activation property. These properties are indicating that the females show higher antigen-presenting activity and mitogenic responses than male. It is also indicated that the females are carrying higher immunoglobulin percentage than males. This antigen-presenting activity is increased antibody production after both primary and secondary antigen stimulation property, and also a higher homograft rejection rate (Fairweather et al., 2008; Gardner et al., 2020; Venter et al., 2020; Ellul et al., 2020).

10.4 OMEGA-3 FATTY ACID SUPPLEMENTATION

The various cells, tissues, receptors, and membranes are having different lipid composition profiling. These profiling of lipid are changes due to the availability of big chain omega-3 fatty acids compounds. These compounds use create changes under the lipid profiling nature, functions, and characters (Burdge & Calder, 2005).

These changes of lipid profiling of cells, tissues, receptors, and membranes have been seen after the supplements of big chain omega-3 fatty acids compounds. The commonly used EPA and DHA which are generally found in fish oil. These big chain omega-3 fatty acids compounds are taking normally in form of capsules. It induces in changes in platelets, plasma lipids, leukocytes, erythrocytes, cardiac tissue, colonic tissue, and most likely in many other cells and tissue types. This supplements capsules having mainly big chain omega-3 fatty acids compounds like omega-6 (Calder & Yaqoob, 2009). These big chain omega-3 fatty acids compounds are also potentially play a role in the management of blood related problems to solve (Calder & Yaqoob, 2009).

10.5 MECHANISMS OF ACTION OF VERY LONG-CHAIN OMEGA-3 FATTY ACIDS (LCOFAS)

The mechanism of big chain omega-3 fatty acids compounds are summarized in form of figure that is shown in Figure 10.1. This big chain omega-3 fatty acids compounds are induced the changes in tissues and associated membranes, due to increase under fluidity and a numbers of physical natures. It plays a significant role for development of tissue, receptors, transporters, and intercellular cell signaling pathways. These positive changes and cells, tissues, and membranes' increase the cells, tissue, membranes, receptors, and other associated functions well. It enhances cell functions through a variety of mechanisms that are shown in Figure 10.1. These include:

- The effects on the improvement of cell singling pathways improve the various cell associated activities such as transcription, translation, DNA changes, and gene expression. These all changes also can be helpful for the receptors concerning activities of-α and γ, and the sterol regulatory element-binding proteins (Calder, 2007).
- These big chain omega-3 fatty acids compounds are induced the alteration in lipid membranes activity like eicosanoids. The eicosanoids are producing various inflammations mediators which induce inflammations. A numbers of anti-inflammatory drugs act on and block the production or release into the body, and decrease the inflammations, platelet aggregation and muscle activity, and urinal functions. These big chain omega-3 fatty acids compounds can be used for the treatment of the above-mentioned diseases (Frisardi et al., 2011).
- These big chain omega-3 fatty acids are compounds are reduced the production of ARA also. It can inhibit the synthesis of ARA. In recently clinical research have been found that big chain omega-3

fatty acids such as EPA and DHA was showed protective effect for damaging of tissues in our body (Freitas et al., 2018).

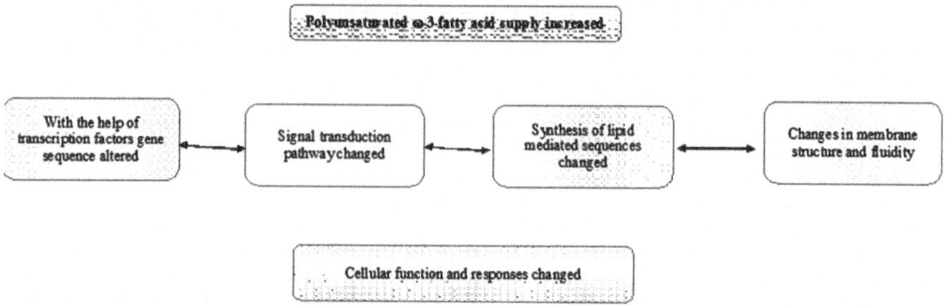

FIGURE 10.1 Cell function influenced by the polyunsaturated omega-3 fatty acids.

10.5.1 AN INCREASED INTAKE OF VERY LONG-CHAIN OMEGA-3 FATTY ACIDS (LCOFAS) IS BENEFICIAL TO HEALTH

Through the above-mentioned diagram big chain omega-3 fatty acids induce some changes on cells their tissues physiological properties. It is associated with many health issues or complications, and these reported studies represented that it may be helpful to control and management of various diseases such as cardiac disease. It can be used for the treatment and management of various cardiac concerning complications (Skulas-Ray et al., 2008). It also can be used for the removal of complications of blood related diseases such as platelet activity, plasma concentration, vascular functions, thrombosis (Feingold, 2021), and inflammation (Jho et al., 2004), etc., research finding also revealed that big chain omega-3 fatty acids remove the problems of critical others diseases and can reduced the death rate in all over the world, if people take this big chain omega-3 fatty acids in their regular diet in different forms (Harris et al., 2008). A number of others research have been reported that big chain omega-3 fatty acids showed good results for the recovery in joint pain, inflammation, and concerning diseases (Marchioli et al., 2002).

In many research under big chain omega-3 fatty acids have been used for the treatment of arthritis, bowel diseases. Under the investigational research study it has been seed big chain omega-3 fatty acids like DHA decreased the inflammations of the bowel, and arthritis.

Research work finding reported that the big chain omega-3 fatty acids compounds can used for the treatment of inflammation others diseases such as eye, brain, and others., it also easily to reach the brain and other parts of tissue due lipid soluble, and it can be play vital role (Echeverría et al., 2017). The under big chain omega-3 fatty acids compounds also increase the memory power, learning ability of child in during growing age (Marchioli et al., 2002; Giacobbe et al., 2020).

10.6 CONCLUSION

The positive results of a decade's worth of study have confirmed the position of -3 PUFAs as a possible contender for preventing or possibly treating autoimmune illnesses including type-1 diabetes, RA, SLE, and MS. Many of the good benefits of -3 PUFAs may undoubtedly be recognized to their anti-inflammatory properties; however, other mechanisms, such as mTOR (mammalian target of rapamycin) activity modulation, may also be at work. The large doses of EPA/DHA or fat-1 gene therapy to generate endogenous-3 PUFAs, it has a significant promise in the clinical treatment of these severe disorders, despite certain technical problems, notably clinical assessment of usefulness and protection.

KEYWORDS

- **autoimmunity**
- **polyunsaturated acids**
- **systemic lupus**

REFERENCES

Asif, M., (2011). Health effects of omega-3, 6, 9 fatty acids: Perilla frutescens is a good example of plant oils. *Oriental Pharmacy & Experimental Medicine, 11*(1), 51–59.

Bahreynian, M., Feizi, A., & Kelishadi, R., (2020). Is fatty acid composition of breast milk different in various populations a systematic review and meta-analysis. *Int. J. Food. Sci. Nutr., 71*, 909–920.

Bali, A., Vlaši, D., Žužul, K., Marinovi, B., & Bukvi, M. Z., (2020). Omega-3 versus omega-6 polyunsaturated fatty acids in the prevention and treatment of inflammatory skin diseases. Int. *J. Mol. Sci., 21*, 741.

Beare-Rogers, J. L., Dieffenbacher, A., & Holm, J. V., (2001). Lexicon of lipid nutrition (IUPAC technical report). *Pure and Applied Chemistry, 73*(4), 685–744.

Burdge, G. C., & Calder, P. C., (2005). Conversion of $\alpha $-linolenic acid to longer-chain polyunsaturated fatty acids in human adults. *Reproduction Nutrition Development, 45*(5), 581–597.

Calder, P. C., & Yaqoob, P., (2009). Omega-3 polyunsaturated fatty acids and human health outcomes. *Biofactors., 35*(3), 266–272.

Calder, P. C., & Yaqoob, P., (2009). Understanding omega-3 polyunsaturated fatty acids. *Postgraduate Medicine, 121*(6), 148–157.

Calder, P. C., (2007). Immunomodulation by omega-3 fatty acids. *Prostaglandins, Leukotrienes and Essential Fatty Acids, 77*, 327–335.

Calder, P. C., (2014). Very long chain omega-3 (n-3) fatty acids and human health. *European Journal of Lipid Science and Technology, 116*(10), 1280–1300.

Downing, L. H., & Mujic, B. K., (2009). Infectious diseases are sleeping monsters: Conventional and culturally adapted new metaphors in a corpus of abstracts on immunology. *Ibérica, Revista de la Asociación Europea de Lenguas Para Fines Específicos., 17*, 61–82.

Doyle, H. A., & Mamula, M. J., (2012). Autoantigenesis: The evolution of protein modifications in autoimmune disease. *Current Opinion in Immunology, 24*(1), 112–118.

Ellul, S., Marx, W., Collier, F., Saffery, R., Tang, M., Burgner, D., Carlin, J., Vuillermin, P., & Ponsonby, A. L., (2020). Barwon infant study investigator group. Plasma metabolomic profiles associated with infant food allergy with further consideration of other early life factors. *Prostaglandins Leukot. Essent. Fatty Acids, 159*, 102099.

Fairweather, D., Frisancho-Kiss, S., & Rose, N. R., (2008). Sex differences in autoimmune disease from a pathological perspective. *The American Journal of Pathology, 173*(3), 600–609.

Feingold, K. R., (2021). *Triglyceride Lowering Drugs*. Endotext *[Internet]*. www.endotext.org (accessed on 20 February 2023).

Freitas, H. R., Isaac, A. R., Malcher-Lopes, R., Diaz, B. L., Trevenzoli, I. H., & De Melo, R. R. A., (2018). Polyunsaturated fatty acids and endocannabinoids in health and disease. *Nutritional Neuroscience, 21*(10), 695–714.

Frisardi, V., Panza, F., Seripa, D., Farooqui, T., & Farooqui, A. A., (2011). Glycerophospholipids and glycerophospholipid-derived lipid mediators: A complex meshwork in Alzheimer's disease pathology. *Progress in Lipid Research, 50*(4), 313–330.

Gardner, K. G., Gebretsadik, T., Hartman, T. J., Rosa, M. J., Tylavsky, F. A., Adgent, M. A., Moore, P. E., Kocak, M., Bush, N. R., Davis, R. L., et al., (2020). Prenatal omega-3 and omega-6 polyunsaturated fatty acids and childhood atopic dermatitis. *J. Allergy. Clin. Immunol. Pract., 8*, 937–944.

Giacobbe, J., Benoiton, B., Zunszain, P., Pariante, C. M., & Borsini, A., (2020). The anti-inflammatory role of omega-3 polyunsaturated fatty acids metabolites in pre-clinical models of psychiatric, neurodegenerative, and neurological disorders. *Front. Psychiatry, 11*, 122.

Godessart, N., & Kunkel, S. L., (2001). Chemokines in autoimmune disease. *Current Opinion in Immunology, 13*(6), 670–675.

Goodnow, C. C., (2007). Multistep pathogenesis of autoimmune disease. *Cell, 130*(1), 25–35.

Goris, A., & Liston, A., (2012). The immunogenetic architecture of autoimmune disease. *Cold Spring Harbor Perspectives in Biology, 4*(3), 7260.

Halken, S., Muraro, A., De Silva, D., Khaleva, E., Angier, E., Arasi, S., Arshad, H., Bahnson, H. T., Beyer, K., Boyle, R., et al., (2021). EAACI guideline: Preventing the development of food allergy in infants and young children (2020 update). *Pediatr. Allergy. Immunol., 32*, 843–858.

Jacobsen, C., Nielsen, N. S., Horn, A. F., & Sørensen, A. D. M., (2013). *Food Enrichment with Omega-3 Fatty Acids* (1st edn., pp. 1–10). Elsevier.

Mackay, I. R., Leskovsek, N. V., & Rose, N. R., (2008). Cell damage and autoimmunity: A critical appraisal. *Journal of Autoimmunity, 30*(1, 2), 5–11.

Marrack, P., Kappler, J., & Kotzin, B. L., (2001). Autoimmune disease: Why and where it occurs. *Nature Medicine, 7*(8), 899–905.

Mathur, D., Rout, S., Mishra, B. K., Rodas, G. L., Vallamkondu, J., Kandimalla, R., & Casanova, B., (2020). *Potential Pathological Biomarkers in Multiple Sclerosis,* 1–66.

Micaleff, M., Munro, I., Phang, M., & Garg, M., (2009). Plasma n-3 polyunsaturated fatty acids are negatively associated with obesity. *Br. J. Nutr., 102*, 1370–1374.

Nitbani, F. O., Tjitda, P. J. P., Nurohmah, B. A., & Wogo, H. E., (2020). Preparation of fatty acid and monoglyceride from vegetable oil. *Journal of Oleo Science, 69*(4), 277–295.

Rosenblum, M. D., Remedios, K. A., & Abbas, A. K., (2015). Mechanisms of human autoimmunity. *The Journal of Clinical Investigation, 125*(6), 2228–2233.

Sartorio, M. U. A., Pendezza, E., Coppola, S., Paparo, L., D'Auria, E., Zuccotti, G. V., & Berni, C. R., (2022). Potential role of omega-3 polyunsaturated fatty acids in pediatric food allergy. *Nutrients, 14*, 152.

Schieffer, D., Naware, S., Bakun, W., & Bamezai, A. K., (2014). Lipid raft-based membrane order is important for antigen-specific clonal expansion of $CD4^+$ T lymphocytes. *BMC Immunology, 15*, 58.

Scrimgeour, C. M., & Harwood, J. L., (2007). Fatty acid and lipid structure. *The Lipid Handbook, 3*, 1–36.

Skulas-Ray, A. C., West, S. G., Davidson, M. H., & Kris-Etherton, P. M., (2008). Omega-3 fatty acid concentrates in the treatment of moderate hypertriglyceridemia. *Expert Opinion on Pharmacotherapy, 9*(7), 1237–1248.

Suseno, S. H., Hayati, S., & Izaki, A. F., (2014). Fatty acid composition of some potential fish oil from production centers in Indonesia. *Oriental Journal of Chemistry, 30*(3), 975.

Thewissen, M., Somers, V., Venken, K., Linsen, L., Van, P. P., Geusens, P., Damoiseaux, J., & Stinissen, P., (2007). Analyses of immunosenescent markers in patients with autoimmune disease. *Clinical Immunology, 123*(2), 209–218.

Venter, C., Agostoni, C., Arshad, S. H., Ben-Abdallah, M., Du Toit, G., Fleischer, D. M., Greenhawt, M., Glueck, D. H., Groetch, M., Lunjani, N., et al., (2020). Dietary factors during pregnancy and atopic outcomes in childhood: A systematic review from the European academy of allergy and clinical immunology. *Pediatr. Allergy Immunol., 31*, 889–912.

PART III

Marine-Derived Sources Used to Boost Immunity

CHAPTER 11

IMMUNOMODULATORY EFFECTS OF EDIBLE MICROALGAE

SHRUTI SRIVASTAVA,[1] MD. AKHLAQUR RAHMAN,[2] and SHANTHY SUNDARAM[1]

[1]*Center of Biotechnology, Nehru Science Center, University of Allahabad, Prayagraj, Uttar Pradesh, India*

[2]*Department of Biotechnology, S.S. Khanna Girls' Degree College, Prayagraj, Uttar Pradesh, India*

ABSTRACT

Microalgae are photosynthetic microorganisms that thrive in different environments. They show enormous diversity with respect to morphology, physiology, and genetic traits. Due to their biochemical and genetic diversity, they serve as an excellent source of new natural bio-products with potential applications in several biotechnological sectors. The different compounds derived from microalgae have shown several properties like anticancer antimicrobial anti-inflammatory and immunomodulatory. In the recent years, compounds stimulating both innate and adaptive immune responses have been used to overcome or fight several pathological conditions such as cancer, etc. Immunomodulators are used in place of classical antibiotic treatment to elevate host defense response. Easy cultivation, high nutritional value and long period use as food supplement are the advantages that made it a valuable source for conducting immunomodulating studies. Microalgae possessing bioactive compounds with immunomodulatory properties are of

Immune-Boosting Nutraceuticals for Better Human Health: Novel Applications. Urmila Jarouliya, PhD, Raj K. Keservani, MPharm, Rajesh K. Kesharwani, PhD, Virendra K. Patel, PhD, and Adi D. Bharti, MPharm (Eds.)

great interest for their possible role as a new class of potential drugs to treat human pathological disorders. High amount of β-carotene and chlorophyll in Spirulina and Chlorella help in providing cellular protection. Chlorella Growth Factor derived from Chlorella help in treating many degenerative diseases. Furthermore, these microalgae derived compounds can promote better health and alleviate the risk of the development of degenerative diseases.

11.1 INTRODUCTION

Microalgae are unicellular algae inhabiting both marine and freshwater habitat. They range from 1 to 50 μm in diameter and are highly diverse with 2,00,000–8,00,000 species (Richmond, 2013). They can also thrive well in extreme environmental conditions such as warm, freezy, high or low salinity and light intensities, etc. (Plaza et al., 2009). Being photosynthetic microorganisms, they contribute to around 40% of global photosynthesis (Moreno-Garrido, 2008). These microorganisms are capable of producing different interesting natural bioproducts with industrial and pharmaceutical applications (Mimouni et al., 2012). Current days advances in the field of genome sequencing, metagenomics, transcriptomics, meta transcriptomics, proteomics, and metabolomics have added in the *in-silico* identification of several gene clusters which are potentially involved in the synthesis of bioactive compounds from a pool of species thereby uncovering various novel drugs from microalgae (Lauritano et al., 2019; Mishra et al., 2019).

Various compounds from microalgae are acting as a promising drug in treating human diseases by activating the immune system in both humans and murine models for example the release of pro and anti-inflammatory mediators (Heo et al., 2012; Kong et al., 2016 and Mosser et al., 2008) and modification of macrophage activation protecting the host from various pathogens and harmful agents. The immune system of the host develops a sophisticated defense mechanisms to protect itself from pathogens and toxic substances which is categorized as innate immune response and adaptive immune response.

Innate immune response which gets hastily activated after a pathogenic invasion, constitutes the first line of host response involving a huge number of molecules and cells. The cells comprise of macrophage, monocyte, neutrophils, basophils, eosinophils, dendritic cells (DCs) and the natural killer (NK) cells (Shishido et al., 2012). Contrary to this adaptive immune response comprises fewer number of cells namely B- and T-lymphocytes and are based on the antigen specific receptors expressed on the surface of these cells (Medzhitov, 2007). Cells of the immune system modulate their activities with the help of mediators. Macrophages and lymphocytes predominantly produce cellular mediators called cytokines that entails interleukins (ILs), tumor necrosis factor (TNF-α) and IFN-γ (Chan, 2019). Inflammatory tissues damage and autoimmune diseases are caused due to uncontrolled responses of the immune system towards pathogenic organisms and toxic substances. These deleterious effects are regulated by balancing the magnitude of co-stimulatory and inhibitory signals (Pardoll, 2012). The primary mediator of the effector functions are the T-cells that exhibit various co-inhibitory receptors like, lymphocyte-activation gene 3 (LAG-3), programed cell death protein 1 (PD-1), and cytotoxic T-lymphocyte-associated protein 4 (CTLA-4) (Pardoll, 2012).

Immunomodulation is a process of stimulating or suppressing the immune response for therapeutic purposes which involves the activation of the immune system to fight diseases such as microbial infections cancer and to reduce inflammatory reactions upon injury and infection (Ben-Aharon et al., 2018). Various microalgae are discovered that show immunomodulation in humans or in other organisms like sheep and mouse. Dried microalgae or their raw extracts show activity against several cells of the immune system (Gennaro & Chiara, 2019). Aqueous polysaccharide extracts from *Chlorella stigmatophora* tested *in vivo* or *in-vitro* mouse models showed activation of the phagocytosis by macrophages from the peritoneal cavity (Guzman et al., 2009). Condensed water-soluble extract of commercially available spray dried Spirulina serve as a food supplement that augments interferon (IFN) production and NK cell cytotoxicity (Hirahashi et al., 2002; Wu et al., 2006).

11.2 SPIRULINA

Spirulina (Arthrospira) is a prokaryotic cyanobacteria belonging to Cyanophyta (Figure 11.1(a)). They evolved more than 3 million years ago, responsible for creating the present oxygenic atmosphere with importance in regulating the planetary biosphere (Romano et al., 2004). It can be found in tropical and subtropical lakes with high pH (between 8 and 11) and elevated carbohydrate and bicarbonate concentrations. For commercial purpose, it can be grown in huge outdoor or greenhouse ponds under controlled conditions. It is considered a blue–green algae as it contains chlorophyll (green) and phycocyanin (blue). Spirulina are also found in very warm water at 32°C to 45°C and can even survive in temperatures of 60°C. It is an autotroph and thus carry out photosynthesis. As it is gram-negative its cell wall is made up of peptidoglycan. The trichome is helical in shape. The largest commercial producer of Spirulina are United States, Thailand, India, Taiwan, China, Pakistan, and Burma. In 1981, FDA by issuing a GRAS (generally recognized as safe) certificate approved Spirulina to be legitimately promoted as a safe food supplement for human beings (Costa & Morais, 2013).

The cell wall of *Spirulina platensis* is made up of complex sugars and proteins which is digested easily. Spirulina is rich in protein content and has high digestibility. They contain substantial amounts of essential polyunsaturated fatty acid (PUFA) as well as phenolic compounds (Borges et al., 2013). Due to its high nutritional value and presence of different bioactive compounds, Spirulina is one of the most consumed microalgae worldwide (Borges et al., 2013) since ancient times it has been used as a source of protein and vitamins. Its protein content is 50 to 70%(w/w) and carbohydrate is 10 to 20%(w/w) of its dry weight with the lipid proportion ranging between 5 and 10% (w/w). Vitamin B1, B2, notably vitamin B12 are all abundant in this food. Apart from this Spirulina has high content of pigments, essential amino acids, essential fatty acids, antioxidants, minerals, and oligo-elements important ones being iron, calcium, magnesium, phosphate, and potassium (Costa andMorais, 2013). β-carotene accounts for approximately 80% ofSpirulina's carotenoids content along with other components like tocopherols, phycocyanin, and phycoerythrin (Priyadarshani & Rath, 2012). Bioactive compounds present in Spirulina are depicted in Table 11.1.

TABLE 11.1 Bioactive Compounds in Edible Microalgae (by Wet Weight)

Bioactive Compounds	Spirulina	Chlorella
Protein	60–70%	55–60%
Lipids	5% (ALA)	>10% (ω-3 PUFAs)
Dietary fiber	2%	>30%
Minerals	10% (P, Mg, K, Ca, Fe, Zn)	>10% (Fe, Ca, Mg, K, Zn)
Vitamins	E, B1, B2, B3, B9	C, E, K1, B12, B1, B2, B6
Pigments	Phycocyanin, carotenoids	1–4% Chlorophylls, carotenoids

Several studies have reported the therapeutic role of Spirulina, which elucidates its application in the cure of hyperlipidemia, HIV, cancer, anemia, obesity, diabetes, hypertension, increased immunological response in safeguarding kidney towards heavy metals and drugs and decrease in level of glucose and lipid in serum (Ambrosi et al., 2008; Colla et al., 2007; Colla et al., 2008; Torres-Duran et al., 2007). In addition to its antioxidant activity Spirulina also exerts immunomodulatory activities by regulating cytokines such as interleukins IL-1β, IL-2, IL-4, IL-6, IL-10, and TNF-α (Mao et al., 2000).

Spirulina causes inhibition of viral replication and also builds up cellular and humoral immunity by inhibiting progression of cancer (Ayehunie et al., 1996). Several studies have depicted that Calcium-Spirulan, a polymerized sugar molecule extracted from Spirulina are safe for human cells but interfere with replication of HIV-1, herpes simplex, human cytomegalovirus (HCMV), influenza A virus, mumps virus and measles virus (Hayashi & Hayashi, 1996). Recently an aqueous extract of *S. platensis* was found to be effective ininhibiting replication of HIV-1 in human T-cells, peripheral blood mononuclear cells and Langerhans cell in an *in-vitro* research (Ayehunie et al., 1996). Even after an infection has occurred, microalgal compounds with antiviral characteristics can immunomodulate. It has been shown in studies that Spirulina constantly improves the function of the immune system in mice, hamsters, chickens, turkeys, cats, and fish. Phycocyanin extracted from Spirulina imitate the erythropoietin hormone (EPO) and help in haematopoiesis. It also modulates the production of WBCs when bone marrow stem cells are damaged.

11.3 CHLORELLA

The term Chlorella is acquired from the Latin word "ella" meaning green and small. It has existed since the pre-Cambrian period about 2.5 billion years ago. It is a eukaryotic genus of green unicellular microalgae belonging to the phylum Chlorophyta (Figure 11.1(b)) (Plaza et al., 2009). It thrives well in both fresh water and salt water. The Japanese discovered Chlorella as a food supplementand currently Japan is the world leader in the consumption of Chlorella microalgae. Chlorella contains chlorophyll proteins polysaccharide vitamin minerals carotenoids and essential amino acid (leucine, lysine, methionine, phenylalanine, threonine, tryptophan, valine, and histidine) and dietary fiber. Its overall composition includes 53% (w/w) protein, 23%(w/w) carbohydrate, 9%(w/w) lipids, 5%(w/w) minerals and oligo-elements (Costa & Morais, 2013). The biomass of Chlorella is also abundant in vitamins of B-complex, notably vitamin B12. Chlorella also contains huge amounts of folate and iron. It can be used as a food without having any risk to human health as it has GRAS certificate issued by FDA (Costa & Morais, 2013; Costa et al., 2006). Chlorella is a very useful component of food and medicine as it has tremendous nutritional and functional characteristics. Polysaccharides (PS) of Chlorella gained much attention because of its diverse beneficial health activities, like immunomodulation, antioxidation, anti-hyperlipidemia, antitumor, neuroprotection, and anti-asthmatic effect.

Chlorella contains several bioactive compounds (Table 11.2) which depicts antitumor, anticoagulant, antibacterial, antioxidant, antihyperlipidemic property along with liver protecting and immunostimulatory effects of enzymatic protein hydrolysate (Plaza et al., 2009; Cha et al., 2010; Kokou et al., 2012; Li et al., 2013; Medina-Jaritz et al., 2013). Various antioxidant compounds identified from Chlorella such as lutein, α-carotene, β-carotene, ascorbic acid and α-tocopherol are important as natural colorants and additives along with having beneficialrole in reducing cancer and treating muscular degeneration (Plaza et al., 2009; Zhao & Sweet, 2008). β-1,3 glucan, an important bioactive compounds of Chlorella acts as an immune stimulator by reducing free radicals and blood cholesterol levels. It also prevents atherosclerosis and hypercholesterolemia as well as tumor formation (Beijerinck, 1890). Under stressful condition it also synthesizes TAG.

Taipei (Taiwan chlorella manufacturing co.) produces a huge amount of Chlorella that is about 4,00,000 tons of biomass annually (Rosch & Posten, 2012). Dr. Martinus Willem Beijerinck in 1890 discovered *Chlorella vulgaris* for the first time (Beijerinck, 1890). Later in the year 1903 another species *Chlorella pyrenoidosa* was identified (Chick H, 1903). Till date more than 20 species of Chlorella with over 100 strains have been characterized (Wu et al., 2001). Presently three verities of Chlorella are found, *including Chlorella vulgaris, Chlorella lobophora and Chlorella sorokiniana* (Krienitz et al., 2004). *Chlorella sorokiniana*, a thermotolerant mutant of *Chlorella pyrenoidosa*, was first identified by Sorokin in 1953 (Sorokin & Myers, 1953; Lizzul et al., 2018).

TABLE 11.2 Essential Bioactive Compounds Isolated from Different Spirulina and Chlorella Species

Microalga	Bioactive Compounds	References
Spirulina sp.	Polysaccharides	Amaro et al. (2013)
Spirulina platensis	Phycocyanin, C-phycocyanin, phenolic acids, tocopherols (vitamin E), neophytadiene, phytol, PUFAs (*n*-3) fatty acids, oleic acid, linolenic acid, palmitoleic acid.	Iba˜nez & Cifuentes (2013); Singh & Dhar (2011); Ayehunie et al. (1996)

TABLE 11.2 *(Continued)*

Microalga	Bioactive Compounds	References
Spirulina fusiformis	Diacylglycerols.	Singh & Dhar (2011)
Chlorella sp.	Carotenoids, sulfated polysaccharides, sterols, PUFAs (*n*-3) fatty acids.	Iba˜nez & Cifuentes (2013)
Chlorella vulgaris	Canthaxanthin, astaxanthin, peptide, oleic acid.	Priyadarshani & Rath (2012); Ayehunie et al. (1996); Ibañez & Cifuentes (2013)
Chlorella minutissima	Eicosapentaenoic acid (EPA).	Singh & Dhar (2011)
Chlorella ellipsoidea	Zeaxanthin, violaxanthin.	Amaro et al. (2013)
Chlorella pyrenoidosa	Lutein, sulfated polysaccharide.	Ayehunie et al. (1996)

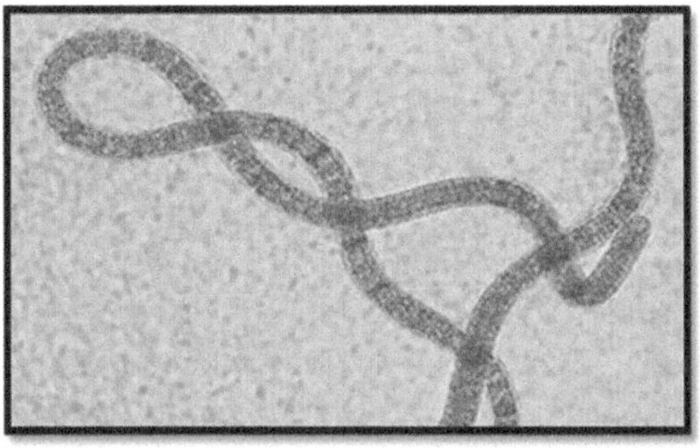

FIGURE 11.1(A) Spirulina.

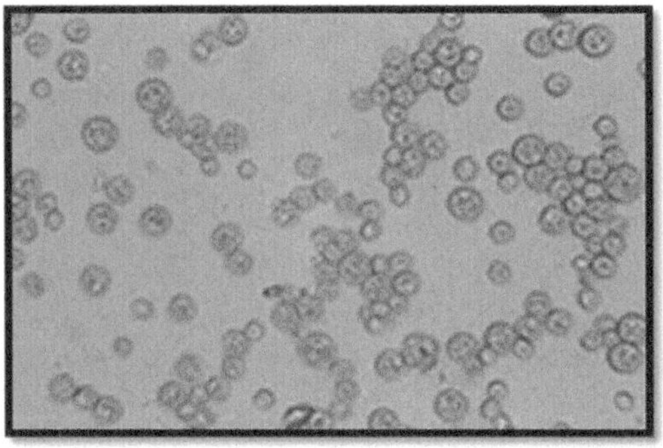

FIGURE 11.1(B) Chlorella.

11.4 COMPOUNDS WITH IMMUNOMODULATORY ACTIVITY IN MICROALGAE

11.4.1 SULFATE POLYSACCHARIDES (PS)

Edible microalgae are very good polysaccharides (PS) producers, which include exocellular polysaccharides (EPS) as well as sulfated polysaccharides (sPS). PS are antiviral agents, health foods, antioxidants, anti-inflammatory, and immunomodulatory compounds, and can also be employed as bone joint lubricants and ships drag-reducing component (Raposo et al., 2013). Sulfated PS from marine sources exhibit immunomodulatory activities (Chen et al., 2019; Liu et al., 2019; Muhamad et al., 2019; Sun et al., 2019; Vishwakarma & Vavilala, 2019; Abdala et al., 2019; Bahramzadeh et al., 2019 and Cui et al., 2019) with potential use in drug development. Sulfated PS are present in various macroalgae and microalgae found insea-water as well as freshwater environment (Cao et al., 2019).

11.4.2 SULFOLIPIDS

In plant and algal chloroplasts Sulfolipids are components of the thylakoid (Frentzen, 2016) and also form the anionic part of the mono- and digalactosyl-diacylglycerols (Kobayashi, 2016). Sulfoquinovose, are the constructing block of sulfolipids, and additionallyanimportant part of the sulfur cycle (Roy et al., 2003). The amount of Sulfolipids produced by photosynthetic organisms is 1,010 tons annually (Speciale et al., 2016). Sulfolipids elicits diverse capabilities to cure and avert human pathological conditions. As an example, it was elucidated that they are a strong inhibitor of glutaminyl cyclase thereby acting against Alzheimer's disease (Michael et al., 2016). Additionallysulfolipids can activate the immune system, and can be used as vaccine adjuvants (Manzo et al., 2017). Manzo et al. generated a synthetic sulfoglycolipid (called β-SQDG18) using microalgae-derived sulfolipids as an adjuvant to stimulate effective activation of DC and enhance immune reaction towards cancer cell (i.e., B16 mouse melanoma cells).β-SQDG18, elevated the proportion of CD83-positive DCs, augmented the release of pro-inflammatory cytokines IL-12 and INF-γ and boosted theexpression of IL-1α, IL-1β, IL-18, and IL-27 on an exposure of 24 hours. Synthetic sulfolipids are capable of lowering tumors growth and instigate multiplication of both lymphocytes and APCs memory, proving to be a better adjuvant in cancer vaccination. This syntheticsulfolipids was patented as Sulfavant (EP3007725A1, WO2014199297A1) (Manzo et al., 2019).

11.4.3 POLYUNSATURATED FATTY ACIDS (PUFAS)

PUFAs are fatty acids with multiple double bonds in their carbon chain. The omega-6 (ω-6) and omega-3 (ω-3) are the two promising series of PUFA. The two different classes were based on the positioning of the first unsaturation on carbon backbone from the methyl, omega-C. Microalgae can produce both omega 6 and omega 3 fatty acids and different microalgal species are marked for their PUFA production. Microalgae are taken into consideration as treasured options of sustainable and environment friendly PUFA producers (Khozin et al., 2016). The omega 6 fatty acids comprise of γ-linoleic acid (GLA) and arachidonic acid (ARA), while the omega 3 fatty acid encompass eicosapentaenoicacid (EPA) and docosahexaenoic acid (DHA) (Cui et al., 2019). Biological functions associated to DHA and EPA, include range of beneficial cardiovascular effects (Adkins & Kelley, 2010), antimicrobial activities (Le & Desbois, 2017), anti-inflammatory modulation (Schulte et al., 2019), antioxidant properties (Singh et al., 2010), beneficial effects on respiratory diseases (Garcia et al., 2015; Miyata & Arita, 2015), antitumoral activity (Nappo et al., 2012), and inflammation mitigation in adipose tissues (Yamada et al., 2017). Han et al. (2017) showed that EPA elevate the expansion of macrophage, stimulate the nitric oxide (NO) and cytokines release such as IL-1β, IL-6, TNF-α, and INF-γ by activating GPR120-mediated Raf kinases mitogen-activated proteinkinases p44/p42 (ERK1/2)-inhibitor of nuclear factor kappa-B kinase subunit β (IKKβ)-nuclear factorkappa B-p65 (NF-kB) signaling pathways.

11.4.4 ASTAXANTHIN (ASX)

Astaxanthin (ASX) is a carotenoid pigment, present in various freshwater and marine microorganisms like bacteria, yeast, fungi, and microalgae, and also in aquatic organisms that sustain on them. It possessantioxidant (Speranza et al., 2012), anti-inflammatory (Galasso et al., 2017), anti-obesity (Yoshida et al., 2010), cardiovascular (Hussein et al., 2005) and anti-proliferativeactivities (Martinez et al., 2018) In the last few years the use of ASX has elevated as a food supplement for human health. The market capacity of ASX was around US $100 million in 2018 (Brendler & Williamson, 2019).

It has become an essential compound in the world market as its safe utilization has been discussed in detail (Novoveská et al., 2019). The astaxanthin derived from the microalga *Haematococcuspluvialis* is approved by the Food and Drug Administration (FDA) and is commercially available. It has enormous utilization in optic and visual health (Brendler & Williamson, 2019), dermatological health and body exercise performance (Davinelli et al., 2018) and recovery (Brendler & Williamson, 2019). It has a vital role in curing Parkinson disease (PD) (Grimmig et al., 2016) and cancer (Martinez et al., 2018). ASX is also reported to amplify immune response in several studies. ASX enhance the cytotoxic activity of the NK cell, as well as increase the overall T- and B-cell populations. ASX is shown to rapidly elevate the lymphocyte expansion in *in-vitro* mice model along with elevation in the amount of IL-2 and INF in *ex vivo* mice model (Lin et al., 2015).

11.5 IMMUNOMODULATION IN SPIRULINA

Aqueous extract of Spirulina shows a significant impact on the immune system stimulating the NK cells, thereby elevating the phagocytic activity of the macrophages. It stimulates the formation of cytokines and antibodies (Schwartz & Shklar, 1987) and has an essential role in the activation and mobilization of B- and T-lymphocytes. This extract suppresses the expression of CD28 and co-stimulation of human T-cells it generates immunosuppressive response similar to that of the drug cyclosporine (Ismail et al., 2009). It is seen that a photosynthetic pigment phycocyanin extracted from Spirulina inhibits the liberation of histamines from mast cells in an allergic response. It repressed tumor cells growth, stimulatedactivity of NK cells and also caused lymphocytes present in the spleen to produce TNF-α (Subhashini et al., 2004). It is indicated by the age-related changes analysis using flow cytometry that Spirulina conserves the function of the intestinal epithelium that provides the first line of defense against infections in mucosa (Sheahan et al., 2008). Feeding Spirulina to mice showed these age-associated differences in the intestinal intraepithelial lymphocytes. Cells having CD45+ act as amarker for IELs. The amount of CD45+CD8+ cells were greater in older mice and smaller in adult mice with CD4+CD8+ cells greater in older mice (Sheahan et al., 2008).

A study carried with a different strain of Spirulina (*Spirulina fusiformis*) showed dose-dependent suppression of humoral immunity, cell mediated immunity (delayed type hypersensitivity) and TNF-α in mice. Spirulina also causes a reduction in the mitogen induced T-lymphocyte expansion which depicts its immunosuppressive potential (Raffaella et al., 2009). It elicits a powerful IL-1β, IL-responses causing short term amplification of the adaptive immune response and also stimulates Th-1 type responsewhich potentially activates cell-mediated immunity (Khan et al., 2005).

The immunostimulatory effects of Spirulina was investigated by the morphologic and morphometric changes in the rat's spleen by performing NBT test of peritoneal macrophages (Khan et al., 2005; Zhang et al., 2001). Spirulina activates the cells of the innate immune system and hence can stimulate the mucosal and systemic immune system components. It was noted by performing cytokine assay test that Spirulina increases the expression of genes encoding the chemokines interleukin (IL-8), MCP-1, MIP-1α, MIP-1β, IP-10; the cytokines TNF-α, IL-1β and the enzyme cyclo-oxygenase-2 (COX-2). The expression of cytokine mRNA was evaluated by RT-PCR and the THP-1 activation was confirmed (Hayashi et al., 1996; Ayehunie et al., 1998 and Hayashi et al., 1996).

Abiliprotein of Spirulina, phycocyanin reduces allergic inflammation by suppressing the IgE antibody and the immune system activity of mucosa against infectious illness is maintained in C3H/HeN and BALB/cA mice (Simpore et al., 2005). Spirulina also protects against hay fever (Mittal et al., 1999). It is also shown that Spirulina reduces inflammation in arthritis patients by stimulating the secretion of IL-2 that regulates inflammatory response (Hosoyamada et al., 1991).

Spirulina has an immunostimulatory effects on TNF-α which plays a crucial part in evoking an autoimmune disease in people who are genetically susceptible. It is shown that Spirulina has a direct effect on myeloid lineages by exposing blood cellstreated withBCG cell wall skeleton to Spirulina. The skeleton of BCG cell wall act as a ligand for TLR-2 and 4 to promote maturation of monocyte or macrophage. Spirulina is beneficial in oral administration as it participates in the Toll mediated signaling in blood cells. NK cells important for NK-mediated IFN-γ production are indirectly affected by Spirulina (Juarez et al., 2009 and Khan et al., 2005). NK cells show a two-fold increase in activity by including Spirulina as dietary supplementation at 10,000 ppm in Cornell k-strain White leghorns and broiler chicks (Khan et al., 2005).

Animal studies have revealed that Spirulina stimulates the immune system through several ways which includes increasing the phagocytosis by macrophages, that cause NK cells accumulation in tissues, enhancing the formation of antibodies and cytokines by the exhilaration and deployment of T-and B-cells (Qureshi & Ali, 1996; Mao et al., 2000; Khan et al., 2005). Spirulina has remarkable immunomodulatory effects. According to Mao et al. (2000), spirulina produced interleukin IL-1, IL-4, and IFN-γ by 2.0, 3.3, and 13.6 times more than baseline levels, respectively. Treating PBMCs with phycocyanin extracted from *Spirulina platensis* caused a substantial elevation in IFN-γ synthesis and reduction in IL-4 production (Sheikh et al., 2014). A study conducted on a mouse affected with systemic candidiasis and breast cancer it is shown that *Spirulina platensis* depicts tumoricidal activity by raising the formation of pro-inflammatory TNF-α and IFN-γ (Shokri et al., 2014).

The immunomodulatory activity of purified phycocyanin from *Spirulina platensis* was examined on murine cell and macrophage cell line J 774A that revealed increased release of TNF-α, IL-1β and IL-6 from J77 4A cells in addition to IL-1β and COX-2 protein expression. *Spirulina platensis* and phycocyanin inhibit the inflammation associated genes expression by lipopolysaccharide (LPS) treated BV-2 microglial cells. Preliminary studies show that in BV-2 microglial cells LPS upregulated the genes for inducible nitric oxide synthase (iNOS), COX-2, TNF-α, and IL-6 (Chen et al., 2012).

Moreover, Spirulina or phycocyanin exposure significantly decreased the expression of the above-mentioned genes. Sin et al., 2014 and Ali et al., 2015 reported that *Spirulina maxima*inhibit TNF-α and IL-6 genes in human cell line CCD-986 sk) (Sin et al., 2014; Ali et al., 2015). It is shown that in both male and female subjects Spirulina supplement increases IL-2 concentration in plasma and decreases IL-6 concentration (Park et al., 2008). *In vivo* and *in vitro* studies showed that *Spirulina fusiformis* has marked immunosuppressive effects (Rasool & Sabina, 2009). Significant suppression of immune reactions, delayed type hypersensitivity response and TNF-α formation was shown in *Spirulina fusiformis* treated mice.

Mitogen-induced proliferation of T-lymphocyte was reduced by *Spirulina fusiformis, in-vitro*. Hence it may serve as an alternative treatment to arthritis where uncontrolled expansion of T-cells take place (Hwang et al., 2013). Food supplement with Spirulina or phycocyanin downregulates the expression of genes related to inflammation and immune system. Spirulina helps in improving anemia and immunosenescence in older people (Selmi et al., 2011). Spirulina supplement contains immunoreactive insulin, it can potentially benefit individuals with diabetes mellitus (Anwer et al., 2012). Spirulina and its bioactive component phycocyanin display immunomodulation by stimulating antibodies formation and up or down regulating genes for IL-1β, IL-2, IL-4, IL-6, IL-10, IFN-γ, IL-17, and TNF-α (Table 11.3).

TABLE 11.3 Potential of Spirulina in Innate and Adaptive Immune Response

Type of Immune Reaction	Mode of Action	References
Innate immune reaction	Activation of chemokines IL-8, IL-1 β, IL-2, TNF-α, NF-kappa B, MCP-1, MIP-1α, MIP-1β, IP-10, enzyme cyclo-oxygenase-2 and THP-1 genes. Enhancement in the antigen-specific and total IgA antibody level in the Peyer's patches, mesenteric lymph nodes and intestinal mucosa also in the spleen cells in mice. Decreasing serum IgG1 and IgE antibody level alleviating allergic response. Toll mediated signaling in blood cells, triggering macrophages, natural killer cells (NK), and neutrophils. NK-mediated IFN production by cooperative presence of IL-12 and IL-18. *In-vitro* induction of caspase dependent apoptosis in HeLa cells; up-regulating pro-apoptotic gene and deactivation of anti-apoptotic gene, help in the tumoral apoptosis signal transduction; caspases 2, 3, 4, 6, 8, 9, and 10 activation.	Zhang et al. (2001); Hayashi et al. (1996); Ayehunie et al. (1998); Mittal et al. (1999); Juarez et al. (2009); Lee & Werth (2004)
Innate immune/ hypersensitive reaction	Modulating the differentiation of Th2 cells mediated by inhibiting the production of IL-4 in patients with allergic rhinitis. IgE antibody suppression. Decrease the level of LTB4 and prostaglandin E2 in the ear inflammation test of mouse induced by arachidonic acid.	Park et al. (2008); Hirahashi et al. (2002); Mao et al. (2000)
Adaptive immune response	Potentiating Th-1 type and cell-mediated response; antigen presenting cells. Regulating T-cell subpopulation in adult and old mice. Changing leukocyte subset expansion and cytokine release response against *Candida albicans* (CA) and tetanus toxoid (TT), *in vitro*. Efficient antiviral properties against herpes simplex virus type 1, cytomegalovirus, influenza virus and human immunodeficiency virus type-1. Enhanced primary humoral immune reaction against sheep's RBCs. In peritoneal macrophages raising the amount of phagocytic cells. Increased mitogen-induced lymphocytes expansion.	Schwartz & Shklar (1987); Sheahan et al. (2008); Raffaella et al. (2009); Khan et al. (2005); Lee & Werth (2004); Shklar & Schwartz (1988); Hayashi et al. (1994)

Monocyte and macrophages are the phagocytic cells in vertebrates that are involved in innate as well as adaptive immunity. They cause engulfment and digestion of pathogen and invigorate lymphocytes and other cells of the immune system. Nitric oxide (NO) and cytokines are the mediators secreted by macrophages in response to the antigen. Cytokines play a vital role in the inflammatory response and include TNF, IL-1, IL-8 (pro-inflammatory cytokine) and IL-10 (anti-inflammatory cytokine) (Batshan et al., 2001). The pro-inflammatory cytokines work against exogenously derived pathogens, whereas anti-inflammatory cytokines decrease inflammatory response and maintain homeostasis. Spirulina amplify the function of macrophages and increases the production of IL-1, but the level of IgG antibody remained unchanged (Raja et al., 2008). IL-1 activates the inflammatory response, increases the production of acute-phase proteins that are vital in the neurological immune response, and has a direct effect on the brain (Mosmann & Coffman, 1989). Spirulina supplemented chickens had increased level of macrophages with increased phagocytic activity and NO synthase activity (Batshan et al., 2001). The same effect of increased phagocytic activity of macrophage was also seen in cats (Qureshi & Ali, 1996) and dogs (Zhang et al., 2001) as well as humans (Hirahashi et al., 2002; Hayashi et al., 2004) fed with *S. platensis* extract. Researchers have shown that the immunomodulatory activity is mediated via the innate immune system using Spirulina either as powder or hot water extract with phycocyanin and water-soluble PS as active components. They cause immunomodulatory effect by increasing the expansion of RBCs, granulocytes, monocytes, and the cells of the fibroblast lineage originated from mice bone marrow (Hayashi et al., 2004).

Immulina, a polysaccharide fraction from Spirulina having high molecular weight stimulates NF-κB and induce IL-1β and TNF-α mRNA expression in THP-1 human monocytes (Pugh & Pasco, 2011). The expression of chemokines like interleukin (IL-8), MCP-1, MIP-1α, MIP-1β, IL-10 and also TNF-α, IL-1β and COX-2 are elevated by its use. Antibodies to CD14 and TLR2 suppress the activation of NF-κB by Immulina but not by antibodies against TLR4.

Administration of *Spirulina fusiformis* to a mice in a dose dependent manner causes a decrease in humoral immune reaction, cell mediated immune reaction, delayed type hypersensitivity response (DTH) and TNF-α. It also suppresses the expansion of T-cell under the influence of a mitogen. This depicted that it has a potential immunosuppressive effect and hence validates its use as a drug (Rasool & Sabina, 2009). Dose dependent treatment of *S. fusiformis* causes a reduction in the level of hemagglutinating antibodies in the serum. It also decreases the release of lymphokines that promote inflammatory reaction. Overproduction of inflammatory mediators has damaging effects, but moderate levels are necessary for the survival of the host. Hence their synthesis is regulated tightly. Reactive oxygen species (ROS), pro-inflammatory cytokines, matrix metalloproteins as well as NO production are induced by TNF-α and thereby it can be used in treating pain, swelling, destruction of joints caused by rheumatoid arthritis (Beutler, 1995). TNF-α is drastically lowered by *Spirulina fusiformis*in arthritis instigated by adjuvantsin mice (Haworth et al., 1991). The spleen cell of mice fed with Spirulina extract produce increased level of IFN-γ which shows a drag towards Th1 type cell mediated immune response. Also, it is seen that regularly consuming extract of Spirulina in hot water for two months by humans causes increased liberation of IFN-γ by NK cells as an IL-12 and IL-18 response (Hirahashi et al., 2002). Spirulina also show immunomodulatory activity by stimulating the secretion of IgA at the mucosal surface, responsible for protection against bacteria and viruses (Kilian & Russell, 1994). It has been observed that treatingwith Spirulina or its fraction for a longer duration increased IgA response. A mice fed with an Immulina mixed extract depicts alteration in different parameters of immune system. The Peyer's patch cell showed a two-fold increase in *ex vivo* formation of IgA and IL-6 as well as IFN-γ production increased to four folds in Immulina treated mice. *Spirulina platensis* consumption by men for a period of more than a year increases the level of S-IgA in saliva. A study showed that a similar type of cell is targeted initially by this extract which mediates local as well as systemic response. Hence it is evident that Spirulina has a profound role in mucosal immunity by increasing the secretion of IgA in human saliva (Ishii et al., 1999). It has been well studied that Spirulina suppresses the histamine liberation from mast cells and thus exhibit anti-inflammatory response (Yang, Lee, and Kim, 1997; Kim et al., 1998). People suffering from allergic rhinitis were supplemented with Spirulina for three months after which the peripheral blood mononuclear cell was withdrawn for checking cytokines IL-2, IFN, and IL-4 levels which regulate IgE-mediated allergic response. It was observed that the level of IL-4 decreased by 32% by Spirulina treatment. A Japanese team showed that the extract of *S. platensis* in hot water increased the formation of IFN-γ and injury of NK cell when given orally in male volunteers (Hirahashi et al., 2002). It is also suggested that Spirulina can also play a role in changing immunity by overcoming nutritional deficiencies.

Phycocyanin, a major constituent of *S. platensis* is a water-soluble protein pigment associated on the outer surface of light harvesting complex in PSII. It has several medicinal as well as pharmaceutical properties such as antioxidant (Romay & Gonzalez, 2000), antitumor and immune boosters (Li et al., 2004). Recently it has been shown that PC induces apoptosis and thereby regress the growth of tumor cells (Reddy et al., 2003). It is already known that Spirulina platensis acts as a good resource of Selenium enrichment (Chen et al., 2006; Chen et al., 2006). The effect of selenium enriched PC (Se-PC) over the severity of anaphylactic reaction and response of circulating antibodies on a model allergen was studied in rats which depicted increased specific IgG response in Se-PC treated mice (Egorova et al., 2006). PC increases the quantities of CD59 expression and apoptotic Fas proteins as well as decrease the replication of HeLa cells (Li et al., 2005). PC possess antitumor activity and thus by combining with the receptors of mitosis depressors present on the cancerous cells they act as blocker of mitosis. Due to this linking and stimulation of protein kinasespromotingcell death the CD59 gene transcription is promoted. Excess CD59 expressionactivates Fas protein present on the HeLa cells which integrate with Fas antigen and in turn activate death domain that conducts proapoptotic signal in tumor cells thereby causing their death. It is suggested that daily consuming little dose of PC may avoid cancer development or its reoccurrence (Boehm et al., 1997). In another study, it is also shown that PC elevates lymphocyte activity (Balachandran et al., 2006) and increases the general immunity of one's body.

11.6 IMMUNOMODULATION IN CHLORELLA

Aqueous soluble polysaccharide from Chlorella show strong immunomodulatory activities in mice, human blood cells and human clinical trials (Qi & Kim, 2017; Tabarsa et al., 2015; Qi & Kim, 2018 and Suárez et al., 2005). TNF-α along with IL-1β and IL-6 produced by activated monocytes and macrophages that stimulates cell proliferation in various types of cells are important initiator of the inflammatory response and mediator (Sibi & Rabina, 2016). Aqueous solution of *Chlorella vulgaris* influence production of TNF, IL-6 and IL-10. Increased levels of TNF-α produces anti-inflammatory IL-6 that hampers TNF-α and help maintain the Th1/Th2 balance (An et al., 2008). The mononuclear cells produce high amounts of these cytokines on stimulation with the lowest concentration of Chlorella. Sibi andRabina described the suppression of cytokines and mediators of inflammatory response by the aqueous extracts of *Chlorella vulgaris (*An et al., 2008).

Chlorella vulgaris supplemented diet improve lymphocyte proliferation and phagocytosis by macrophages, stimulates cytokine expression, increases NK cells mediated cytotoxicity and reduces changes in the spleen's histology. Hence *Chlorella vulgaris* supplementation can effectively cure immunological disorders and may prove advantageous for patients undergoing immunosuppressive therapies.

Researches have shown that *Chlorella vulgaris* and its extracts help in improving the immune functions (Merchant et al., 2000 and An et al., 2008) by regulating lipid metabolism or tumors (Shibata et al., 2001). Oral administration of extract from *Chlorella vulgaris* elevates the IL-2 mRNA level in the spleen and macrophages of mice (Hasegawa et al., 2000). Hasegawa et al. (1999) described antigen specific response of *Chlorella vulgaris*hot water extract on mice which exhibited decreased IgE release and IL-6 mRNA expression participating in type 2 helper T-cell response along with elevated level of IL-12 and g-interferon mRNA, raising type 1 helper T-cell response and decreasing the type 2 helper T-cell response. Thus, it shows that hot-water Chlorella extract supplement proves to be beneficial in overcoming allergic reactions involving type 2 helper T-cell response. Polysaccharide of Chlorella soluble in hot water can be applicable as an agent to stimulate anti-microbial activity. In humans four week supplementation of Chlorella increases the secretion of salivary SIgA thereby increasing the mucosal immune function that performs a critical role as the first line of defense against pathogenic microbial invasion (Lamm et al., 1995). It is also suggested in a study that a short-term supplementation of Chlorella increases activity of NK cell and facilitates production of IFN-γ, interleukin-12, and interleukin-1β (IL-1β) (Abel et al., 2018 and Kwak et al., 2012).

Immunomodulatory properties of polysaccharide from Chlorella is broadly studied under *in vitro* and *in vivo* conditions. Yang et al. (2006) showed in a study that polysaccharide from *C. pyrenoidosa* remarkably invigorate the index of phagocytosis, activate T-cell via delayed-type hypersensitivity, and raise the level of antibody titer. Another study revealed that the raw polysaccharide and its fractions (F1, F2, and F3) extracted from *C. vulgaris* show immunomodulatory activities thereby stimulating the multiplication of macrophages. NO, prostaglandin E2, and pro-inflammatory cytokine release were elevated, as were the expressions of related mRNAs. Purified fractions outperform crude polysaccharide in terms of immunomodulation. The majority of investigations on Chlorella polysaccharide immunomodulating activity focused on immunostimulatory effects, but there were few outliers. Guzmán et al. (Guzmán et al., 2003) discovered that the crude sulfated PS isolated from *C. stigmatophora* have immunosuppressive activity, loweringthe phagocytic index *in vitro* and *in vivo* which may be due to the negative charge of the sulfate group. Primarily the identification of a particular receptors and the various pathways of signal transduction mediate macrophage activation by polysaccharide.

The *C. pyrenoidosa* water-soluble polysaccharide (CWSP) can bind to TLR4 and cause macrophage cells to produce interleukin (IL)-1β. This involves the addition of a phosphate group on JNK/p38. It can also positively regulate the expression of surface HLA molecules (HLA-DA, -DB, -DC, etc.), CD80 and CD86 on macrophages. In addition, CWSPcan assist in the formation of tumor necrosis factor (TNF)-α and IL-1β in J774A.1 macrophages and C57BL/6J mice (Hsu et al., 2010). Qi & Kim (Qi & Kim, 2018) demonstrated that polysaccharide of *C. ellipsoide* activates nuclear factor-kappa-B (NF-κB) and mitogen-activated protein kinase (MAPK) signaling pathways in RAW 264.7 cells, resulting in increased iNOS and mRNA expression of cytokines (IL-1β, IL-6, IL-10, and IL-12). PS immunomodulatory actions are regulated by structural characteristics such as molecular weight, monosaccharide compositions, glycosidic linkages, and functional groups such as sulfate and acetyl groups. According to other research, the immunomodulatory action of Chlorella PS is proportional to their molecular weight. The structural relation of Chlorella polysaccharide with immunomodulatoryactivity needs to be further investigated (Yuan et al., 2020; Ferreira et al., 2015; Xu et al., 2014; Zhang et al., 2016 and Perera et al., 2018).

Chlorella is considered as a powerful alga against cancer. For the study, laboratory mice were fed chlorella for 10 days before being implanted with three distinct cancers. It was reported by Moss that about 70% of the mice supplemented with chlorella were free from cancer whereas the untreated mice passed away of it in about 20 days. In another study patients suffering from glioblastoma were treated with powdered and liquid chlorella, accompanying chemotherapy and radiation therapy. Though patients of glioblastoma show 10% survival rate but Chlorella treatment increased it to 40%. This study proved that Chlorella can play a major role in effectively curing cancer. It functions as a beneficial nutritional and detoxifying food.

To analyze the effect of orallyfed Chlorella protein hydrolysate (Cv-PH) on innate and adaptive immune response in undernourished mice, an experimental study was conducted (Morris et al., 2007). It was depicted that the

Cv-PH treated starved mice showed increased: (a) hemopoiesis; (b) activation of macrophage and its phagocytic activity; and (c) enhancement of humoral and cell mediated immunity. Several physiological roles played by chlorella and its hydrophilic extracts include improving immune system, hypoglycemic effects, lowering higher lipid state in fat fed animals, etc. Moreover lipid soluble extract of chlorella (LEC) has less physiological role. LEC blocks NO synthesis induced by LPS by blocking the expression of iNOS mRNA. Chlorella powder shows dose dependent inhibition of peptidase cathepsin S, thromboxane A(2) synthase, and cyclooxygenase-2. It also inhibits TNF-α converting enzyme, protein tyrosine phosphatase (SHP-2), calpain, protein kinases, and protein tyrosine phosphatases (Cheng et al., 2008). Chlorella contributes to the anti-inflammatory and antithrombotic effects by inhibiting the action of cyclooxygenase-2 and thromboxane synthase.

11.7 CONCLUSION

Microalgae being a rich source of bioactive compounds have been proposed to be used as a food supplement. Recently the immunomodulatory effects of different active extracts, fractions, and pure compounds from edible microalgae are reported. In order to recognize the optimum active concentration, its mode of action and potential use further analysis needs to be performed. Spirulina has different immunomodulating and anti-inflammatory activities. It triggers the antibodies formation and regulate the cytokines gene expression. The immunomodulatory, antioxidativeand anti-inflammatory properties are interlinked in Spirulina. It inhibits inflammatory responses via antioxidant and anti-inflammatory mechanism. Phycocyanin and β-carotene of Spirulina are essential for its antioxidant, immunomodulatory, and anti-inflammatory response. Chlorella supplemented diet for humans and mammals are related to antioxidant, antidiabetic, immunomodulatory, antihypertensive, and antihyperlipidemic reactions. Its PS can be developed into functional foods and as potential drugs.

The discovery of huge amount of new compounds from edible microalgae can serve in the development of new potential drugs for immunotherapy. Adequate characterization of all immunomodulatory compounds from different species of edible microalgae have not yet been completed, and hence they can serve as a potential source for new immunomodulatory molecules.

KEYWORDS

- **anti-inflammatory**
- **antimicrobial**
- **bioactive compounds**
- **chlorella**
- **immunomodulatory**
- **microalgae**

REFERENCES

Abdala, D. R. T., Casas, A. V., Arrojo, A. M. A., Cardenas, C., Dobretsov, S., & Figueroa, F. L., (2019). Immunomodulatory and antioxidant activities of sulfated polysaccharides from *Laminaria ochroleuca, Porphyra umbilicalis*, and *Gelidium corneum. Mar. Biotechnol., 21*, 577–587.

Abel, A. M., Yang, C., Thakar, M. S., & Malarkannan, S., (2018). Natural killer cells: Development, maturation, and clinical utilization. *Front. Immunol., 9*, 1869.

Adkins, Y., & Kelley, D. S., (2010). Mechanisms underlying the cardioprotective effects of omega-3 polyunsaturated fatty acids. *J. Nutr. Biochem., 21*, 781–792.

Al-Batshan, H. A., Al-Mufarrej, S. I., Al-Homaidan, A. A., & Qureshi, M. A., (2001). Enhancement of chicken macrophage phagocytic function and nitrite production by dietary *Spirulina platensis. Immunopharmacol. Immunotoxicol., 23*, 281–289.

Albracht-Schulte, K., Gonzalez, S., Jackson, A., Wilson, S., Ramalingam, L., Kalupahana, N. S., & Moustaid-Moussa, N., (2019). Eicosapentaenoic acid improves hepatic metabolism and reduces inflammation in dependent of obesity in high-fat-fed mice and in HepG2 cells. *Nutrients, 11*, 599.

Ali, E. A., Barakat, B. M., & Hassan, R., (2015). Antioxidant and angiostatic effect of *Spirulina platensis* suspension in complete Freund's adjuvant-induced arthritis in rats. *PLoS One, 10*(4), e0121523.

Amaro, H. M., Barros, R., Guedes, A. C., Sousa-Pinto, I., & Malcata, F. X., (2013). Microalgal compounds modulate carcinogenesis in the gastrointestinal tract. *Trends in Biotechnology, 31*(2), 92–98.

Ambrosi, M. A., Reinehr, C. O., Bertolin, T. E., et al., (2008). Propriedadesde sa´ude de *Spirulina* spp. *Revistade Ci^enciasFarmac^euticasB´asica e Aplicada, 29*(2), 109–117.

An, H. J., Rim, H. K., Lee, J. H., Seo, M. J., Hong, J. W., Kim, N. H., & Kim, H. M., (2008). Effect of *Chlorella vulgaris* on immune-enhancement and cytokine production *in vivo* and *in vitro. Food Sci. Biotechnol., 17*, 953–958.

Anwer, R., Khursheed, S., & Fatma, T., (2012). Detection of immunoactive insulin in *Spirulina. J. Appl. Phycol., 24*, 583–591.

Ayehunie, S., Belay, A., Baba, T. W., & Ruprecht, R. M., (1998). Inhibition of HIV-1 *Spirulina platensis* (*Arthrospira platensis*). JAIDS: *J Acquir Immune Def Synd. Human Retrovirol., 18*(11), 7–12.

Ayehunie, S., Belay, A., Hu, Y., Baba, T., & Ruprecht, R., (1996). *Inhibition of HIV-1 Replication by an Aqueous Extract of Spirulina Platensis (Arthrospira Platensis).* 7th IAAA Conference, Knysna, South Africa.

Bahramzadeh, S., Tabarsa, M., You, S., Li, C., & Bita, S., (2019). Purification, structural analysis a mechanism of murine macrophage cell activation by sulfated polysaccharides from *Cystoseira indica*. *Carbohydr. Polym., 205*, 261–270.

Balachandran, P., Nirmal, D., Pugh, G. M., & David, S., (2006). Toll-like receptor 2-dependent activation of monocytes by *Spirulina* polysaccharide and its immune enhancing action in mice. *Int. Immunopharmacol., 6*, 1808–1814.

Beijerinck, M. W., (1890). Culturversuche mit zoochlorellen, lichen-engonidien und anderen niederen algen. *Botanische Zeitung, 47*, 725–739.

Ben-Aharon, O., Magnezi, R., Leshno, M., & Goldstein, D. A., (2018). Association of immunotherapy wit durable survival as defined by value frameworks for cancer care. *JAMA Oncol., 4*, 326 332.

Beutler, B., (1995). TNF, immunity and inflammatory disease lesions of the past decade. J. *Investig. Med., 43*, 227–235.

Boehm, U., Klamp, T., Groot, M., & Howard, J. C., (1997). Cellular responses to interferon-gamma. *Ann. Rev. Immunol., 15*, 749–795.

Borges, J. A., Rosa, G. M., Meza, L. H. R., Henrard, A. A., Souza, M. R. A. Z., & Costa, J. A. V., (2013). *Spirulina* sp. LEB-18 culture using effluent from the anaerobic digestion. *Brazilian Journal of Chemical Engineering, 30*(2), 277–287.

Brendler, T., & Williamson, E. M., (2019). Astaxanthin: How much is too much? A safety reviews. *Phytother. Res., 33*, 3090–3111.

Cao, S., He, X., Qin, L., He, M., Yang, Y., Liu, Z., & Mao, W., (2019). Anticoagulant and antithrombotic properties *in vitro* and *in vivo* of a novel sulfated polysaccharide from marine green alga *Monostroma nitidum*. *Mar. Drugs, 17*, 247.

Cha, K. H., Kang, S. W., Kim, C. Y., Um, B. H., Na, Y. R., & Pan, C. H., (2010). Effect of pressurized liquids on extraction of antioxidants from chlorella vulgaris. *Journal of Agricultural and Food Chemistry, 58*(8), 4756–4761.

Chan, A. H., & Schroder, K., (2019). Inflammasome signaling and regulation of interleukin-1 family cytokines. *J. Exp. Med., 217*, e20190314.

Chen, J. C., Liu, K. S., Yang, T. J., et al., (2012). *Spirulina* and C-phycocyanin reduce cytotoxicity and inflammation-related genes expression of microglial cells. *Nutr. Neurosci., 15*(6), 252–256.

Chen, T., Zheng, W., Wong, Y. S., Yan, F., & Bai, Y., (2006). Accumulation of selenium in mixotrophic culture of *Spirulina platensis* on glucose. *Bioresour. Technol., 97*, 2260–2265.

Chen, T., Zheng, W., Yang, F., Bai, Y., & Wong, Y. S., (2006). Mixotrophic culture of high selenium-enriched *Spirulina platensis* on acetate and the enhanced production of photosynthetic pigments. *Enzyme Microb. Technol., 39*, 103–107.

Chen, X., Song, L., Wang, H., Liu, S., Yu, H., Wang, X., Li, R., et al., (2019). Partial characterization, the immune modulation and anticancer activities of sulfated polysaccharides from filamentous microalgae *Tribonema* sp. *Molecules, 24*, 322.

Cheng, F. C., Feng, J. J., Chen, K. H., Imanishi, H., Fujishima, M., Takekoshi, H., Naoki, Y., & Shimoda, M., (2008). *Chlorella* powder inhibits the activities of peptidase cathepsin S, PLA2, cyclooxygenase-2, thromboxane synthase, tyrosine phosphatases, tumor necrosis factor-alpha converting enzyme, calpain and kinases. *Int. J. Food Sci. Nutr., 23*, 1–10.

Chick, H., (1903). A study of a unicellular green alga, occurring in polluted water, with especial reference to its nitrogenous metabolism. *Proc. Royal Soc. Biol. Sci. Ser. B, 71*, 458–476.

Colla, L. M., Muccillo-Baisch, A. L., & Vieira, C. J. A., (2008). *Spirulina platensis* effects on the levels of total cholesterol, HDL and triacylglycerols in rabbits fed with a hypercholesterolemic diet. *Brazilian Archives of Biology and Technology, 51*(2), 405–411.

Colla, L. M., Reinehr, C. O., Reichert, C., & Costa, J. A. V., (2007). Production of biomass and nutraceutical compounds by *Spirulina platensis* under different temperature and nitrogen regimes. *Bioresource Technology, 98*(7), 1489–1493.

Costa, J. A. C., & Morais, M. G., (2013). Microalgae for food production. In: Soccol, C. R., Pandey, A., &. Larroche, C., (eds.), *Fermentation Process Engineering in the Food Industry* (p. 486). Taylor & Francis.

Costa, J. A. C., & Morais, M. G., (2013). Microalgae for food production. In: Soccol, C. R., Pandey, A., & Larroche, C., (eds.), *Fermentation Process Engineering in the Food Industry* (p. 486). Taylor & Francis.

Costa, J. A. V., Radmann, E. M., Cerqueira, V. S., Santos, G. C., & Calheiros, M. N., (2006). Perfil de ´acidos graxos das microalgas *Chlorella vulgaris* e *Chlorella minutissima* cultivadas em diferentes condicoes. *Alimentos e Nutric¸˜ao Araraquara, 17*(4), 429–436.

Cui, J. F., Ye, H., Zhu, Y. J., Li, Y. P., Wang, J. F., & Wang, P., (2019). Characterization and hypoglycemic activity of a Rhamnan-type sulfated polysaccharide derivative. *Mar. Drugs, 17*, 21.

Cui, Y., Thomas-Hall, S. R., & Schenk, P. M., (2019). *Phaeodactylum tricornutum* microalgae as a rich source of omega-3 oil: Progress in lipid induction techniques towards industry adoption. *Food Chem., 297*, 124937.

Davinelli, S., Nielsen, M. E., & Scapagnini, G., (2018). Astaxanthin in skin health, repair, and disease: A comprehensive review. *Nutrients, 10*, 522.

Egorova, E. A., Gmoshinskiĭ, I. V., Zorin, S. N., & Mazo, V. K., (2006). Studies of immunomodulation caused by selenium-enriched phycocyanin. *Vopr Pitan, 75*, 19–21.

El Sheikh, S. M., Shalaby, M. A. M., Hafez, R. A., et al., (2014). The immunomodulatory effects of probiotic bacteria on peripheral blood mononuclear cells (PBMCS) of allergic patients. *Am. J. Immunol., 10*(3), 116–130.

Ferreira, S. S., Passos, C. P., Madureira, P., Vilanova, M., & Coimbra, M. A., (2015). Structure function relationships of immunostimulatory polysaccharides: A review. *Carbohydr. Polym., 132*, 378–396.

Frentzen, M., (2004). Phosphatidylglycerol and sulfoquinovosyl diacylglycerol: Anionic membrane lipids and phosphate regulation. *Curr. Opin. Plant Biol., 7*, 270–276.

Galasso, C., Corinaldesi, C., & Sansone, C., (2017). Carotenoids from marine organisms: Biological functions and industrial applications. *Antioxidants, 6*, 96.

Garcia De, A. M., Leal, S., Caralt, B., Roca, O., Sabater, J., & Masclans, J. R., (2015). The role of omega-3polyunsaturated fatty acids in the treatment of patients with acute respiratory distress syndrome: A clinical review. *BioMed Res. Int.*, 653750.

Grimmig, B., Morganti, J., Nash, K., & Bickford, P. C., (2016). Immunomodulators as therapeutic agents in mitigating the progression of Parkinson's disease. *Brain Sci., 6*, 41.

Guzmán, S., Gato, A., Lamela, M., & Calleja, J. M., (2003). Anti-inflammatory and immunomodulatory activities of polysaccharide from *Chlorella stigmatophora* and *Phaeodactylum tricornutum*. *Phyther. Res. 17*, 665–670.

Guzman, S., Gato, A., Lamela, M., Freire-Garabal, M., & Calleja, J. M., (2003). Anti-inflammatory and immunomodulatory activities of polysaccharide from *Chlorella stigmatophora* and *Phaeodactylum tricornutum*. *Phytother. Res., 17*, 665–670.

Han, L., Song, S., Niu, Y., Meng, M., & Wang, C., (2017). Eicosapentaenoic acid (EPA) induced macrophages activation through GPR120-Mediated Raf-ERK1/2-IKKbeta-NF-kappaB p65 signaling pathways. *Nutrients, 9*, 397.

Hasegawa, T., Ito, K., Ueno, S., Kumamoto, S., Ando, Y., Yamada, A., Nomoto, K., & Yasunobu, Y., (1999). Oral administration of hot water extracts of *Chlorella vulgaris* reduces IgE production against milk casein in mice. *Int. J. Immunopharmacol., 21*, 311–323.

Hasegawa, T., Noda, K., Kumamoto, S., Ando, Y., Yamada, A., & Yoshikai, Y., (2000). Biological activity of microalgae can be enhanced by manipulating the cultivation temperature and irradiance. *Int. J. Immunopharmacol., 22*, 877–885.

Haworth, C., Brennan, F. M., Chantry, D., Turner, M., Maini, R. N., & Feldman, M., (1991). Expression of granulocyte-macrophage colony-stimulating factor in rheumatoid arthritis: Regulation by tumor necrosis factor alpha. *Eur. J. Immunol., 21*, 2575–2579.

Hayashi, O., Ishii, K., Kawamura, C., Yen, H. S., YeBao, N., et al., (2004). Enhancement of mucosal immune functions by dietary *Spirulina platensis* in human and animals. *Nutr. Sci., 7*, 31–34.

Hayashi, O., Katoh, T., & Okuwaki, Y., (1994). Enhancement of antibody production in mice by dietary *Spirulina platensis*. *J. Nutr. Sci. Vitaminol. (Tokyo), 40*(5), 431–441.

Hayashi, T., & Hayashi, K., (1996). Calcium spirulan, an inhibitor of enveloped virus replication, from a blue-green alga *Spirulina platensis*. *J. Nat. Prod., 59*, 83–87.

Hayashi, T., Hayashi, K., Maeda, M., & Kojima, I. A., (1996). Natural sulphated polysaccharide, calcium spirulan, isolated from *Spirulina platensis*: *In vitro* and *ex vivo* evaluation of anti-herpes simplex virus and anti-human immunodeficiency virus activities. *AIDS Res. Hum. Retroviruses, 12*(15), 1463–1471.

Hayashi, T., Hayashi, K., Maeda, M., & Kojima, I., (1996). Calcium spirulan, an inhibitor of enveloped virus replication, from a blue-green alga *Spirulina platensis*. *J Nat Prod., 59*(1), 83–87.

Heo, S. J., Yoon, W. J., Kim, K. N., Oh, C., Choi, Y. U., Yoon, K. T., Kang, D. H., et al., (2012). Anti-inflammatory effect of fucoxanthin derivatives isolated from *Sargassum* siliqua strum in lipopolysaccharide-stimulated RAW 264.7 macrophage. *Food Chem. Toxicol., 50*, 3336–3342.

Hielscher-Michael, S., Griehl, C., Buchholz, M., Demuth, H. U., Arnold, N., & Wessjohann, L. A., (2016). Natural products from microalgae with potential against Alzheimer's disease: Sulfolipids are potent glutaminyl cyclase inhibitors. *Mar. Drugs, 14*, 203.

Hirahashi, T., Matsumoto, M., Hazeki, K., Saeki, Y., Ui, M., & Seya, T., (2002). Activation of the human innate immune system by *Spirulina:* Augmentation of interferon production and NK cytotoxicity by oral administration of hot water extract of *Spirulina platensis*. *Int. Immunopharmacol., 2*, 423–434.

Hirahashi, T., Matsumoto, M., Hazeki, K., Saeki, Y., Ui, M., & Seya, T., (2002). Activation of the human innate immune system by *Spirulina*: Augmentation of interferon production and NK cytotoxicity by oral administration of hot water extract of *Spirulina platensis*. *Int. Immunopharmacol., 2*, 423–434.

Hirahashi, T., Matsumoto, M., Hazeki, K., Saeki, Y., Ui, M., & Seya, T., (2002). Activation of the human innate immune system by *Spirulina*: Augmentation of interferon production and NK cytotoxicity by oral administration of hot water extract of *Spirulina platensis*. *Int. Immunopharmacol., 2*(4), 423–434.

Hosoyamada, Y., Takai, T., & Kato, T., (1991). Effects of water-soluble and insoluble fractions of *Spirulina* on serum lipid components and glucose tolerance in rats. *Journal of Japanese Soc. Nutr. Food Sci., 44*, 273–277.

Hsu, H. Y., Jeyashoke, N., Yeh, C. H., Song, Y. J., Hua, K. F., & Chao, L. K., (2010). Immunostimulatory bioactivity of algal polysaccharides from *Chlorella pyrenoidosa* activates macrophages via toll-like receptor 4. *J. Agric. Food Chem., 58*, 927–936.

Hussein, G., Nakamura, M., Zhao, Q., Iguchi, T., Goto, H., Sankawa, U., & Watanabe, H., (2005). Antihypertensive and neuroprotective effects of astaxanthin in experimental animals. *Biol. Pharm. Bull., 28*, 47–52.

Hwang, J. H., Chen, J. C., & Chan, Y. C., (2013). Effects of C-phycocyanin and *Spirulina* on salicylate-induced tinnitus, expression of NMDA receptor and inflammatory genes. *PLoS One, 8*(3), e58215.

Iba˜nez, E., & Cifuentes, A., (2013). Benefits of using algae as natural sources of functional ingredients. *Journal of the Science of Food and Agriculture, 93*(4), 703–709.

Ishii, K., Katoch, T., Okuwaki, Y., & Hayashi, O., (1999). Influence of dietary *Spirulina platensis* on IgA level in human saliva. *J. Kagawa Nutr. Univ., 30*, 27–33.

Ismail, M. F., Ali, D. A., Fernando, A., et al., (2009). Chemoprevention of rat liver toxicity and carcinogenesis by *Spirulina. Int. J. Biol Sci., 5*(4), 377–387.

Juarez, M. A., Mascher, D., Torres, P. V., & Farias, J. M., (2009). Effects of *Spirulina* on vascular reactivity. *J. Med. Food, 12*(1), 15–20.

Khan, M., Shobha, J. C., Mohan, I. K., Naidu, M. U., Sundaram, C., & Singh, S., (2005). Protective effect of *Spirulina* against doxorubicin-induced cardiotoxicity. *Phytother Res., 19*(12), 1030–1037.

Khan, Z., Bhadouria, P., & Bisen, P. S., (2005). Nutritional and therapeutic potential of *Spirulina. Curr. Pharm. Biotechnol., 6*(5), 373–379.

Khan, Z., Bhadouria, P., & Bisen, P. S., (2005). Nutritional and therapeutic potential of *Spirulina. Curr. Pharm. Biotechnol., 6*(5), 373–379.

Khozin-Goldberg, I., Leu, S., & Boussiba, S., (2016). Microalgae as a source for VLC-PUFA production. *Subcell Biochem., 86*, 471–510.

Kilian, M., & Russell, M. W., (1994). Function of mucosal immunoglobulins. In: Ogra, P. L., Mestecky, J., Lamm, M. E., Strober, W., McGhee, J. R., & Bienenstock, J., (eds.), *Handbook of Mucosal Immunology* (pp. 127–137). Academic, New York.

Kim, H. M., Lee, E. H., Cho, H. H., & Moon, Y. H., (1998). Inhibitory effect of mast cell-mediated immediate-type allergic reactions in rats by *Spirulina. Biochem. Pharmacol., 55*, 1071–1076.

Kobayashi, K., (2016). Role of membrane glycerolipids in photosynthesis, thylakoid biogenesis and chloroplast development. *J. Plant Res., 129*, 565–580.

Kokou, F., Makridis, P., Kentouri, M., & Divanach, P., (2012). Antibacterial activity in microalgae cultures. *Aquaculture Research, 43*(10), 1520–1527.

Kong, Z. L., Kao, N. J., Hu, J. Y., & Wu, C. S., (2016). Fucoxanthin-rich brown algae extract decreases inflammation and attenuates colitis-associated colon cancer in mice. *J. Food Nutr. Res., 4*, 137–147.

Krienitz, L., Hegewald, E. H., Hepperle, D., Huss, V. A. R., Rohr, T., & Wolf, M., (2004). Phylogenetic relationship of *Chlorella* and *Parachlorella gen.* Nov. (*Chloropyta, Trebouxiophyceae*). *Phycologia, 43*, 529–542.

Kwak, J. H., Baek, S. H., Woo, Y., Han, J. K., Kim, B. G., & Kim, O. Y., (2012). Beneficial immunostimulatory effect of short-term *Chlorella* supplementation: Enhancement of natural killer cell activity and early inflammatory response (randomized, double-blinded, placebo-controlled trial). *Nutr. J., 11*, 53.

Lamm, M. E., Nedrud, J. G., Kaetzel, C. S., & Mazanec, M. B., (1995). IgA and mucosal defense. *Acta Pathol. Microbiol. Immunol. Scand., 103*, 241–246.

Lauritano, C., Ferrante, M. I., & Rogato, A., (2019). Marine natural products from microalgae: An – omics overview. *Mar. Drugs, 17*, 269.

Le, P. N. T., & Desbois, A. P., (2017). Antibacterial effect of eicosapentaenoic acid against bacillus cereus and *Staphylococcus aureus*: Killing kinetics, selection for resistance, and potential cellular target. *Mar. Drugs, 15*, 334.

Lee, A. N., & Werth, V. P., (2004). Activation of autoimmunity following use of immunostimulatory herbal supplements. *Arch Dermatol., 140*(6), 723–727.

Li, B., Zhang, X. C., Gao, M. H., & Yu, H., (2004). Study on the anti-tumor immune activity of phycocyanin and polysaccharides from *Spirulina platensis. J. Ocean Univ. China, 34*, 396–402.

Li, B., Zhang, X., Gao, C. X., et al., (2005). Effects of CD59 on antitumoral activities of phycocyanin from *Spirulina platensis. Biomed Pharmacother., 59*, 551–560.

Li, L., Li, W., Kim, Y. H., & Lee, Y. W., (2013). *Chlorella vulgaris* extract ameliorates carbon tetrachloride-induced acute hepatic injury in mice. *Experimental and Toxicologic Pathology, 65*(1, 2), 73–80.

Lin, K. H., Lin, K. C., Lu, W. J., Thomas, P. A., Jayakumar, T., & Sheu, J. R., (2015). Astaxanthin, a carotenoid, stimulates immune responses by enhancing IFN – and Il-2 secretion in primary cultured lymphocytes *in vitro* and *ex vivo. Int. J. Mol. Sci., 17*, 44.

Liu, Y., Ma, N., Sun, X., Duan, M., Luo, T., Jiang, P., Jiang, G., et al., (2019). Effect of intake pattern of sulfated polysaccharides on its biological activity in high fat diet-fed mice. *Int. J. Biol. Macromol., 132*, 9–16.

Lizzul, A. M., Lekuona-Amundarain, A., Purton, S., & Campos, L. C., (2018). Characterization of *Chlorella sorokiniana*, UTEX 1230. *Biology, 7*, 25.

Manzo, E., Cutignano, A., Pagano, D., Gallo, C., Barra, G., Nuzzo, G., Sansone, C., Ianora, A., Urbanek, K., Fenoglio, D., et al., (2017). A new marine-derived sulfoglycolipid triggers dendritic cell activation and immune adjuvant response. *Sci. Rep., 7*, 6286.

Manzo, E., Gallo, C., Fioretto, L., Nuzzo, G., Barra, G., Pagano, D., Krauss, I. R., Paduano, L., Ziaco, M., DellaGreca, M., et al., (2019). Diasteroselective colloidal self-assembly affects the immunological response of the molecular adjuvant sulfavant. *ACS Omega, 4*, 7807–7814.

Mao, T. K., Van De, W. J., & Gershwin, M. E., (2000). Effect of *Spirulina* on the secretion of cytokines from peripheral blood mononuclear cells. *J. Med. Food, 3*(3), 135–140.

Mao, T. K., Van De, W. J., & Gershwin, M. E., (2000). Effect of *Spirulina* on the secretion of cytokines from peripheral blood mononuclear cells. *J. Med. Food, 3*(3), 135–140.

Martinez, A. K. A., Lauritano, C., Romano, G., & Ianora, A., (2018). Marine microalgae with anti-cancer properties. *Mar. Drugs, 16*, 165.

Medina-Jaritz, N. B., Carmona-Ugalde, L. F., Lopez-Cedillo, J. C., & Ruiloba-DeLeon, F. S. L., (2013). Antibacterial activity of methanolic extracts from *Dunaliella salina* and *Chlorella vulgaris. The FASEB Journal, 27*, abstract 1167.5.

Medzhitov, R., (2007). Recognition of microorganisms and activation of the immune response. *Nature, 449*, 819–826.

Merchant, R. E., Carmack, C. A., & Wise, C. M., (2000). Nutritional supplementation with *Chlorella pyrenoidosa* for patients with fibromyalgia syndrome: A pilot study. *Phytother. Res., 14,* 167–173.

Mimouni, V., Ulmann, L., Pasquet, V., Mathieu, M., Picot, L., Bougaran, G., Cadoret, J. P., et al., (2012). The potential of microalgae for the production of bioactive molecules of pharmaceutical interest. *Curr. Pharm. Biotechnol., 13,* 2733–2750.

Mishra, A., Medhi, K., Malaviya, P., & Thakur, I. S., (2019). Omics approaches for microalgal applications: Prospects and challenges. *Bioresour. Technol., 29,* 121890.

Mittal, A., Kumar, P. V., Banerjee, S., Rao, A. R., & Kumar, A., (1999). Modulatory potential of *Spirulina* fusiformis on carcinogen metabolizing enzymes in Swiss albino mice. *Phytother Res., 13*(2), 111–114.

Miyata, J., & Arita, M., (2015). Role of omega-3 fatty acids and their metabolites in asthma and allergic diseases. *Allergol. Int., 64,* 27–34.

Moreno-Garrido, I., (2008). Microalgae immobilization: Current techniques and uses. *Bioresource Technology, 99*(10), 3949–3964.

Morris, H. J., Carrillo, O., Almarales, A., et al., (2007). Immunostimulant activity of an enzymatic proteinhydrolysate from green microalga *Chlorella vulgaris* on undernourished mice. *Enzyme Microb. Technol., 40,* 456–460.

Mosmann, T. R., & Coffman, R. L., (1989). Th1 and Th2 cells: Different patterns of lymphokine secretion lead to different functional properties. *Annu. Rev. Immunol., 7,* 145–173.

Mosser, D. M., & Edwards, J. P., (2008). Exploring the full spectrum of macrophage activation. *Nat. Rev. Immunol., 8,* 958–969.

Muhamad, I. I., Zulkifli, N., Selvakumaran, S. A., & Lazim, N. A. M., (2019). Bioactive algal-derived polysaccharides: Multi-functionalization, therapeutic potential and biomedical applications. *Curr. Pharm. Des., 25,* 1147–1162.

Nappo, M., Berkov, S., Massucco, C., Di Maria, V., Bastida, J., Codina, C., Avila, C., et al., (2012). Apoptotic activity of the marine diatom *Cocconeis scutellum* and eicosapentaenoic acid in BT20 cells. *Pharm. Biol., 50,* 529–535.

Novoveská, L., Ross, M. E., Stanley, M. S., Pradelles, R., Wasiolek, V., & Sassi, J. F., (2019). Microalgal carotenoids: A review of production, current markets, regulations, and future direction. *Mar. Drugs, 17,* 240.

Pardoll, D. M., (2012). The blockade of immune checkpoints in cancer immunotherapy. *Nat. Rev. Cancer, 12,* 252–264.

Park, H. J., Lee, Y. J., Ryu, H. K., et al., (2008). A randomized double-blind, placebo-controlled study to establish the effects of *Spirulina* in elderly Koreans. *Ann. Nutr. Metab., 52*(4), 322–328.

Park, H. J., Lee, Y. J., Ryu, H. K., Kim, M. H., Chung, H. W., & Kim, W. A. Y., (2008). Randomized double-blind, placebo-controlled study to establish the effects of *Spirulina* in elderly Koreans. *Annal. Nutr. Metabol., 52*(4), 322–329.

Perera, N., Yang, F. L., Chern, J., Chiu, H. W., Hsieh, C. Y., Li, L. H., Zhang, Y. L., Hua, K. F., & Wu, S. H., (2018). Carboxylic and: O-acetylmoieties are essential for the immunostimulatory activity of glucuronoxylomannan: A novel TLR4 specific immunostimulator from *Auricularia auricula-judae. Chem. Commun., 54,* 6995–6998.

Plaza, M., Herrero, M., Alejandro, C. A., & Ib´a˜nez, E., (2009). Innovative natural functional ingredients from microalgae. *Journal of Agricultural and Food Chemistry, 57*(16), 7159–7170.

Plaza, M., Herrero, M., Cifuentes, A., & Ibanez, E., (2009). Innovative natural functional ingredients from microalgae. *J. Agric. Food Chem., 57*, 7159–7170.

Priyadarshani, I., & Rath, B., (2012). Commercial and industrial applications of micro algae—A review. *Journal of Algal Biomass Utilization, 3*(4), 89–100.

Pugh, N., & Pasco, D. S., (2001). Characterization of human monocyte activation by a water soluble preparation of *Aphanizomenon flos-aquae*. *Phytomedicine, 8*, 445–453.

Qi, J., & Kim, S. M., (2017). Characterization and immunomodulatory activities of polysaccharides extracted from green alga *Chlorella ellipsoidea*. *Int. J. Biol. Macromol., 95*, 106–114.

Qi, J., & Kim, S. M., (2018). Effects of the molecular weight and protein and sulfate content of *Chlorella ellipsoidea* polysaccharides on their immunomodulatory activity. *Int. J. Biol. Macromol., 107*, 70–77.

Qi, J., & Kim, S. M., (2018). Effects of the molecular weight and protein and sulfate content of *Chlorella ellipsoidea* polysaccharides on their immunomodulatory activity. *Int. J. Biol. Macromol. 107*, 70–77.

Qureshi, M. A., & Ali, R. A., (1996). *Spirulina platensis* exposure enhances macrophage phagocytic function in cats. *Immunopharmacol. Immunotoxicol., 18*, 457–463.

Qureshi, M. A., & Ali, R. A., (1996). *Spirulina platensis* exposure enhances macrophage phagocytic function in cats. *Immunopharmacol. Immunotoxicol., 18*, 457–463.

R¨osch, C., & Posten, C., (2012). Challenges and perspectives of microalgae production. *Technikfolgenabsch¨atzung—Theorieund Praxis, 21*(1).

Raffaella, R., Marzia, B. G., Laura, P., & Elena, M., (2009). Role of the p53/p21 system in the response of human colon carcinoma cells to doxorubicin. *Int. J. Bio Sci., 5*(4), 377–387.

Raja, R., Hemaiswarya, S., Ashok, K. N., Sridhar, S., & Rengasamy, R., (2008). A perspective on the biotechnological potential of microalgae. *Crit. Rev. Microbiol., 34*, 77–88.

Raposo, M. F., De Morais, R. M., & Bernardo De, M. A. M., (2013). Bioactivity and applications of sulphated polysaccharides from marine microalgae. *Mar. Drugs, 11*, 233–252.

Rasool, M., & Sabina, E. P., (2009). Appraisal of immunomodulatory potential of *Spirulina fusiformis*": An *in vivo* and *in vitro* study. *J. Nat. Med., 63*(2), 169–175.

Rasool, M., & Sabina, E. V., (2009). Appraisal of immunomodulatory potential of *Spirulina fusiformis*": An *in vivo* and *in vitro* study. *J. Nat. Med., 63*, 169–175.

Reddy, M. C., Subhashini, J., Mahipal, S. V., Bhat, V. B., Srinivas, R. P., Kiranmai, G., Madyastha, K. M., & Reddanna, P., (2003). C-Phycocyanin, a selective cyclooxygenase-2 inhibitor, induces apoptosis in lipopolysaccharide-stimulated RAW 264.7 macrophages. *Biochem. Biophys. Res. Commun., 304*, 385–392.

Riccio, G., & Lauritano, C. (2019). Microalgae with immunomodulatory activities. *Marine Drugs, 18*(1), 2.

Richmond, A., & Hu, Q., (2013). *Handbook of Microalgal Culture: Applied Phycology and Biotechnology.* Wiley-Blackwel: Hoboken, NJ, USA.

Romano, I., Bellitti, M. R., Nicolaus, B., et al., (2000). Lipid profile: A useful chemotaxonomic marker for classification of a new cyanobacterium in *Spirulina* genus. *Phytochemistry, 54*(3), 289–294.

Romay, C., & Gonzalez, R., (2000). Phycocyanin is an antioxidant protector of human erythrocytes against by peroxyl radicals. *J. Pharm. Pharmacol., 52*, 367–368.

Roy, A. B., Hewlins, M. J., Ellis, A. J., Harwood, J. L., & White, G. F., (2003). Glycolytic breakdown of sulfoquinovose in bacteria: A missing link in the sulfur cycle. *Appl. Environ. Microbiol., 69*, 6434–6441.

Rupasinghe, H. P., Erkan, N., & Yasmin, A., (2010). Antioxidant protection of eicosapentaenoic acid and fish oil oxidation by polyphenolic-enriched apple skin extract. *J. Agric. Food Chem., 58*, 1233–1239.

Schwartz, J., & Shklar, G., (1987). Regression of experimental hamster cancer by beta carotene and algae extracts. *J. Oral Maxillofac Surg., 45*(6), 510–515.

Selmi, C., Leung, P. S., Fischer, L., et al., (2011). The effects of *Spirulina* on anemia and immune function in senior citizens. *Cell Mol. Immunol., 8*(3), 248–254.

Sheahan, S., Bellamy, C. O., Harland, S. N., et al., (2008). TGF beta induces apoptosis and EMT in primary mouse hepatocytes independently of p53, p21Cip1 or Rb status. *BMC Cancer, 8*, 191–201.

Shibata, S., Oda, K. N., Onodera-Masuoka, N., Matsubara, S., Kikuchi-Hayakawa, H., & Ishikawa, F., (2001). Hypocholesterolemic effect of indigestible fraction of *Chlorella regularis* in cholesterol-fed rats. *J. Nutr. Sci. Vitaminol., 47*, 373–377.

Shishido, S. N., Varahan, S., Yuan, K., Li, X. D., & Fleming, S. D., (2012). Humoral innate immune response and disease. *Clin. Immunol., 144*, 142–158.

Shklar, G., & Schwartz, J., (1988). Tumor necrosis factor in experimental cancer regression with alpha-tocopherol, beta-carotene, canthaxanthin and algae extract. *Eur. J. Cancer Clin. Oncol.*, (5), 839–850.

Shokri, H., Khosravi, A., & Taghavi, M., (2014). Efficacy of *Spirulina platensis* on immune functions in cancer mice with systemic candidiasis. *J. Mycol. Res., 1*(1), 7–13.

Sibi, G., & Rabina, S., (2016). Inhibition of pro-inflammatory mediators and cytokines by *Chlorella vulgaris* extracts. *Pharmacogn. Res., 8.*

Simpore, J., Zongo, F., Kabore, F., Dansou, D., Bere, A., & Nikiema, J. B., (2005). Nutrition rehabilitation of HIV-infected and HIV-negative undernourished children utilizing *Spirulina. Annal. Nut. Metabolism, 49*(6), 373–380.

Sin, J. B., Choi, W. Y., & Lee, H. Y., (2014). Comparison of anti-inflammatory activity of *Spirulina maxima* extract by ultrasonication and water extraction process *Spirulina maxima. J. Korean Soc. Food Nutr., 43*(12), 1852–1857.

Singh, N. K., & Dhar, D. W., (2011). Microalgae as second-generation biofuel: A review. *Agronomy for Sustainable Development, 31*(4), 605–629.

Sorokin, C., & Myers, J., (1953). A high-temperature strain of *Chlorella. Science, 117*, 330–331.

Speciale, G., Jin, Y., Davies, G. J., Williams, S. J., & Goddard-Borger, E. D., (2016). YihQ is a sulfoquinovosidase that cleaves sulfoquinovosyl diacylglyceride sulfolipids. *Nat. Chem. Biol., 12*, 215–217.

Speranza, L., Pesce, M., Patruno, A., Franceschelli, S., De Lutiis, M. A., Grilli, A., & Felaco, M., (2012). Astaxanthin treatment reduced oxidative induced pro-inflammatory cytokines secretion in U937: SHP-1 as a novel biological target. *Mar. Drugs, 10*, 890–899.

Spolaore, P., Joannis-Cassan, C., Duran, E., &. Isambert, A., (2006). Commercial applications of microalgae. *Journal of Bioscience and Bioengineering, 101*(2), 8796.

Suárez, E. R., Kralovec, J. A., Noseda, M. D., Ewart, H. S., Barrow, C. J., Lumsden, M. D., & Grindley, T. B., (2005). Isolation, characterization and structural determination of a unique type of arabinogalactan from an immunostimulatory extract of *Chlorella pyrenoidosa. Carbohydr. Res., 340*, 1489–1498.

Subhashini, J., Mahipal, S. V., Reddy, M. C., Mallikarjuna, R. M., Rachamallu, A., & Reddanna, P., (2004). Molecular mechanisms in C-Phycocyanin induced apoptosis in human chronic myeloid leukemia cell line-K562. *Biochem. Pharmacol., 68*(3), 453–462.

Sun, C., Liu, M., Sun, P., Yang, M., Yates, E. A., Guo, Z., & Fernig, D. G., (2019). Sulfated polysaccharides interact with fibroblast growth factors and protect from denaturation. *FEBS Open Bio., 9*, 1477–1487.

Tabarsa, M., Shin, I. S., Lee, J. H., Surayot, U., Park, W., & You, S., (2015). An immune-enhancing water-soluble β-glucan from *Chlorella vulgaris* and structural characteristics. *Food Sci. Biotechnol., 24*, 1933–1941.

Torres-Duran, P. V., Ferreira-Hermosillo, A., & Juarez-Oropeza, M. A., (2007). Anti-hyperlipemic and antihypertensive effects of *Spirulina maxima* in an open sample of Mexican population: A preliminary report. *Lipids in Health and Disease*, 6, Article33.

Vishwakarma, J., & Vavilala, S. L., (2019). Evaluating the antibacterial and antibiofilm potential of sulphated polysaccharides extracted from green algae *Chlamydomonas reinhardtii. J. Appl. Microbiol., 127*, 1004–1017.

Wu, H. L., Hseu, R. S., & Lin, L. P., (2001). Identification of *Chlorella* spp. isolates using ribosomal DNA sequences. *Bot. Bull. Acad. Sin., 42*, 115–121.

Wu, Q., Liu, L., Miron, A., Klimova, B., Wan, D., & Kuca, K., (2016). The antioxidant, immunomodulatory, and anti-inflammatory activities of *Spirulina:* An overview. *Arch. Toxicol., 90*, 1817–1840.

Xu, X. F., Yan, H. D., Tang, J., Chen, J., & Zhang, X. W., (2014). Polysaccharides in *Lentinus edodes*: Isolation, structure, immunomodulating activity and future prospective. *Crit. Rev. Food Sci. Nutr., 54*, 474–487.

Yamada, H., Umemoto, T., Kakei, M., Momomura, S. I., Kawakami, M., Ishikawa, S. E., & Hara, K., (2017). Eicosapentaenoic acid shows anti-inflammatory effect via GPR120 in 3T3-L1 adipocytes and attenuates adipose tissue inflammation in diet-induced obese mice. *Nutr. Metab., 14*, 33.

Yang, F. M., Shi, Y., Sheng, J. C., & Hu, Q. H., (2006). *In vivo* immunomodulatory activity of polysaccharides derived from *Chlorella pyrenoidosa. Eur. Food Res. Technol., 224*, 225–228.

Yang, H. N., Lee, E. H., & Kim, H. M., (1997). *Spirulina platensis* inhibits anaphylactic reaction. *Life Sci., 61*, 1237–1244.

Yoshida, H., Yanai, H., Ito, K., Tomono, Y., Koikeda, T., Tsukahara, H., & Tada, N., (2010). Administration of natural astaxanthin increases serum HDL-cholesterol and adiponectin in subjects with mild hyperlipidemia. *Atherosclerosis, 209*, 520–523.

Yuan, Q. X., Zhang, X. D., Ma, M. Y., Long, T., Xiao, C. L., Zhang, J., Liu, J. K., & Zhao, L. Y., (2020). Immunoenhancing glucuronoxylomannan from *Tremella aurantialba* Bandoniet Zang and its low-molecular-weight fractions by radical depolymerization: Properties, structures and effects on macrophages. *Carbohydr. Polym., 238*, 116–184.

Zhang, H. Q., Lin, A. P., & Sun, Y., (2001). Chemo and radio protective effects of polysaccharide of *Spirulina platensis* on haematopoietic stem cells in dog and mice. *Acta Pharmacol. Sin., 22*(12), 1121–1124.

Zhang, H. Q., Lin, A. P., Sun, Y., & Deng, Y. M., (2001). Chemo and radio protective effects of polysaccharide of *Spirulina platensis* on hemopoietic system of mice and dogs. *Acta Pharmacol. Sin., 22,* 1121–1124.

Zhang, X. R., Qi, C. H., Guo, Y., Zhou, W. X., & Zhang, Y. X., (2016). Toll-like receptor 4-relatedimmunostimulatory polysaccharides: Primary structure, activity relationships, and possible interaction models. *Carbohydr. Polym., 149*, 186–206.

Zhao, L., & Sweet, B. V., (2008). Lutein and zeaxanthin for macular degeneration. *The American Journal of Health-System Pharmacy, 65*(13), 1232–1238.

CHAPTER 12

ANTIOXIDANT ACTION OF GLUTATHIONE AGAINST HEAVY METAL TOXICITY IN FISHES

ANURAG SAMSON LALL,[1] JYOTI VANDANA MANI,[1]
SHANTI BHUSHAN MISHRA,[2] and AMIT KUMAR SINGH[3]

[1]*Department of Chemistry, Sam Higginbottom University of pAgriculture, Technology, and Sciences, Prayagraj, Uttar Pradesh, India*

[2]*Department of Pharmacognosy, United Institute of Pharmacy, Prayagraj, Uttar Pradesh, India*

[3]*Department of Pharmaceutics, Uniteed Institute of Pharmacy, Prayagraj, Uttar Pradesh, India*

ABSTRACT

The purpose of the present chapter is to provide a comprehensive overview of the mechanism of heavy metal induced toxicity, generation of ROS and the antioxidant defense of glutathione (GSH) and associated enzymes such as glutathione peroxidase (GPx), glutathione sulfotransferase (GST) and glutathione reductase (GR) in fishes. ROS are endogenously formed during cellular aerobic respiration in the mitochondrial electron transport chain (ETC). However, the heavy metals with variable oxidation states may participate in Fenton and Haber–Weiss reactions, enhancing the production of ROS. Glutathione is abundantly found intracellularly in fishes and is one of the major components of cellular defense as it provides a way for

Immune-Boosting Nutraceuticals for Better Human Health: Novel Applications. Urmila Jarouliya, PhD, Raj K. Keservani, MPharm, Rajesh K. Kesharwani, PhD, Virendra K. Patel, PhD, and Adi D. Bharti, MPharm (Eds.)

detoxification and elimination of intracellular toxicants and free radicals. Therefore, it is necessary to compile a comprehensive picture of the heavy metal induced toxicity and its mitigation via the antioxidant glutathione in fishes.

12.1 INTRODUCTION

Industrial revolution has enormously improved the quality of life, economy, transportation, and communication; however, it is only at the expense of environmental degradation. The use of contemporary techniques has dramatically increased the production and hence the emission of synthetic chemicals. Surface water being the ultimate destination of several anthropogenic emissions, its contamination poses a substantial threat to the aquatic ecosystems.

Contaminants like pesticides, insecticides, oils, polychlorinated biphenyls (PCBs), polycyclic aromatic hydrocarbons (PAHs), heavy metals and other xenobiotics are frequently found in surface water (Javed & Usmani, 2019). Heavy metals are ubiquitous pollutants which are non-biodegradable and are notorious for their environmental persistence compared to most organic contaminants that quickly break down to carbon dioxide and water (Jozefczak et al., 2012). Diverse agricultural, industrial, and daily life anthropogenic activities are responsible for worldwide occurrence of zinc (Zn), lead (Pb), chromium (Cr), nickel (Ni), arsenic (As), copper (Cu) and other heavy metals in aquatic environments (Eroglu et al., 2015). The cumulative anthropogenic heavy metal discharge is huge, and consequently it has influenced different environmental processes in various ecosystems.

With almost 20,000 species, fishes constitute a major class of vertebrates that are directly affected by heavy metal contamination of water resources. Heavy metals can easily accumulate in various fish tissues including liver, kidneys, and gills (Jezierska & Witeska, 2006). Since fishes occupy the top level of the food chain in aquatic ecosystems, the bioaccumulation of metals may also cause their overall environmental biomagnification. Thus, for assessment of metal contamination in aquatic systems, fishes are regarded as suitable and effective bio-indicator (Sevcikova et al., 2011).

Metal bioaccumulation causes several toxic effects including alterations in enzyme activity, disabilities in locomotion, swimming, respiration, reproduction, and impacts embryonic development (Atchison et al., 1987). Heavy metals toxicity is mainly caused by excess generation of certain reactive species like superoxide radical ($\dot{O}_2^-$), hydrogen peroxide (H_2O_2), hydroxyl radical ($\dot{O}H$), etc., which are collectively termed as ROS that can damage various cell constituents including cell membrane, proteins, lipids, carbohydrates, and DNA (Lushchak, 2016).

Antioxidants such as superoxide dismutase (SOD), catalase (CAT), glutathione (GSH) and its associated enzymes like glutathione peroxidase (GP_x), glutathione reductase (GR) and glutathione-S-transferase (GST) have significant contribution in defense mechanism against ROS (Sevcikova et al., 2011). Usually there exists an equilibrium between the ROS production and antioxidant response. Heavy metal buildup, on the other hand, disrupts this balance creating stress (Srikanth et al., 2013).

Fishes can survive in a diverse variety of conditions with varying temperatures, salinities, and oxygen levels, which significantly influences the rate of ROS production. Thus, the activity of GSH and associated enzymes in fishes substantially differs from that of mammals (Wilhelm-Filho et al., 2007). This review chapter focuses on the heavy metal toxicity, ROS production and the defensive action of GSH and associated enzymes in fishes.

12.2 HEAVY METALS AND THEIR BIOACCUMULATION

Metals having with at least 5 times higher atomic density than water constitute an important group of contaminants called heavy metals. The significance of these toxicants is revealed by the fact that about 13 of them have been listed in the priority pollutant list of the United States Environmental Protection Agency (USEPA) which includes Ag, As, Be, Cr, Cd, Cu, Hg, Ni, Pb, Sb, Se, Tl, and Zn, due to their toxicity and relative abundance (Shofstahl & Hardy, 1990; Novotny, 1995). Since these elements are non-biodegradable, they are usually associated with prolonged environmental persistence. Several heavy metals like Fe, Co, Cu, Mn, Mo, and Zn are required in minimal quantities by living organisms. However, these are toxic at higher concentrations. In contrast, metals like As, Hg, Cd, and Pb do not have any vital or beneficial effect for living beings and their buildup in tissues may cause severe ailments.

Passive metal uptake results in heavy metal bioaccumulation in various tissues of the concerned organism. The accumulated metal concentration is frequently determined not only by the species under investigation but also varies from individual to individual. The dermal features of the aquatic species, the prevalence of metal binding ligands, the porosity of the tissues, and the pace of metabolism all influence the entire process.

The overall pattern of accumulation might be linear or exponential, contingent on the uptake rate, the diffusion gradients, and the magnitude of metal loss. The kinetics of metal uptake in fishes is most commonly explained using an exponential model. This is usually observed when the initial uptake rate exceeds the metal loss. As the external adsorption sites are occupied, the overall accumulation rate gradually decreases, and the frequency of metal loss becomes considerably relevant. Eventually, as the rate of metal uptake and loss become equal, an equilibrium state is achieved and the adsorption follows first order kinetics according to following equation:

$$C_t = C_e\,(1 - e^{\gamma t})$$

where; C_t and C_e are metal concentrations in the tissue at time 't' and equilibrium respectively. γ being the coefficient of elimination and is related to the biological half-life $t_{1/2}$ as follows (Javed & Usmani, 2019):

$$\gamma = 0.693/t_{1/2}$$

When the rate of metal removal is negligibly slow, the accumulation of metal keeps on increasing with exposure and the metal uptake follows a linear pattern.

The magnitude of metal buildup in various tissues is frequently employed as a bio-indicator of metal contamination in aquatic ecosystem. Generally, the heavy metal accumulation in fishes is multidirectional and occurs in all vital organs of fish, including gills, liver, kidneys, and muscles, consequently knocking down the locomotive, respiratory, reproductive as well as endocrine system.

12.3 ENDOGENOUS GENERATION OF REACTIVE OXYGEN SPECIES (ROS)

The term "reactive oxygen species (ROS)" is frequently used to describe highly reactive oxygen-containing chemical species that are responsible for several important biochemical and physiological roles in an organism. The term collectively includes free radicals like superoxide radical ($\dot{O}_2^-$), hydroxyl radical ($\dot{O}H$), alkoxyl radical ($\dot{O}R$), peroxyl radical ($\dot{O}R_2$) as well as other non-radical species such as H_2O_2, peroxynitrite ($ONOO^-$), hypochlorous acid (HOCl), etc. Although minimal ROS concentrations are vital for healthy cellular activity, higher levels may irreversibly deteriorate macromolecules, including lipids, proteins, and nucleic acids, preventing them from functioning normally (Li et al., 2016). ROS are usually formed as by-products during cellular aerobic respiration by the mitochondrial electron transport chain (ETC) (Lushchak, 2016). All aerobic organisms depend on oxidative phosphorylation for production of energy in the form of adenosine triphosphate (ATP). The process involves the transfer of electrons to molecular oxygen through a series of mitochondrial complexes with subsequent translocation of protons across the inner mitochondrial membrane, activating the enzyme ATP synthase for production of ATP. Figure 12.1 displays an overview of the mitochondrial ETC which consists of 4 protein complexes, coenzyme Q (CoQ), cytochrome c (Cyt c) as electron carriers and the enzyme ATP synthase.

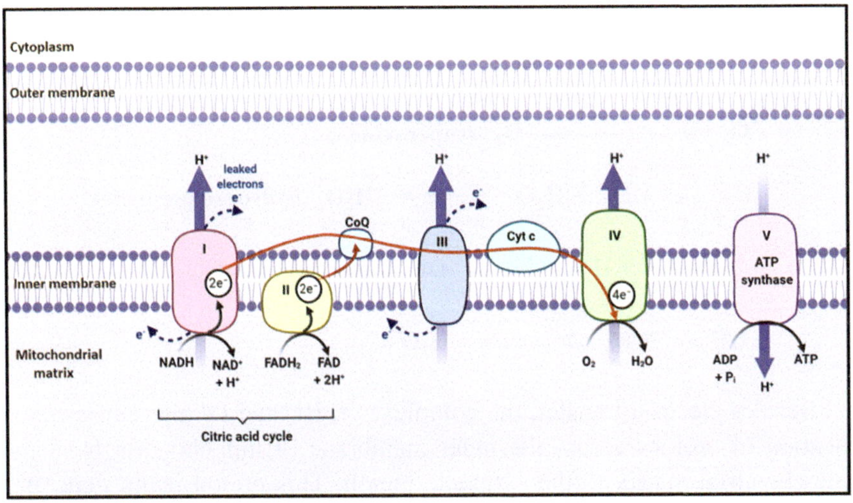

FIGURE 12.1 Mitochondrial electron transfer chain.

Source: Adapted from Sapkota 2021.

The reactions occurring in mitochondrial ETC have been depicted below:

Complex I: $NADH + H^+ + CoQ \longrightarrow NAD^+ + CoQH_2$

Complex II: $Succinate + FADH_2 + CoQ \longrightarrow Fumerate + FAD^+ + CoQH_2$

Complex III: $CoQH_2 + 2\ Cyt\ c\ (Fe^{3+}) \longrightarrow CoQ + 2\ Cyt\ c\ (Fe^{2+}) + 2H^+$

Complex IV:
$$\left\{ \begin{array}{l} Cyt\ c\ (Fe^{2+}) \longrightarrow Cyt\ c\ (Fe^{3+}) + e^- \ [x\ 4] \\ 4H^+ + O_2 + 4e^- \longrightarrow 2H_2O \end{array} \right\}$$

The reduced forms of nicotinamide adenine dinucleotide (NADH) and flavin adenine dinucleotide ($FADH_2$) act as electron donors, while the molecular oxygen serves as the final electron acceptor. Complexes I and II consist of enzymes NADH dehydrogenase and succinate dehydrogenase respectively and catalyze the transfer of two electrons each to the CoQ. Meanwhile, NADH and $FADH_2$ are oxidized to form NAD^+ and FAD, respectively and are given back to the citric acid cycle. Similarly in complex III, the cytochrome reductase helps in transfer of electrons from reduced $CoQH_2$ to Cyt c where the Fe^{3+} ion in the heme group of Cyt c reduces to form Fe^{2+} (ferrous ion) while $CoQH_2$ is oxidized back to CoQ. Finally in complex IV, the cytochrome oxidase assists in reduction of molecular oxygen via transfer of electrons from Cyt c to form water. The overall reduction of oxygen is usually a four-step process with different reactive intermediates involved in each step as shown below (Mahboob, 2013).

$$O_2 + e^- \longrightarrow \dot{O}_2^- \text{ (superoxide ion)}$$

$$\dot{O}_2^- + e^- + 2H^+ + H_2O \longrightarrow H_2O_2 \text{ (hydrogen peroxide)}$$

$$H_2O_2 + e^- + H^+ \longrightarrow \dot{O}H \text{ (hydroxyl radical)}$$

$$\dot{O}H + e^- + H^+ \longrightarrow H_2O$$

Besides electron transfer, the complexes I, III, and IV also cause translocation of protons across the inner membrane of mitochondria from the mitochondrial matrix to the cytosolic region. This proton pump generates an electrochemical gradient which enables the protons to cross the proton channel of the enzyme ATP synthase and pass into the matrix, thus activating the enzyme to synthesize ATP using ADP (Ahmad et al., 2018). In addition to all this process, a partial leakage of electrons occurs frequently at complexes I and III, which causes reduction of oxygen resulting in formation of superoxide ion in the mitochondrial matrix as well as the inter membrane space. The superoxide ion thus formed serves as a precursor for other ROS.

12.4 HEAVY METAL INDUCED TOXICITY

Diverse mechanistic pathways are responsible for heavy metal toxicity. Metals may easily bind with specific protein sites by replacing certain other essential metals, disrupting the overall functioning of cellular proteins and impairing the cell physiology. Sulfuryl (-SH) group of proteins is the characteristic moiety which is targeted by the metals. As the metal binds to the protein structure, the folding and refolding of 3-dimensional protein structure is sterically disrupted (Jan et al., 2015). Besides this, heavy metals with variable oxidation states are also accountable for the generation of ROS by participating in Fenton and Haber–Weiss reactions (Liang et al., 2013; Rubino et al., 2015). Metals (M) like Fe, Cu, Ni, Co, etc., with more than one valence states may catalytically convert endogenous H_2O_2 and superoxide ion to hydroxyl radical, hydroxide ion and oxygen molecule as depicted below:

$$M^{(n+1)+} + \dot{O}_2^{-} \longrightarrow M^{n+} + O_2$$

$$M^{n+} + H_2O_2 \longrightarrow M^{(n+1)+} + \overline{O}H + \dot{O}H \quad \textbf{(Fenton reaction)}$$

$$\textbf{Overall: } \dot{O}_2^{-} + H_2O_2 \longrightarrow \dot{O}H + \overline{O}H + O_2 \quad \textbf{(Haber - Weiss reaction)}$$

Hydroxyl radicals thus formed are deleterious for biological structures and biomolecules. The biomembranes usually contain polyunsaturated fatty acid (PUFA) in the lipid bilayer structure. The C=C bonds in PUFA render the lipid bilayer highly susceptible to oxidation. The oxidative disintegration of lipids caused by hydroxyl radical and other ROS exposure is referred to as lipid peroxidation (Javed & Usmani, 2019). The reactions below depict the sequence of reactions describing the process.

The initiation step involves the generation of lipid radicals during the attack of $\dot{O}H$ radical over lipid structure. These lipid radicals easily undergo oxidation to form lipid peroxyl radicals which may further produce lipid peroxide in the propagation step.

In addition to lipid peroxidation, the hydroxyl radical also attacks and damages the side chains of various amino acids like cysteine, arginine, histidine, etc. Similarly, the sugar moieties and nitrogenous bases present in DNA are also susceptible to $\dot{O}H$ attack (Jaishankar et al., 2014; Jan et al., 2015). $\dot{O}H$ radicals often cause breaking of DNA strands on account of proton abstraction from sugar moieties. Moreover, the oxidation of nitrogen bases in nucleic acids generates (C5-OH and C6-OH) pyrimidine adduct radicals and (C4-OH, C5-OH, and C8-OH) purine adduct radicals. These adducts are usually responsible for the formation of pre-mutagenic DNA lesions (Jan et al., 2015). It is well documented that attack of ȮH radicals and other ROS over DNA often produces 8-oxoguanine. The quantity of 8-oxoguanine formed may be readily analyzed by high pressure liquid chromatography (HPLC) and it reflects the extent of DNA deterioration (Lushchak, 2016).

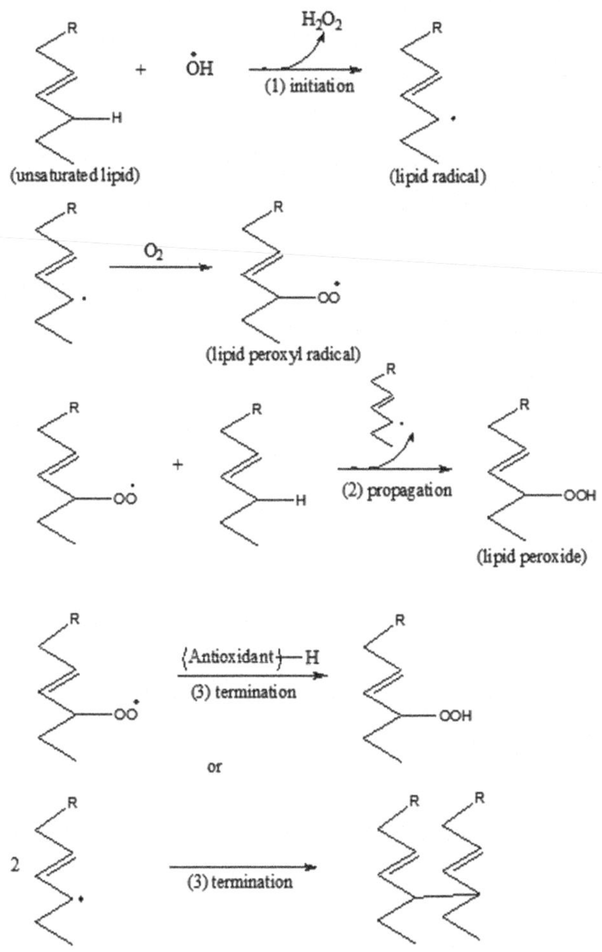

12.5 THE ANTIOXIDANT DEFENSE

Fishes are aerobic life forms, and in order to comply with their energy needs, they depend on the reduction of molecular oxygen. During this process, the high energy phosphate group is attached to ADP facilitating its conversion to ATP, while ROS are formed as byproducts. ROS play essential roles in a number of cellular processes including intracellular signaling, homeostasis, growth, and cell division. However, formation of excess ROS due to certain environmental conditions, stress factors, or the presence of toxicants often proves detrimental to the health of an organism. The excessive ROS generation stimulates an endogenous protection mechanism in fishes which facilitates the formation of antioxidants in hepato-pancreas and other tissues. These antioxidants act as ROS scavengers and thus either prevent or slow down the deterioration caused by free radicals (Mahboob, 2013; Eroglu et al., 2015).

Depending on the action towards ROS, the antioxidants may be classified as primary and secondary antioxidants. Amongst the primary antioxidants, the SOD (EC 1.15.1.1) directly acts on superoxide anion and converts it to molecular oxygen. Similarly, the catalase (CAT, EC 1.11.1.6) and the GPx (EC 1.11.1.9) directly reduce H_2O_2 to water. In contrast, the secondary antioxidants like GSH and NADPH assist the activity of primary oxidants and ensure their uninterrupted performance (Lushchak, 2016).

12.6 GLUTATHIONE AND ASSOCIATED ENZYMES

GSH is a water soluble tripeptide (γ-Glu-Cys-Gly) consisting of glutamine, cysteine, and glycine.

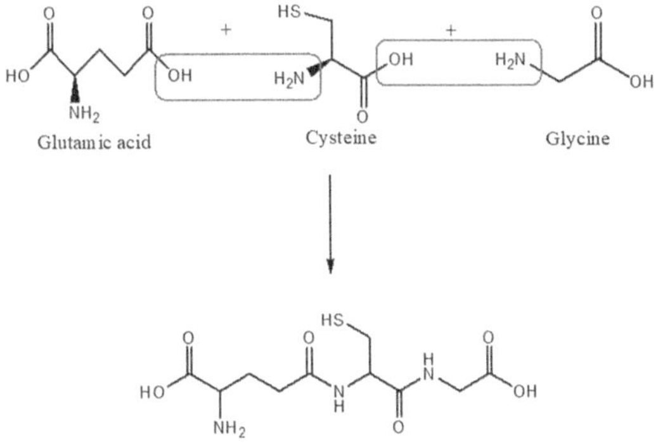

Glutathione

It has relatively low mass and is abundantly found intracellularly in fishes (Kovacevic et al., 2008; Srikanth et al., 2013). Usually, the cytosolic concentration of GSH in most cells lies within the range 1–10 mM (Forman et al., 2009). It is undoubtedly one of the most important antioxidants responsible for defense against ROS and toxicants. The action of GSH is supported by several additional associated enzymes including the GPx (EC 1.11.1.9), GR (EC 1.6.4.2) and GST (EC 2.5.1.18) (Srikanth et al., 2013).

12.6.1 GLUTATHIONE (GSH)-MEDIATED DETOXIFICATION

The major action of GSH and associated enzymes is the detoxification of ROS and electrophilic xenobiotics. The GSH peroxidase (GP_x) effectively reduces H_2O_2 produced during cellular respiration. Although CAT in the peroxisome reduces H_2O_2, GSH is the key antioxidants against H_2O_2 within the mitochondria (Lu, 2009). Similarly, it also acts on lipid peroxides, reducing them to corresponding alcohol and water (Srikanth et al., 2013). The thio (-SH) group of cysteine moiety within the GSH molecule has remarkable affinity towards different metal ions and other electrophiles. Metals like Pb, Cd, Hg, Zn, Cu, etc., specifically bind with the -SH group of GSH and thus form GSH-conjugates having stoichiometry $[(GS^-)_n\,(M^{n+})]$ (Rubino, 2015). Therefore, on exposure to excess metal ion concentration, subjected to the exposure conditions, fishes exhibit either decreased GSH levels due to generation of GSH-conjugates or increased GSH concentration in different tissues on account of increased secretion of GSH in defensive response (Srikanth et al., 2013).

The interaction of GSH with ROS and electrophiles converts its reduced form to the oxidized-disulfide (GSSG) form via a sulfinic acid intermediate (GS-OH) (Rubino, 2015). Since, the reduced form of GSH has a vital role in detoxification and protection of cells, its conversion into oxidized form (GSSG) is often indicative of cellular toxicity.

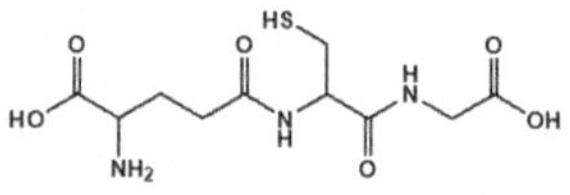

Glutathione (GSH)

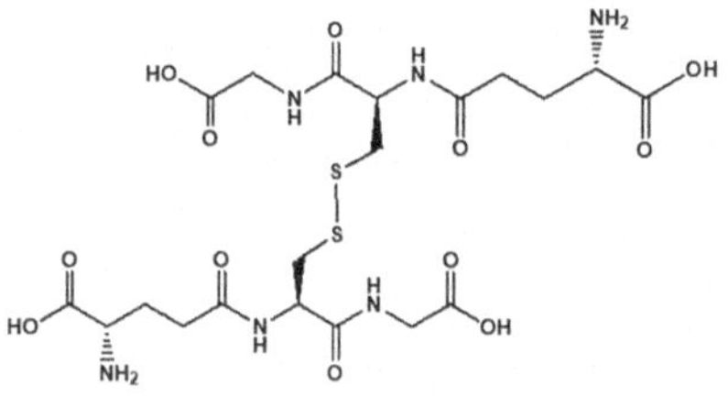

Glutathione disulphide (GSSG)

Usually, in the absence of stress, the molar ratio of GSH: GSSG often exceeds 100:1. Any decrease in this ratio clearly suggests the presence of toxicants causing stress (Zitika et al., 2012). The transformation of GSH to GSSG is reversible. GSSG is reduced back to GSH under the influence of GR. Meanwhile during this process, NADPH molecule is oxidized to NADP, which later reduces back to NADP under the action of glucose-6-phosphate-dehydrogenase (G6PDH) (Lushchak, 2016). Thus GR is significant in regulating the intracellular concentration of reduced form GSH. Moreover, under the influence of the cystolic enzyme GSH-S-transferase (GST), the electrophilic xenobiotics and radicals easily conjugate with the electron rich -SH group of GSH. These polar conjugates are further easily eliminated from the cell (Lu, 2009). The GSH/GSSG redox pair and the function of GSH related enzymes are summarized as follows.

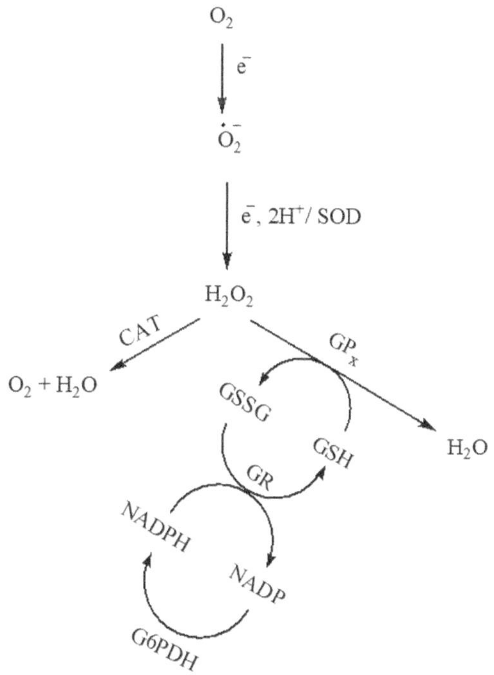

12.6.2 GLUTATHIONE (GSH) AS A CYSTEINE RESERVOIR

Cysteine is an important amino acid containing (-SH) thio group. It is essential in a variety of biological functions, including maintenance of redox balance and contaminant elimination, as well as the production of different proteins, iron-sulfur clusters, coenzyme A, taurine, etc. (Daher et al., 2020). It is also responsible for producing disulfide bridges which stabilize the three-dimensional folding of protein structure (Combs & DeNicola, 2019). The nucleophilic nature of thio group in cysteine provides it a unique ability to bind different electrophiles specially metal ions.

The intracellular concentration (1–10 mM) of GSH is much higher as compared to its extracellular concentration (1–10 μM). Thus, the intracellular redox balance is effectively maintained, however the extracellular oxidative damage is counteracted by cysteine (Hall et al., 2013). Under extracellular conditions, cysteine easily undergoes oxidation to form S-S bonding with another cysteine molecule to form cystine.

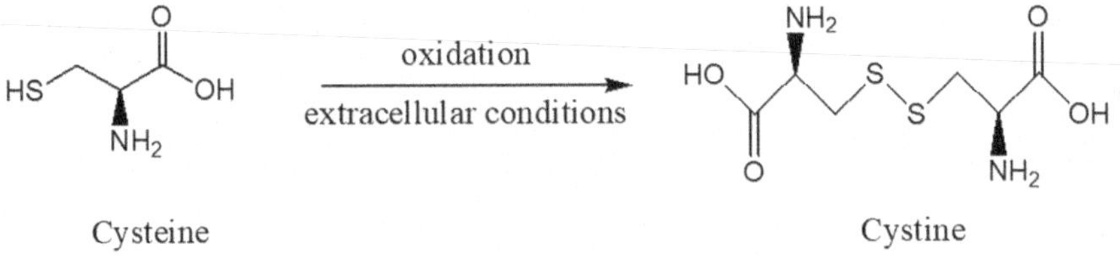

GSH is biosynthesized via trans-sulfuration pathway in several steps by successive combination of cysteine, glutamic acid and glycine (Hall et al., 2013). The initial step involves the formation of γ-glutamyl linkage in cysteine. In the presence of γ-glutamyl cysteine synthetase (GCS), the NH_2 group of cysteine binds with the γ-carboxyl group of glutamic acid. γ-glutamyl cysteine thus formed combines with glycine to form GSH under the action of the enzyme GSH synthetase (Forman et al., 2009). During each of these steps, one molecule of ATP hydrolyzes to form ADP. Figure 12.2 depicts the schematic representation for interconversion between cysteine and GSH.

The thio group of free cysteine while binding to metals, may easily get oxidized, which in turn, induces the reduced metal ion to undergo Fenton reaction producing deleterious $\dot{O}H$ radicals. Thus GSH biosynthesis through cysteine has a protective action in maintaining low levels of free intracellular cysteine (Jozefczak et al., 2012). In fact GSH serves as a consistent reservoir of cysteine (Hall et al., 2013). The γ-glutamyl linkage in GSH has a unique effect of rendering it inert towards the peptidase activity (Lu, 2009; Rubino et al., 2015). As a result, GSH is quite resilient in the intracellular environment and mainly undergoes metabolism extracellularly under the action of enzyme γ-glutamyl transferase (GGT) by transferring the γ-glutamyl moiety to an amino acid, forming corresponding γ-glutamyl amino acid and the cysteinyl–glycine conjugate as depicted in Figure 12.2. The cysteinyl–glycine conjugate thus formed is further dissociated by dipeptidase (DP) yielding back molecules of cysteine and glycine.

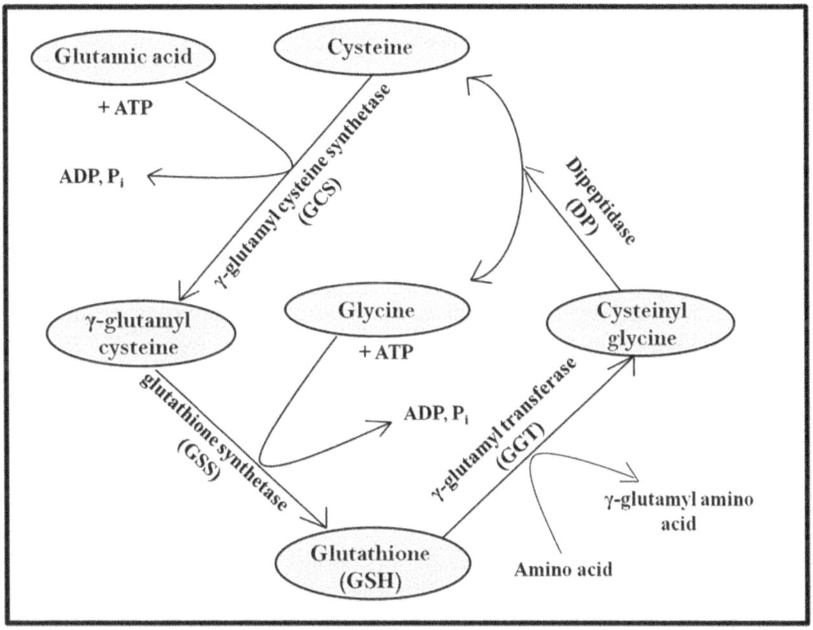

FIGURE 12.2 The interconversion between cysteine and glutathione.

12.6.3 ALTERATIONS IN GSH ACTIVITY

Several factors influence the impact of heavy metal exposure on GSH activity, including the fish species, the tissue being analyzed, the contaminant metal ion, its concentration as well as the exposure conditions (Atli & Canli, 2008). In a study conducted in Obidos lagoon, Portugal, the metal contamination was shown to increase GSH concentration in *Liza aurata* (Pereira et al., 2010). Similarly, exposure of *Channa punctatus* to excessive levels of Cu, Cd, Ni, and Fe reportedly caused an increase in GSH levels (Pandey et al., 2008). On the other hand, in another investigation, Ahmad et al. (2005) revealed decreased GSH concentration in gills and kidney tissues of *Anguilla anguilla* on exposure to Cu at 0.05 and 1.58 ppm concentrations. In a similar way, Monteiro et al. (2010) found that exposure to $HgCl_2$ caused an increase in GSH activity in liver and gill tissues of *Brycon amazonicus*.

Generally, the elevation in GSH activity may be attributed to increased GSH biosynthesis as a stress induced response to combat heavy metal toxicity. Similarly, reduction in GSH levels may be linked with increased formation of GSH-metal conjugates as well as oxidation of GSH to GSSG caused by ROS (Srikanth et al., 2013).

12.7 CONCLUSION

The aim of this review is to present a detailed picture of the toxicity mechanism associated with heavy metal buildup in fish tissues and the defense response provided by the GSH and related enzymes. Despite the availability of literature regarding GSH and its role in stress tolerance, a comprehensive mechanistic compilation of metal toxicity and GSH mediated antioxidant defense mechanism in fishes is much needed. ROS are frequently generated as byproducts of cellular respiration in the mitochondrial ETC. However, by participating in Fenton and Haber-Weiss reaction, heavy metals may cause the overproduction of ROS, which damages various cell components. The intracellular environment within the fish tissues is maintained in a reduced state by GSH. Metals ions may either bind to GSH generating GSH-conjugates or cause ROS-induced oxidation of the -SH group forming GSSG. Moreover, the GSH-associated enzymes GP_x and GST assist in eliminating peroxides and GSH-conjugates, respectively. Similarly, GR AIDS in reducing back GSSG to GSH. The concentration ratio GSH: GSSG is often employed as a bio-indicator of contaminant-induced stress. GSH biosynthesis sustains lower levels of free cysteine, thus further protecting the cellular constituents from deleterious $\dot{O}H$ radicals.

KEYWORDS

- **antioxidant defense**
- **fish**
- **glutathione**
- **heavy metals**
- **ROS**

REFERENCES

Ahmad, I., Oliveira, M., Pacheco, M., & Santos, M. A., (2005). *Anguilla anguilla* L. oxidative stress biomarkers responses to copper exposure with or without β-naphthoflavone pre-exposure. *Chemosphere, 61*(2), 267–275.

Ahmad, M., Wolberg, A., & Kahwaji, C. I., (2018). Biochemistry, Electron Transport Chain. In: *StatPearls*. StatPearls Publishing, Treasure Island (FL); PMID: 30252361.

Atchison, G. J., Henry, M. G., & Sandheinrich, M. B., (1987). Effects of metals on fish behavior: A review. *Environmental Biology of Fishes, 18*(1), 11–25.

Atli, G., & Canli, M., (2008). Responses of metallothionein and reduced glutathione in a freshwater fish *Oreochromis niloticus* following metal exposures. *Environmental Toxicology and Pharmacology, 25*(1), 33–38.

Combs, J. A., & DeNicola, G. M., (2019). The non-essential amino acid cysteine becomes essential for tumor proliferation and survival. *Cancers, 11*(5), 678.

Daher, B., Vučetić, M., & Pouysségur, J., (2020). Cysteine depletion, a key action to challenge cancer cells to ferroptotic cell death. *Frontiers in Oncology, 10*, 723.

Eroglu, A., Dogan, Z., Kanak, E. G., Atli, G. Ü. L. İ. Z. A. R., & Canli, M. U. S. T. A. F. A., (2015). Effects of heavy metals (Cd, Cu, Cr, Pb, Zn) on fish glutathione metabolism. *Environmental Science and Pollution Research, 22*(5), 3229–3237.

Forman, H. J., Zhang, H., & Rinna, A., (2009). Glutathione: Overview of its protective roles, measurement, and biosynthesis. *Molecular Aspects of Medicine, 30*(1, 2), 1–12.

Hall, M. N., Niedzwiecki, M., Liu, X., Harper, K. N., Alam, S., Slavkovich, V., & Gamble, M. V., (2013). Chronic arsenic exposure and blood glutathione and glutathione disulfide concentrations in Bangladeshi adults. *Environmental Health Perspectives, 121*(9), 1068–1074.

Jaishankar, M., Tseten, T., Anbalagan, N., Mathew, B. B., & Beeregowda, K. N., (2014). Toxicity, mechanism and health effects of some heavy metals. *Interdisciplinary Toxicology, 7*(2), 60–72.

Jan, A. T., Azam, M., Siddiqui, K., Ali, A., Choi, I., & Haq, Q. M., (2015). Heavy metals and human health: Mechanistic insight into toxicity and counter defense system of antioxidants. *International Journal of Molecular Sciences, 16*(12), 29592–29630.

Javed, M., & Usmani, N., (2019). An overview of the adverse effects of heavy metal contamination on fish health. *Proceedings of the National Academy of Sciences, India Section B: Biological Sciences, 89*(2), 389–403.

Jezierska, B., & Witeska, M., (2006). The metal uptake and accumulation in fish living in polluted waters. In: *Soil and Water Pollution Monitoring, Protection and Remediation* (pp. 107–114). Springer, Dordrecht.

Jozefczak, M., Remans, T., Vangronsveld, J., & Cuypers, A., (2012). Glutathione is a key player in metal-induced oxidative stress defenses. *International Journal of Molecular Sciences, 13*(3), 3145–3175.

Kovačević, T. B., Borković, S. S., Pavlović, S. Z., Despotović, S. G., & Saičić, Z. S., (2008). Glutathione as a suitable biomarker in hepatopancreas, gills and muscle of three freshwater crayfish species. *Archives of Biological Sciences, 60*(1), 59–66.

Li, R., Jia, Z., & Trush, M. A., (2016). Defining ROS in biology and medicine. *Reactive Oxygen Species (Apex, NC), 1*(1), 9.

Liang, S. S., Shiue, Y. L., Kuo, C. J., Guo, S. E., Liao, W. T., & Tsai, E. M., (2013). Online monitoring oxidative products and metabolites of nicotine by free radicals' generation with Fenton reaction in tandem mass spectrometry. *The Scientific World Journal, 2013.*

Lu, S. C., (2009). Regulation of glutathione synthesis. *Molecular Aspects of Medicine, 30*(1–2), 42–59.

Lushchak, V. I., (2016). Contaminant-induced oxidative stress in fish: A mechanistic approach. *Fish Physiology and Biochemistry, 42*(2), 711–747.

Mahboob, S., (2013). Environmental pollution of heavy metals as a cause of oxidative stress in fish: A review. *Life Sci. J., 10*, 336–347.

Monteiro, D. A., Rantin, F. T., & Kalinin, A. L., (2010). Inorganic mercury exposure: Toxicological effects, oxidative stress biomarkers and bioaccumulation in the tropical freshwater fish matrinxã, *Brycon amazonicus* (Spix and Agassiz, 1829). *Ecotoxicology, 19*(1), 105–123.

Novotny, V., (1995). Diffuse sources of pollution by toxic metals and impact on receiving waters. In: *Heavy Metals* (pp. 33–52). Springer, Berlin, Heidelberg.

Pandey, S., Parvez, S., Ansari, R. A., Ali, M., Kaur, M., Hayat, F., & Raisuddin, S., (2008). Effects of exposure to multiple trace metals on biochemical, histological and ultrastructural features of gills of a freshwater fish, *Channa punctata* Bloch. *Chemico-Biological Interactions, 174*(3), 183–192.

Pereira, P., De Pablo, H., Vale, C., & Pacheco, M., (2010). Combined use of environmental data and biomarkers in fish (*Liza aurata*) inhabiting a eutrophic and metal-contaminated coastal system–Gills reflect environmental contamination. *Marine Environmental Research, 69*(2), 53–62.

Rubino, F. M., (2015). Toxicity of glutathione-binding metals: A review of targets and mechanisms. *Toxics, 3*(1), 20–62.

Sapkota, A. (2021). Electron transport chain: Definition, Components, Steps, FAQs. Microbe Notes. https://microbenotes.com/electron-transport-chain/

Sevcikova, M., Modra, H., Slaninova, A., & Svobodova, Z., (2011). Metals as a cause of oxidative stress in fish: A review. *Vet. Med., 56*(11), 537–546.

Shofstahl, J. H., & Hardy, J. K., (1990). Method for the determination of the priority pollutant metals by HPLC. *Journal of Chromatographic Science, 28*(5), 225–229.

Srikanth, K., Pereira, E., Duarte, A. C., & Ahmad, I., (2013). Glutathione and its dependent enzymes' modulatory responses to toxic metals and metalloids in fish—A review. *Environmental Science and Pollution Research, 20*(4), 2133–2149.

Wilhelm, F. D., (2007). Reactive oxygen species, antioxidants and fish mitochondria. *Front Biosci., 12*, 1229–1237.

Zitka, O., Skalickova, S., Gumulec, J., Masarik, M., Adam, V., Hubalek, J., & Kizek, R., (2012). Redox status expressed as GSH: GSSG ratio as a marker for oxidative stress in paediatric tumor patients. *Oncology Letters, 4*(6), 1247–1253.

CHAPTER 13

CORAL REEF: A SOURCE OF VALUE-ADDED COMPOUNDS BENEATH THE WATER

SHANTI BHUSHAN MISHRA[1] and SHRADHANJALI SINGH[2]

[1]*Department of Pharmacognosy, United Institute of Pharmacy, Prayagraj, Uttar Pradesh, India*

[2]*Department of Pharmaceutical Chemistry, United Institute of Pharmacy, Prayagraj, Uttar Pradesh, India*

ABSTRACT

Over the past decade, marine pharmacognosy and biotechnology have been applied in the fields of human disease and public health, seafood safety, and new material development. Dozens of promising marine life products are being further developed, including cancer therapies. Coral reefs are the ocean's most diverse and complex ecosystems in the oceans and are close relatives of sea anemones and jellyfish. According to the National Ocean Services based at USA, corals have been classified as an important source of new drugs to cure Alzheimer's, cancer, viruses, arthritis, heart disease, and bacterial infections among other serious illnesses. However, the promising importance of coral reefs as a huge resource of lifesaving and life-enhancing products is still not properly empathized by the public and policymakers. In spite of these findings, the therapeutic use of coral reefs is considered unexplored. This chapter illustrates about various compounds obtained from coral reefs and their associated biological activities.

Immune-Boosting Nutraceuticals for Better Human Health: Novel Applications. Urmila Jarouliya, PhD, Raj K. Keservani, MPharm, Rajesh K. Kesharwani, PhD, Virendra K. Patel, PhD, and Adi D. Bharti, MPharm (Eds.)

13.1 INTRODUCTION

Anonymity lies in the seas and oceans on account of large expansion. In view of exploring marine ecosystem, a new approach in technology plays a vital role. There are quite a lot of living and non-living substances that can be perilous to human despite of numerous bio-active compound present which can be helpful to the human race (Edwin et al., 2014). Corals are considered as minute creatures called polyps, which can be put together along with seabed. Considering habitat of astonishing environment there are rich sources suggesting profusion of life.

Coral reefs were come into picture since 500 million years ago considering warm tropical climates (Hoegh-Guldberg et al., 2017). Considering around 25% of diversified ocean coral reefs envelopes < 0.2% of the ocean base. Identification of around 5,000 kinds of reef fish, and additional 2,500 species of corals has been performed. Suggesting intensified prolific ecosystem, algae, and sea grasses were two important types of plants. With respect to uphold enormously fertile submarine world, still to be explored, the plants and animals exert intellectually to maintain balance to sustain numerous medicinal treasures. Suggesting amid the biodiversity, the coral system in the world, has the highest level. Several kinds of plants and animals were present in the ecosystem (Grimsditch, 2004).

With respect to investigating the assorted life forms, 40,000 years ago early aboriginals first considered it and studied it. Amongst seven natural wonders of the world, the reef extends for 2,300 km considering size, beauty, complexity, and rarity. Declaration as a world heritage site and Marine National Park for it was considered by governing body of Great Barrier Reef Marine Park Authority (GBRMPA, 2009).

13.1.1 STRUCTURE AND FUNCTION OF CORALS

13.1.1.1 STRUCTURE

A coral reef is made up of calcium carbonate, or limestone. It is absorbed by water by coral polyps and colonies of coralline algae. Most of the underlying base of the reef is dead, containing sheet of coral skeletons.

Considering the extensive base structure, the living reef is put together on the top, suggesting minute coral polyps considering new level (Ian Whittaker et al., 2016). Surrounding the polyp, calcium carbonate forms skeleton. Several colored tentacles of the coral polyps serve the beautiful color of corals and the zooxanthellae algae survived in the tissues of numerous species. Amid the structural features of corals, they are considered as a primordial group comprising a simple, radially structured body by way of a single aperture that serves as a mouth and anus. Two layers of cells suggest being composed in the body, differentiating by a jelly-like sheet with the absence of internal organs. Amidst the food component, corals acquires particular vicious cells named nematocysts on the retractable tentacles (Figure 13.1). Oceanic upwelling incidentally experiences cold weather, which brings nutrients to float up to the surface. This facilitates the development of plankton (Coral Reef Alliance, 2020). Being small carnivores, including coral polyps apart from plankton-rich waters, tropical oceans do not give plenteous feeding floors. These small creatures will flourish and thus making amazing reefs. The survival of microscopic algae in large numbers in the tissues had solved the mystery. Within hermatypic corals, these algae, named zooxanthellae, live symbiotically (Cox, 1986).

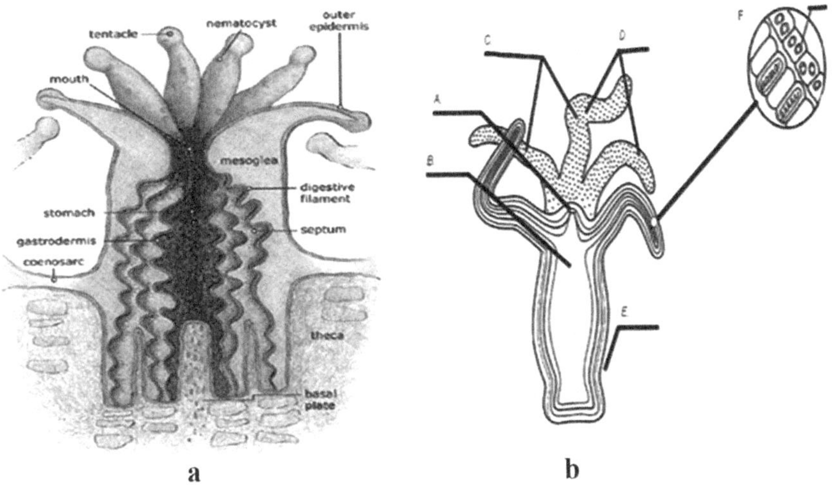

FIGURE 13.1 (a) Diagrams of coral polyp; and (b) coral showing different body parts: (A) mouth (taken food and get rid of waste); (B) stomach (digest the captured prey); (C) nematocysts or stinging cells (help the polyp capture food and protect the polyp from predators); (D) tentacles (act as the animal's primary ways of defense); (E) epidermis (outer tissue layer of the polyp); and (F) zooxanthellae (make food for the polyp through photosynthesis).

Source: a) Reprinted from National Ocean Service, 2020.

Coral produces nitrogen wastes which were absorbed by Zooxanthellae. Along with the nutrients, sunlight, and carbon dioxide has been used by the algae to compose sugars during the process of photosynthesis. Amid providing and metabolizing wastes products by the algae, some of this food is absorbed by the coral polyp. Around 98% of the food for the coral produces sugar by zooxanthellae (National Ocean Service, 2020). With respect to a sunny place, the algae provide a safe place to live for corals. This type of linking is called mutualism. Therefore, the zooxanthellae by way of their brilliant red, orange, and brown, provide corals devoid of doing any work, clean, and well-fed environment.

Considering the classification, although the looks of coral polyps were similar to that of plant, in reality, it is an animal classified into the Phylum Cnidaria (Coelenterata). Phylum Cnidaria is further subdivided into three classes: the hydrozoans (Class Hydrozoa); the jellyfish (Class Scyphozoa); and the corals and sea anemones (Class Anthozoa) (Darwin, 1962). Corals have been considered as the living creature composed of various chemical entities; hence new therapeutic drugs and industrial processes considers to have potential application.

Bioactivities against certain human ailments were explained by greater than 100 new chemicals already been accomplished by corals. Investigation as potential pharmaceutical products with therapeutic benefits for humankind has led to development in the field of research. Amongst the explorations, a group of chemicals is more effective alongside breast, lung, and blood cancers as compared to current drugs, and an additional moiety which is toxic to a malarial parasite (National Research Council, 1999).

13.2 CHEMICAL COMPOUNDS AND BIOACTIVITIES SHOWN BY CORALS

In general, sea sponges and ascidians have the highest levels of bioactive chemicals of all well-known animals. Most of the new classes of chemical compounds and their structures discovered in the last few decades come from sponges. Some of these chemicals and the bioactivities associated with these one is outlined in subsections.

13.2.1 ECTEINASCIDIN-743

A marine tetra hydro isoquinoline alkaloid, Ecteinascidin-743 (ET-743), isolated from *Ecteinascidia turbinata*, that has shown very potent anti-tumor activity in preclinical phases. Ecteinascidin 743 may kill cancer cells by destruct their DNA and stop them from dividing. Clinical trials have shown that this compound has anti-cancer action in patients with advanced sarcoma and osteosarcomas. Studies showed that this chemical stops the formation of microtubules in cancer cells which are required for normal mitotic cell division and growth of tissues. If tumor cells cannot undergo mitosis, then the development of the tumor is prevented (Erba et al., 2001).

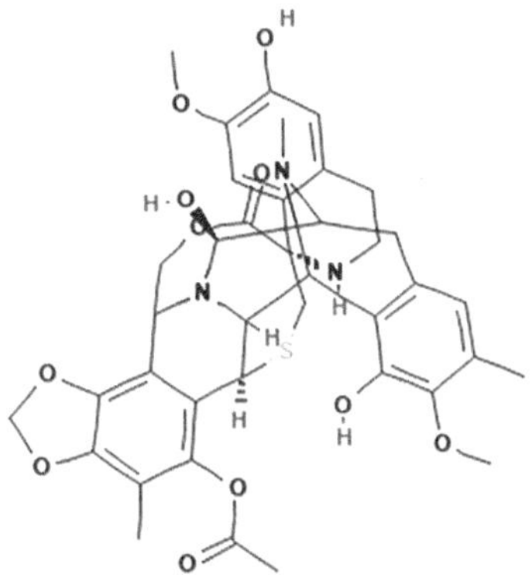

Chemical structure of Ecteinascidin-743.

13.2.2 STYLISSADINE A AND B

It is obtained from the sponge of great barrier reef, i.e., *Stylissa flabellate*. This chemical compound acts against pain receptor named P2X7 hence known as P2X7 antagonists. They could offer new treatment options for inflammatory diseases such as rheumatoid arthritis, osteoarthritis (OA), and chronic obstructive pulmonary diseases (COPD) (Buchanan et al., 2007).

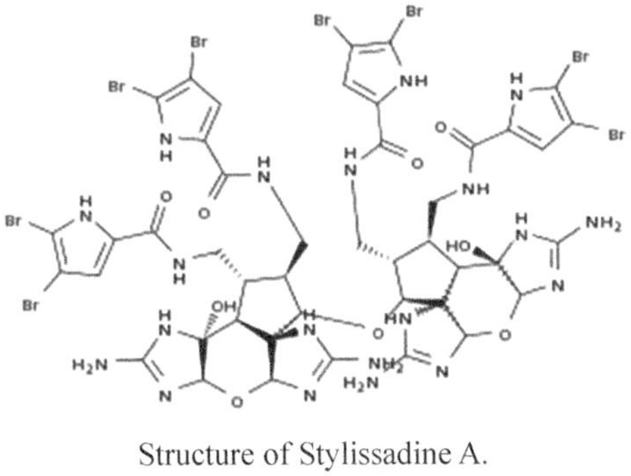

Structure of Stylissadine A.

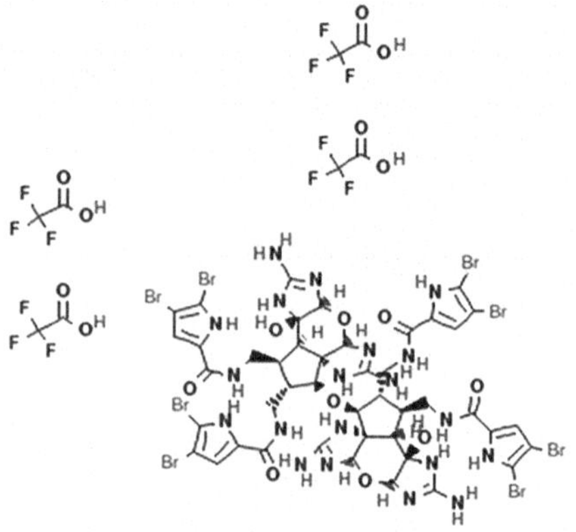

Structure of Stylissadine B.

13.2.3 PSAMMAPLIN A AND BISAPRASIN 11′-SULFATE

It is obtained from sponge named as *Aplysinella rhax*, from the Great Barrier Reef. This compound has been successfully tested at preclinical levels for controlling cardiovascular function, blood flow, and metabolism. Chemically, Psammaplin A, is a symmetrical bromotyrosine disulfide. In 1991, Jimenez & Crews isolated psammaplins A, B, C, and D and prepsammaplin A from *Psammaplysilla purpurea*. Bisaprasin and psammaplins A and D have been reported for the prevention of growth of Gram-positive bacterium *Staphylococcus aureus*. Researches on the chemical components of *Aplysinella rhax* extract, that inhibited the binding of [3H]1,3-dipropyl-8-cyclopentylxanthine ([3H]DPCPX) on adenosine A1 receptors, results in the isolation of two new derivatives of the psammaplin family, psammaplin A 11′-sulfate, bisaprasin 11′-sulfate, along with psammaplin A (Ngoc et al., 2000).

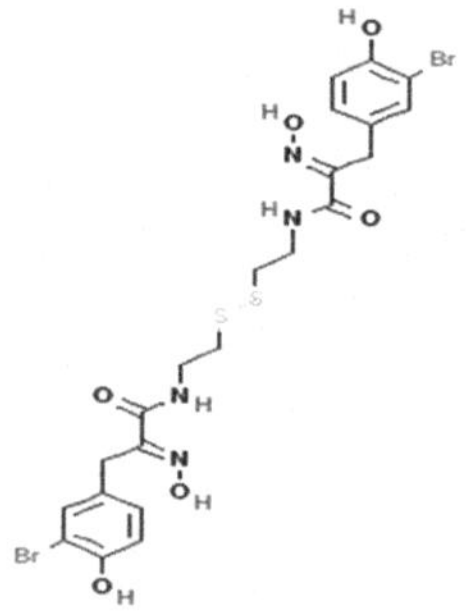

Structure of Psammaplin A.

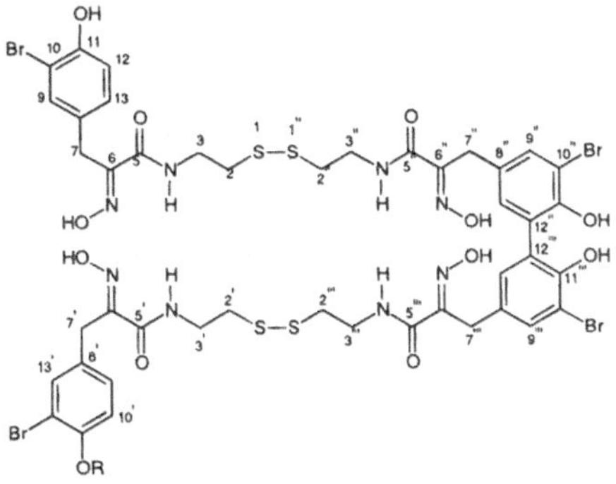

R = H Bisaprasin
R = SO_3Na Bisaprasin 11′-sulfate

Structure of Bisaprasin 11′-sulfate.

13.2.4 ADOCIASULFATE 1,7,8,9

It is obtained from a sponge known as *Adocia aculeata,* from largest coral system Great Barrier Reef. This compound acts as an inhibitor of protein pump that is important for bone resorption cell activity. Adocia sulfate 1, adocia sulfate 7, and adocia sulfate 8, separated from an extract of the sponge *Adocia* sp. are inhibitors of proton pump in chicken bone-derived membrane vesicles. Structural elucidation by 2D-NMR spectroscopy showed that these are novel hexaprenoid hydroquinone sulfates (John et al., 1999).

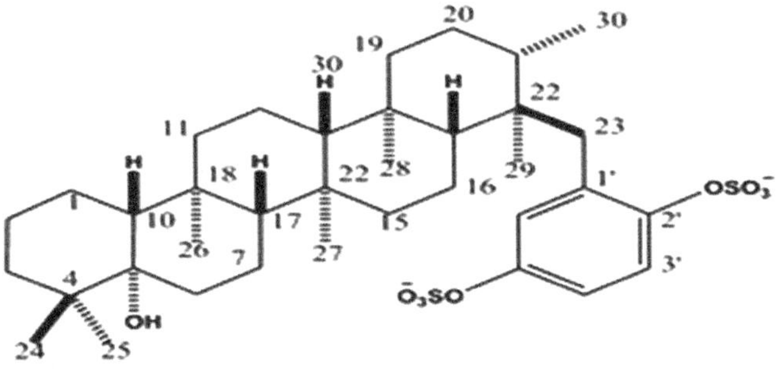

Structure of Adociasulfate 8.

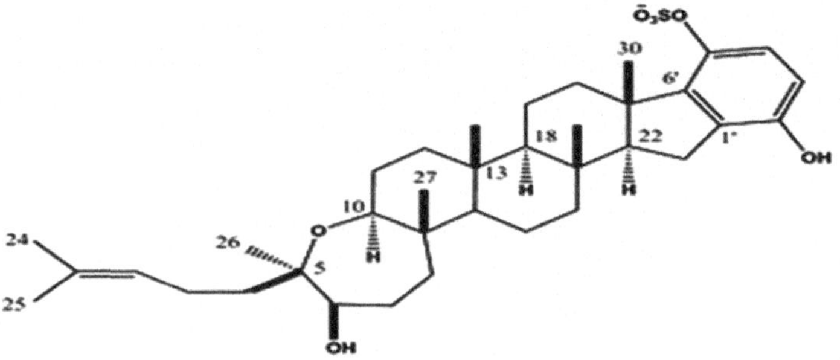

Structure of Adociasulfate 9.

13.2.5 DYSINOSIN A

It is isolated from *Citronia astra,* a new species of sponge, from the outer Great Barrier Reef system. This chemical entity is a theoretically perfect drug due to its small chemical structure. This new natural product dysinosin A has been obtained from sponge of the family *Dysideidae* found near Lizard Island, North Queensland, Australia. Dysinosin A is a powerful inhibitor of the factor VIIa blood coagulation cascade and serine protease thrombin (Zhang et al., 2017).

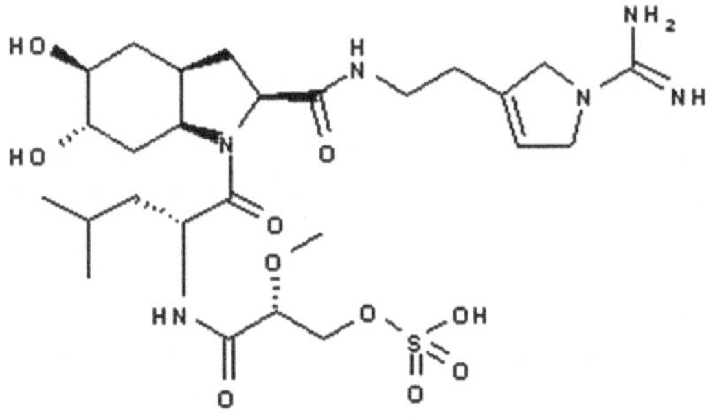

Structure of Dysinosin A.

13.2.6 AXINELLAMINES B, C, AND D

It is isolated from a new species of *Axinella* sp. *1333*., A sponge that produces a gastric bactericide that may be effective in treating and preventing gastric and duodenal ulcers and stomach cancer. Baran and coworkers have completed the racemic synthesis of Axinellamines A and B as well as the synthesis of postulated species 1,9-dideoxy-pre-Axinellamine. Both syntheses proceed through a highly functionalized cyclopentene intermediate which dictates the stereochemistry of the final product. Baran's synthesis is an achievement that marks the first synthesis of these marine natural products however several steps suffer from poor diastereo-selectivities and low yields (Yamaguchi et al., 2008; O'Malley et al., 2007).

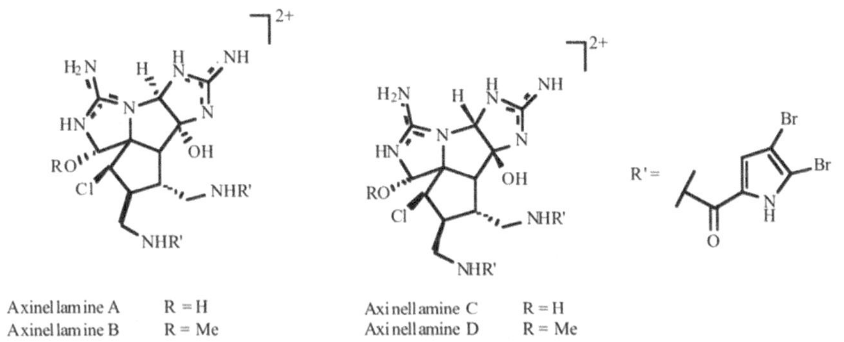

Structure of Axinellamine.

13.2.7 EXIGUAQUINOL

It is derived from the *Neopetrosia exigua, a* species of sponge found in tropical Australasia. This compound shows corresponding functions to the chemicals from *Axinella sponge*. Exiguaquinol was isolated from the marine sponge, *Neopetrosia exigua*, which contains five fused rings, four contiguous stereogenic centers, an aryl sulfate and an alkyl sulfonate in its structure. Exiguaquinol is a significant lead compound for H. pylori infection without disruption of beneficial gut bacteria. However, due to the paucity of substance isolated from remote natural sources, limited studies have been carrying out on exiguaquinol and synthesis is the best way to access larger quantities of exiguaquinol for testing (Qaralleh, 2016; Schwarzwalder et al., 2016).

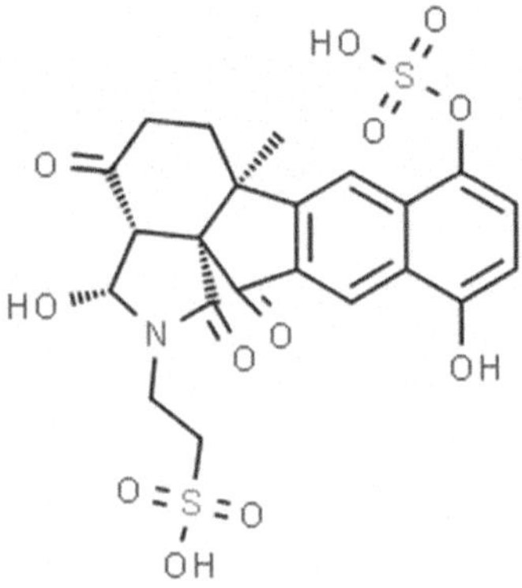

Structure of Exiguaquinol.

13.2.8 FLEXIBILIN D AND FLEXIBILIQUINONE

It was isolated from the soft corals, *Sinularia flexibilis* and *Sinularia gaweli* respectively. The metabolites of Flexibiliquinone and Flexibilin D have possessed the property to inhibit the deposition of the pro-inflammatory iNOS and COX-2 derived proteins of the lipopolysaccharide (LPS) stimulated RAW264.7 macrophage cells (Hu et al., 2013; Lin et al., 2013).

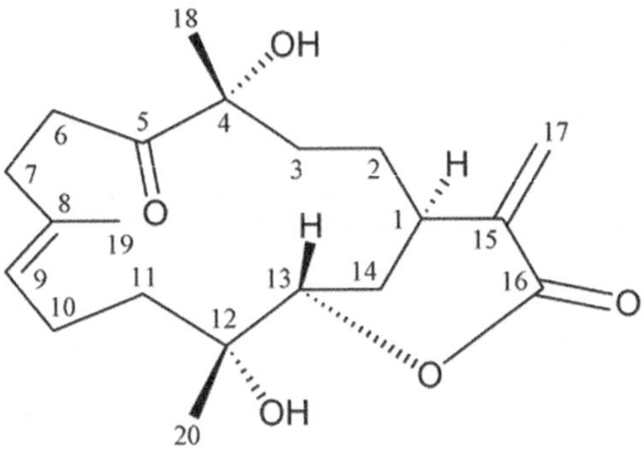

Structure of Flexibilin D.

13.2.9 11-EPI-SINULARIOLIDE ACETATE AND SINULARIN

The Cembrane type compound 11-*epi*-sinulariolide acetate was isolated and purified from soft coral *Sinularia querciformis*. 11-*epi*-sinulariolide acetate significantly inhibited the nitric oxide (NO) synthase induced by expression of the proinflammatory proteins and cyclooxygenase-2 in LPS stimulated murine macrophages. Because of its decreasing the expression of pro-inflammatory proteins and disease progression in adjuvant induced arthritis rats model, this marine chemical Ya-s11 may act as a valuable therapeutic compound for the treatment of RA (Lin et al., 2013; Huang et al., 2012).

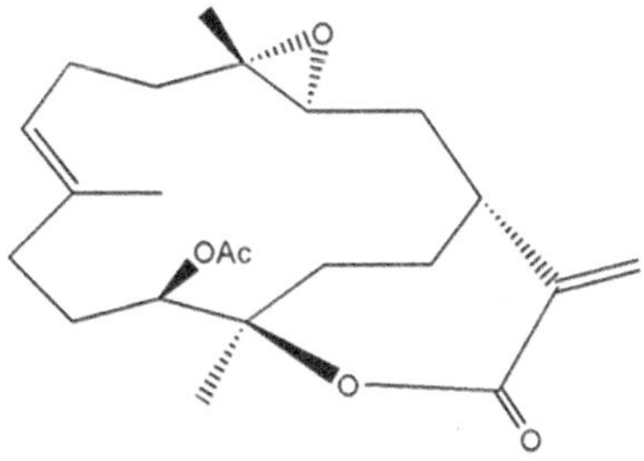

Structure of 11-epi-Sinulariolide.

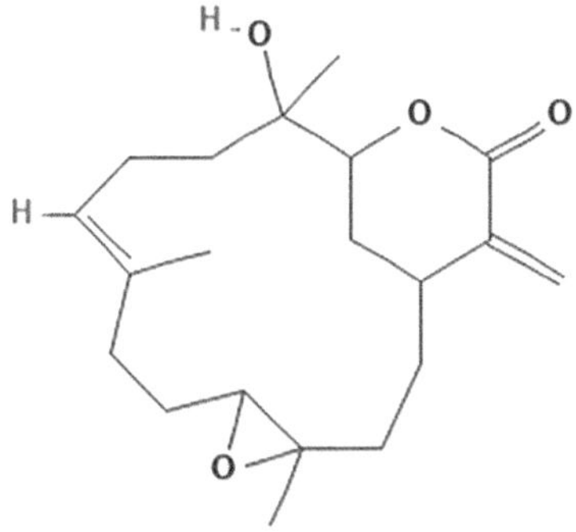

Structure of Sinularin.

13.2.10 5α,8α-EPIDIOXYSTEROL

A new chemical entity 5α,8α-epidioxysterol, i.e., yalongsterol A, accompanied by two known similar steroids 5α,8α-epidioxy-24-methyl-cholesta-6,24(28)-dien-3β-ol and (22E,24S)-5α,8α-epidioxy-24-methyl-cholesta-6,22-dien-3β-ol, were isolated from the soft coral *Sinularia sp.* collected from South China Sea. Their structural elucidation was done by extensive spectroscopic analysis and obtained spectral data compared with available literature (Yang et al., 2019).

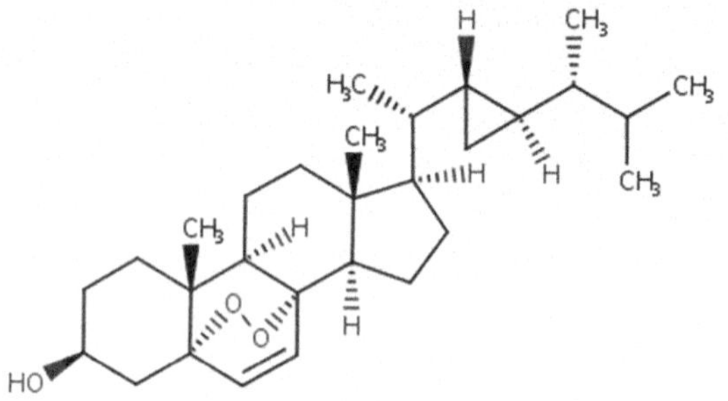

Chemical structure of 5α,8α-Epidioxysterol.

13.2.11 9,11-SECOSTEROL

9,11-secosterol, 24-nor-9,11-seco-11-acetoxy-3 beta,6 alpha-dihydroxy-cholest-7,22(E)-dien-9-one, showed growth inhibitory (IC_{50} below 10 μM and cytotoxic activities against Ehrlich ascites tumor cells, human leukemia K562, and human cervical cancer HeLa, *in vitro*. During inhibition of cell proliferation, 9,11-secosterol causes restoration of HeLa and K562 cells in the metaphase stage of mitosis. Therefore, abnormal mitosis may play a central role in the cytotoxicity of 9,11-secosterol in these cell lines (Lopp et al., 1994).

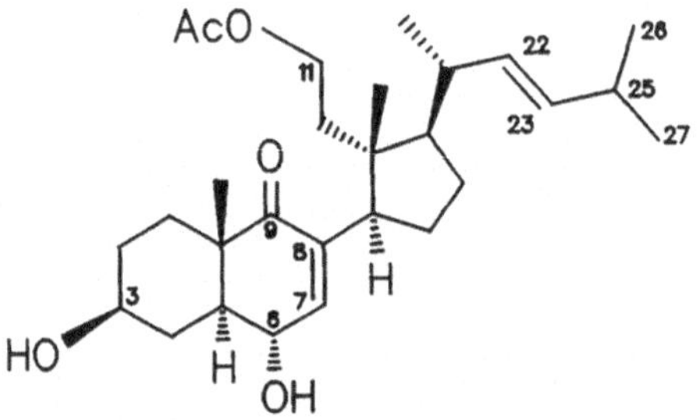

Chemical structure of 9,11-secosterol.

13.2.12 5-EPISINULEPTOLIDE ACETATE (5EPA)

5-Episinuleptolide acetate (5EPA), an anticancer agent of norcembranoidal diterpene category recently isolated from the Formosan soft coral *Sinularia* sp., showed potent activity against cancer cell lines K562, Molt 4 and HL

60. The anti-proliferative assay and annexin V-FITC/propidium iodide (PI) apoptotic assay, stated that the HL 60 cell line is more sensible towards 5EPA. This diterpenoid led to the activation of caspases-3, -8, and -9 as well as PARP cleavage. It also induced the formation of ROS, accumulation of calcium and disruption of mitochondrial membrane potential (Huang et al., 2013).

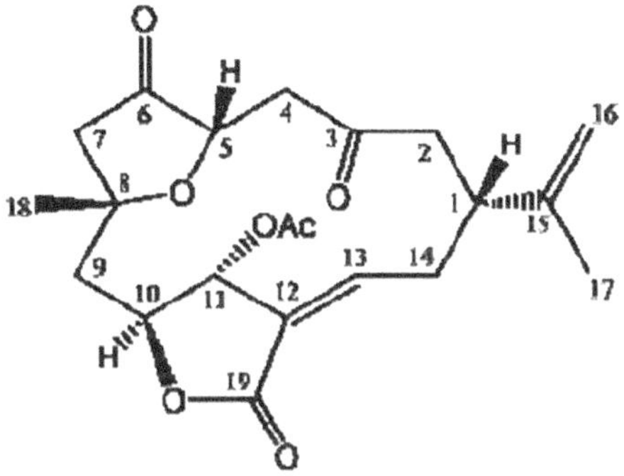

Chemical structure of 5-Epsinuleptolide acetate (5EPA).

13.2.13 SCLEROSTEROIDS

There are five recently investigated pregnane-type steroids, sclerosteroids J to N, and a diterpenoid with a new chemo variety 3-methyl-5-(10′-acetoxy-2′,6′,10′-trimethylundecyl)-2-penten-5-olide, have been isolated from a soft coral *Scleronephthya gracillimum*. Sclerosteroids show cytotoxicity against A549, MDA-MB-231 and HepG2, cancer cell lines and also substantially inhibit the accumulation of the pro-inflammatory inducible nitric oxide synthase (iNOS) protein and could effectively reduce the accumulation of cyclooxygenase 2 protein in LPS stimulated RAW264.7 macrophage cells. Nine new compounds of steroidal category, sclerosteroids A–I were isolated from the *Scleronephthya gracillimum which is an* azooxanthellate type soft coral. The chemical arrangement of sugar moieties in steroidal glycosides was determined by high performance liquid chromatography analysis of the *o*-tolylthiocarbamate derivatives of the liberated sugars from hydrolysis of these steroidal glycosides. Anti-inflammatory and cytotoxic activities of these steroidal compounds were measured using *in vitro* methods (Fang et al., 2012, 2013).

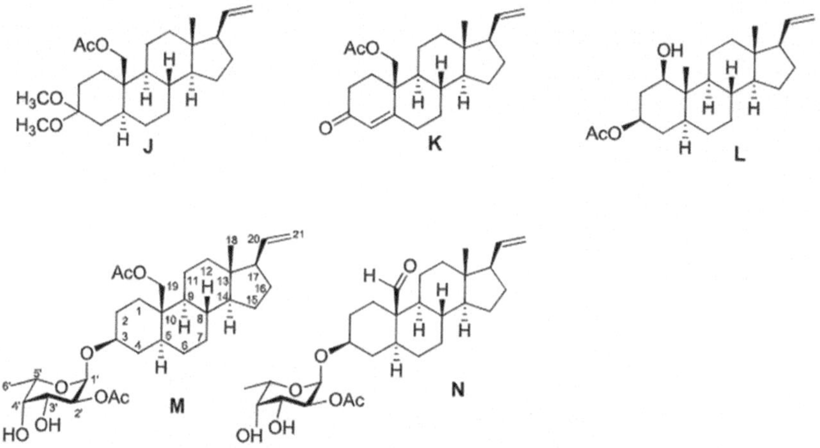

Chemical structure of sclerosteroids (J-N).

13.2.14 PARAMINABIC ACID

Three new compounds of steroidal carboxylic acids class, paraminabic acids A–C have been extracting out from a formosan type soft coral *Paraminabea acronocephala*. It shows strong cytotoxicity against the cancer cell lines, i.e., MDA-MB-231, Hep3B, MCF-7, and A-549 (Chao et al., 2013).

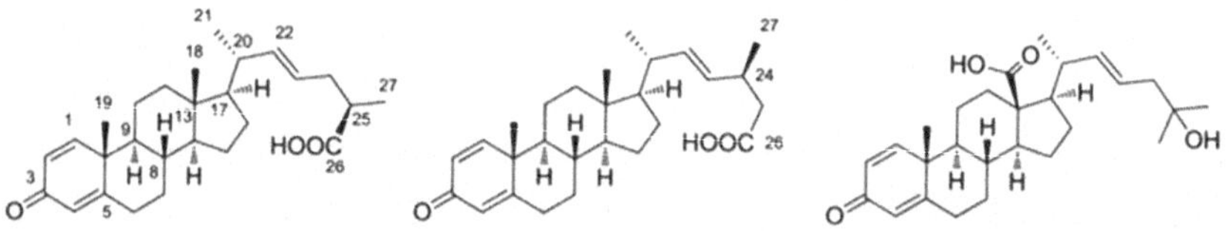

paraminabic acid A paraminabic acid B paraminabic acid C

Chemical structure of Paraminabic acids (A–C).

13.2.15 CAPNELLENES

The isolation of acetylated capnellenes was performed by acetone extraction of fresh soft coral *Capnella imbricata.* Four isolated compounds were named as 3β-acetoxy-Δ9(12)-capnellene-8β,10α-diol, 5α-acetoxy-Δ9(12)-capnellene-8β,10α-diol, 3β,14-diacetoxy-Δ9(12)-capnellene-8β,10α-diol, and 2β,5α-diacetoxy-Δ9(12)-capnellene-8β,10α-diol. From the acetone extract and methylene chloride extracts of the *Capnella imbricata*, 8β-acetoxy-Δ9(12)-capnellene-10α-ol, Δ9(12)-capnellene-10α-ol-8-one, Δ9(12)-capnellene-8β,15-diol, Δ9(12)-capnellene-8β,10α, 13β-triol, Δ9(10)-capnellene-12-ol-8-one, 8β,10α-diacetoxy-Δ9(12)-capnellene, and 8β-acetoxy-Δ9(12)-capnellene, were purified. Capnellenes are major active compounds which shows cytotoxic, antimicrobial, and anti-inflammatory activities. Capnellene-8β,10α-diol and its acylated derivatives show significant cytotoxicity against cervical epitheloid carcinoma (HeLa, medulloblastoma (Daoy), oral epidermoid (KB), and colon adenocarcinoma (WiDr) human tumor cell lines tested using *in vitro* methods (Shen et al., 2008). The diol was also useful against breast and colon cancer cell lines, human leukemia and renal leiomyoblastoma. Recent researches shows that Capnellene derivatives could be useful for the treatment of neuropathic pain. Neuropathic pain is assessed by damage to central or peripheral nerves that results in pathological nociceptive transmission, a neuronal process that responds to virulent stimuli (Jean et al., 2009; Chang et al., 2008).

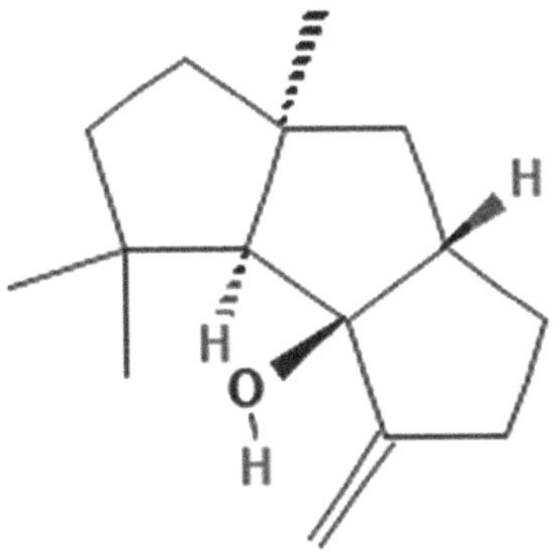

Chemical structure of Capnellene.

13.2.16 LEMNALOL

An ylangene-type sesquiterpenoid named as Lemnalol (8-isopropyl-5-methyl-4-methylene-decahydro-1,5-cyclo-naphthalen-3-ol) is isolated from Japanese soft coral Lemnalia tenuis and Fosmasan soft corals Lemnalia cervicorni. Some studies have reported that lemnalol progressively inhibits the expression of iNOS and cyclooxygenase 2 in LPS-stimulated macrophages, with significant antinociception effects in carrageenan-induced inflammation and chronic constriction injury-induced neuropathy in experimental animals (Lee et al., 2013; Kikuchi et al., 1983).

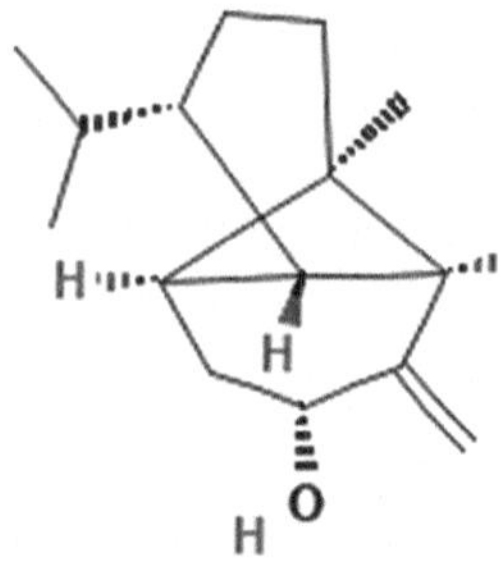

Chemical structure of Lemnalol.

13.2.17 ECHINOCLERODANE A

Echinoclerodane A is the initial clerodane-type derivative extracted from the marine creatures belonging to the phylum *Cnidaria* and this chemical element showed cytotoxicity and anti-inflammatory activity. A novel clerodane-type diterpene, echinoclerodane A was extracted from a Formosan gorgonian coral Echinomuricea sp. Echinoclerodane A showed moderate cytotoxicity towards HL-60, DLD-1, MOLT-4, and LoVo tumor cells and inhibitory effects on superoxide anion formation and elastase release by human neutrophils (Cheng et al., 2012).

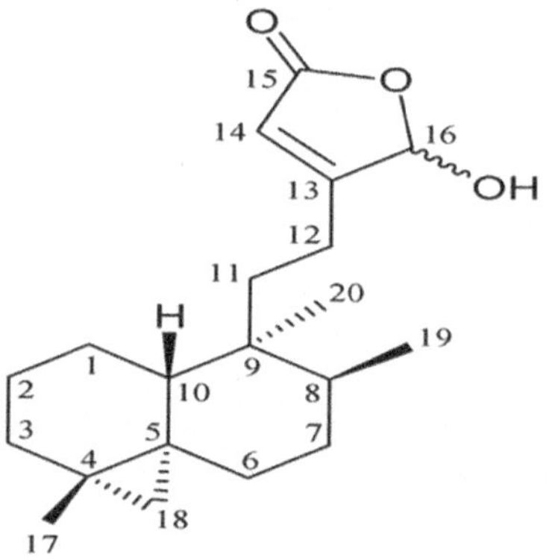

Chemical structure of Echinoclerodane A.

13.2.18 NEBROSTEROIDS Q, R, AND S

A novel 19-oxygenated steroid, nebrosteroid Q and two new 19-norergosterols, nebrosteroids R and S having cytotoxic activity were extracted from the soft coral *Nephthea chabrolii* in San-Hsian-Tai. Their chemical structures have been elucidated by spectral analysis, and their anticancer property against certain cancer cells and anti-viral activity against human cytomegalovirus (HCMV) were determined *in vitro* (Wang et al., 2013).

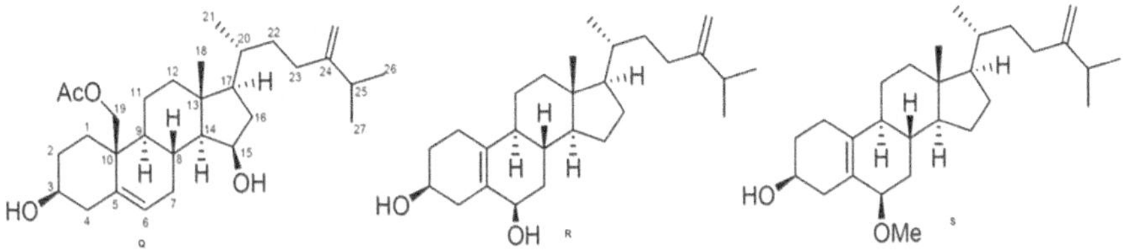

Chemical structure of Nebrosteroids (Q-S).

13.2.19 CYTARABINE

It is an antineoplastic compound that owes its existence to the marine sponge *Cryptotheca crypta* which is common on coral reefs in the Caribbean Sea and inhibits DNA synthesis. Its effects are peculiar to the S phase of the cell cycle to pause usual cell development and division. Cytarabine is metabolized to its triphosphate form (cytosine arabinoside triphosphate) intracellularly. It also known as cytosine arabinoside (ara-C), is a chemotherapy medication used to treat acute lymphocytic leukemia, myeloid leukemia, chronic myelogenous leukemia, and non-Hodgkin's lymphoma. It also has antiviral and immunosuppressive properties (Aronson, 2016; Novotny & Rauko, 2009).

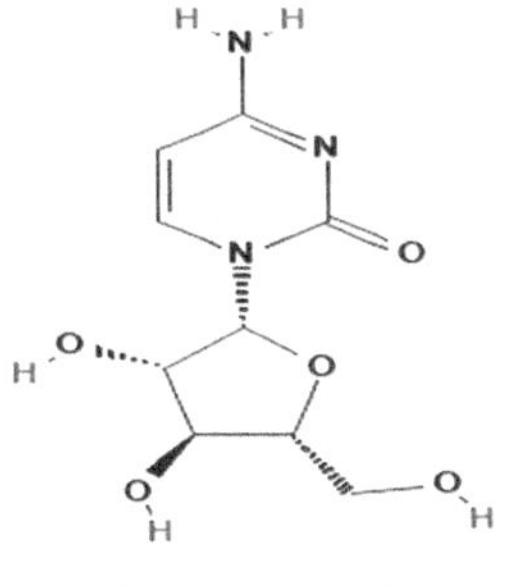

Chemical structure of Cytarabine.

13.2.20 MANZAMINES

Manzamines are complex *β*-carboline alkaloids isolated from Indonesian-Pacific marine sponges and are characterized by having a complicated and new nitrogen-containing polycyclic system. Higa et al. (1986) first time reported manzamine A which was isolated from the Okinawan marine sponge of the genus *Haliclona.* Till date, there are 16 species which belongs to 8 families of marine sponges have been confirmed to yield *β*-carboline manzamine and manzamine-related alkaloids. The occurrence of manzamine alkaloids in a variety of unrelated sponges has led to the assumption of a probable microbial origin for the biosynthesis of these compounds. Manzamine A was exposed to have potent anti-malarial activity in an assay against rodent malaria parasite *Plasmodium berghei in vivo*. Kara et al. at National University of Singapore (NUS) tested (–)-8-hydroxymanzamine A, manzamine A, and manzamine F, along with the existing antimalarial drugs, artemisinin, and chloroquine, with mice infected with *Plasmodium berghei.* It is fascinating to note the difference in the effects and structures of the manzamines tested. It appears that the functionality on the 8-membered ring of manzamines is important for activity (Ang et al., 2000; Hamann, 2007; Radwan et al., 2012).

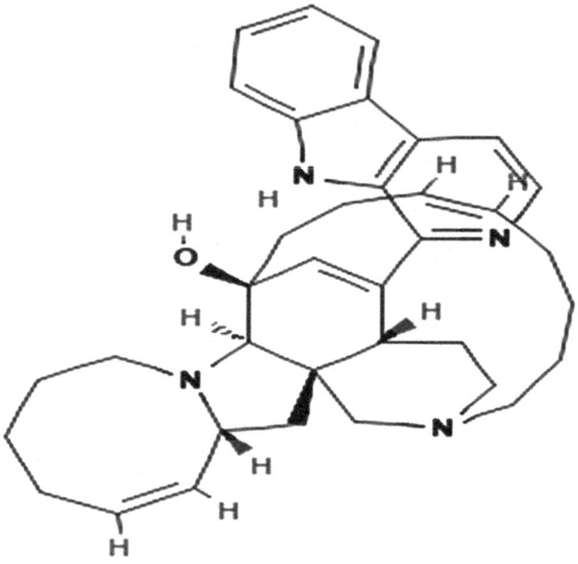

Chemical structure of Manzamine A.

13.3 BIOACTIVITIES IN CORALS

13.3.1 ANTI-INFLAMMATORY ACTIVITY

Inflammation is a physiological defensive mechanism that is stimulated by the immune system upon encountering unfamiliar invaders in an attempt to eliminate the adverse stimuli and begin recovery of the organism.

Diterpenes are one precise class of compounds isolated from corals that have been exhibit therapeutic benefits, particularly anti-inflammatory. Three diterpenoids isolated from *Eunicea fusca*, a gorgonian coral indigenous to Florida, have been found to have anti-inflammatory effects superior to indomethacin (Cooper et al., 2014), a commonly used anti-inflammatory medication. Another diterpene isolated from the same coral species, fucoside E, exhibits the same antimicrobial activity as well as anti-inflammatory activity. Sinularin, a compound, hinders pathways that increase the severity of neuroinflammation. One specific clinical relevance that has proven to show potential in relation to coral natural products is rheumatoid arthritis. 11-epi-sinulariolide acetate (Ya-s11), a cembrane-type compound, have been isolated from the soft coral, *Sinularia querciformis*, and evaluated for its *in vitro* anti-inflammatory activity as well as in adjuvant-induced arthritis in rats (Lin et al., 2013). AIA is the distinct murine model equivalent to rheumatoid arthritis used in experiments to check treatment options. Ya-s11 had exhibited the strong inhibition of the production of proinflammatory proteins inducible nitric oxide synthase (iNOS) and COX-2 in murine macrophages. In the adjuvant induced arthritic (AIA) mice, Ya-s11 completely reduced the symptoms of arthritis by inhibiting migration of inflammatory mediators to joints and preventing bone destruction (Chen et al., 2017).

13.3.2 ANTICANCER ACTIVITY

Sinularia, a natural metabolite from which an agglutinin called sinularian was isolated and found to agglutinate experimental targets, e.g., rabbit erythrocytes and murine leukemia cell lines, but not human or sheep erythrocytes. *Sinularia crassa* were found to downregulate expression of proinflammatory proteins and showed anticancer activity against human liver cancer cells viz. HepG2 (Cooper et al., 2014). Moreover, numerous compounds have also been isolated from corals of this genus Sinularia, including a new nitrogenous diterpenoids, novel, and known lobanes, and cembranes. The lobanes and cembranes had been tested for cytotoxic activity against three human cancer cell lines viz. H460, SF268 and MCF7 and exhibited a 50% inhibition of tumor growth with no selectivity. One more compound, 13-acetoxysarcocrassolide, separated from the soft coral *Sarco-phyton crassocaule* applied for cytotoxic activity against bladder cancer cells (Wright et al., 2012).

Three new steroidal compounds, nebrosteroids Q, R, and S, from the soft coral *Nephthea chabrolii* were isolated and confirmed for its antiviral and anticancer activity. Researches shows that these three steroids showed cytotoxicity against the P-388 (mouse lymphocytic leukemia) cell line. The anticancer properties of coral derived molecules are particularly attractive, moreover, molecules isolated from other soft corals like *Lobophytum michaelae,* Sarcophyton, and *Nephthea chabrolii* also possess anticancer properties.

13.3.3 NEUROPROTECTIVE ACTIVITY

One of the most likely treatment methods that can be added to the advantages of soft corals is its ability to reduce neurological deficits. 11-dehydrosinularolide, a natural metabolite from corals can be used to treat Parkinson's disease. Austrasulfone, which is isolated from the soft coral *Cladiella australis*, established for neuroprotective characteristics via anti-inflammatory pathways (reduction of COX-2 and iNOS) (Cooper et al., 2014). It also showed efficacy in reducing neuropathic pain and retards the progression of multiple sclerosis and atherosclerosis models in rodents. One more molecule, capnellene which is isolated from a soft coral, *Capnella imbricata* exhibits anti-inflammatory and antinociceptive actions by reducing iNOS and COX-2 (Jean et al., 2009).

13.4 SIGNIFICANCE OF CORALS

Coral reefs are the most biologically diverse and precious ecosystems on the earth, which provides valuable and important ecosystem benefits. Coral ecosystems are a food source for millions of people; provide shelter, protect shorelines from storms and erosion, maintain aquatic ecosystem for economically important fish species; providing job and income to local economy through recreation, fishing, and tourism; and are the source of new drugs and hotspots of oceanic biodiversity.

13.4.1 BIODIVERSITY

Biodiversity is the assortment of live creatures that could be found in a particular spot, local region, ecosystem, planet, and so on. Coral reefs are believed to have the most noteworthy biodiversity of any environment on the earth even more than a evergreen rainforest. Occurrence not more than 1% of the seabeds, coral reefs are places to more than 25% of oceanic creatures. Coral reefs are the ultimate nursery, spawning, breeding, hatching, and feeding grounds for several organisms. As far as biodiversity, the diversity of species existing on a coral reef is more noticeable than in any other shallow-water oceanic environment, and it is one of most diverse biosystem on the planet, yet coral reefs covers less than one tenth of 1% of the ocean base. Coral reefs provides shelter to more than 800 hard coral reef species and about 4,000 species of fishes. Elkhorn coral (*Acropora palmata*) and Staghorn coral (*Acropora cervicornis*), are the prime species living on present Caribbean reefs. They are mostly fast growing and able to manage the damage caused by hurricanes and by hasty change in sea level. Caribbean reefs also house lettuce coral (*Agaricia lamarki*), brain coral (*Colpophyllia natans*), scleractinian coral (*Montastrea annularis*), star coral (*Montastrea faveolata*), greater star coral (*Montastrea cavernosa*) and shet coral (*Agaricia lamarcki*). Coral reefs are patchy habitats but they give tremendous breeding grounds for a lot number of species. Moray eels (*Gymnothorax moringa*), Sea fans (*Gorgonia*), crustaceans and conch, long spine sea urchins (*Diadema antillarum*), and other molluscs are considered in corals (Barlow, 1981).

13.4.2 COASTAL PROTECTION

Coral reefs are viable natural barrier from waves which has permitted a number of individuals to settle on tropical shores and reef islands. Future sea level rise estimates are often viewed as the most significant threat to these lowlands. However, the interaction between more frequent high tide flooding in the sea and wave climate on reef flats is likely to be an important mechanism controlling coastal flooding and erosion on tropical beaches in the future. The three-dimensional structure of coral communities on coral reefs plays an important role in dispersing wave energy and protecting tropical shorelines. Strong coral reefs have uneven surfaces and complex structures that severely degrade the power of incoming waves; this protects seashores from waves, currents, and storms, serving to avoid loss of property, life, erosion, and damage (Harris et al., 2018).

13.4.3 FISHERIES

Coral reefs are important for the fisheries industry of the Caribbean and the world. Coral reefs provides a platform for about a quarter of sea fishes. Because of this large number of fish, fishing is the primary industry for the people of the Caribbean. Caribbean fisheries are an indispensable source of food and a source of income for local people as well as national and international fishing sectors. Reefs are also the main source of protein, in the form of shellfish, fishes, and molluscs, to more than 1 billion people in the tropics (Hinrichsen, 1997). The fish that live and grow on coral reefs are an important source of food for over a billion people globally. Nearly half of all jointly managed fisheries in the USA rely on coral reefs and connected habitats for a part of their developmental stages of life cycles. As per the records of NOAA National Marine Fisheries Service, the yearly commercial price of US fisheries from coral reefs to be more than $100 million. Globally, one estimate suggests that that out of the total net profit of $29.8 billion provided by coral reefs, the fisheries profit is $5.7 billion. The value of coral reef fisheries in Southeastern Asia only is US $2.4 billion annually. These figures don't take into account the value of deep-sea coral reef, which are themselves host to many commercially valued species and varieties and thus represents additional fisheries value (NOAA, 2019).

13.4.4 MEDICINE

Numerous species and verities of coral reefs produce chemicals for protection or attack, specifically the moderate-moving or immobile species such as nudibranchs and sponges.

The discovery of potential new pharmaceuticals, called bioprospecting, has been popular in terrestrial ecosystem for decades. Though, bioprospecting in marine environment is relatively new and has lots of potential. Aquatic creatures those having habitat in coral reef ecosystems are indispensable sources of new medicines to treat cancer, induce labor; treating human bacterial infections, arthritis, heart disease, asthma, viruses, ulcers, and other diseases; in addition to sources of nutritional supplements, cosmetics, and enzymes. The drugs and other potentially useful chemical entities recognized from coral reef ecosystems known as the medicine mines of the 21st century and the list of official and potential novel drugs continue to grow.

13.4.5 TOURISM AND RECREATION

Coral reefs are a significant focus of tourism in tropical areas. Scuba diving and snorkeling are two activities for tourists who visit and explore coral reefs. In areas where coral reefs are protected, these two activities provide a large income to local companies operating dive and snorkel shops. Cesar, Burke, and Pet-Soedeit estimated that coral reefs provide nearly US $9.6 billion net benefits annually from tourism globally. Well managed coral reef-based tourism can also provide noteworthy substitute or additional sources of income generation to coastal communities in developing nations. Local economies are being stronger through receiving billions of dollars from these tourists through aquatic tours, sports fishing trips, restaurants, hotels, etc., located near reef ecosystems. Within the Caribbean, the skeletons of corals and algae provide sediments that create beach sand. The clear, white sand is one of the essential draws for tourists to the Caribbean (Cesar et al., 2003).

13.5 CONCLUSION

Since corals are stable animals, many of them have developed chemicals to defend themselves from predators. Researchers are trying to explore the beneficial potential of these chemicals. Flora and fauna on Coral reefs are the vital sources of novel medicines at aquatic ecosystem being developed to treat rheumatoid arthritis, cytotoxicity, and cancer, Alzheimer's disease, numerous bacterial infections, viral infections, cardio vascular disease, and other metabolic disorders. In an upcoming era, coral reef ecosystems could represent an important source of nutraceuticals, cosmeceuticals, medical treatments, biopesticides, and other useful commercial products.

KEYWORDS

- **anticancer**
- **biodiversity**
- **coral reef**
- **marine bio-prospects**
- **Marine drugs**

REFERENCES

Ang, K. K., Holmes, M. J., Higa, T., et al., (2000). *In vivo* antimalarial activity of the beta-carboline alkaloid manzamine A. *Antimicrobial Agents and Chemotherapy, 44*, 1645–1649.

Aronson, J. K., (2016). *Meyler's Side Effects of Drugs: The International Encyclopedia of Adverse Drug Reactions and Interactions*, 798, 799. Elsevier B.V.

Barlow, G. W., (1981). Patterns of parental investment, dispersal and size among coral-reef fishes. *Environmental Biology of Fishes, 6*, 65–85.

Buchanan, M. S., Carroll, A. R., Addepalli, R., et al., (2007). Natural products, stylissadines A and B, specific antagonists of the P2X7 receptor, an important inflammatory target. *Journal of Organic Chemistry, 72*, 2309–2317.

Cesar, H., Burke, L., & Pet-Soedeit, L., (2003). The economics of coral reef degradation. Zeist, Netherlands: Cesar environmental economics consulting (CEEC) and WWF-Netherlands. assets.panda.org/downloads/cesardegradationreport100203.pdf (accessed on 20 February 2023).

Chang, C. H., Wen, Z. H., Wang, S. K., & Duh, C. Y., (2008). Capnellenes from the formosan soft coral *Capnella imbricata*. *Journal of Natural Product, 71*, 619–621.

Chao, C. H., Wu, Y. C., Wen, Z. H., & Sheu, J. H., (2013). Steroidal carboxylic acids from soft coral *Paraminabea acronocephala*. *Marine Drugs, 11*, 136–145.

Chen, L., Deng, H., Cui, H., et al., (2017). Inflammatory responses and inflammation-associated diseases in organs. *Oncotarget., 9*, 7204–7218.

Cheng, C. H., Chung, H. M., Hwang, T. L., et al., (2012). Echinoclerodane A: A new bioactive clerodane-type diterpenoid from a gorgonian coral *Echinomuricea* sp. *Molecules (Basel, Switzerland), 17*, 9443–9450.

Cooper, E. L., Hirabayashi, K., Strychar, K. B., & Sammarco, P. W., (2014). Corals and their potential applications to integrative medicine. *Evidence-Based Complementary and Alternative Medicine: ECAM*, 184959. https://doi.org/10.1155/2014/184959.

Coral reef Alliance, (2020). *Saving the World's Coral Reefs*. https://coral.org/coral-reefs-101/coral-reef-ecology/coral-polyps/ (accessed on 20 February 2023).

Cox, E. F., (1986). The effects of a selective corallivore on growth rates and competition for space between two species of Hawaiian corals. *Journal of Experimental Marine Biology and Ecology, 101*, 161–174.

Darwin, C., (1962). *The Structure and Distribution of Coral Reefs*. Berkeley, CA: University of California press.

Erba, E., Bergamaschi, D., Bassano, L., et al., (2001). Ecteinascidin-743 (ET-743), a natural marine compound, with a unique mechanism of action. *European Journal of Cancer, 37*, 97–105.

Fang, H. Y., Hsu, C. H., Chao, C. H., et al., (2013). Cytotoxic and anti-inflammatory metabolites from the soft coral *Scleronephthya gracillimum*. *Marine Drugs, 11*, 1853–1865.

Great Barrier Reef Marine Park Authority, (2009). *Structure, Classification and Function of Corals.* https://www.qm.qld.gov.au/microsites/biodiscovery/03sponges-and-corals/structure-classification-function.html (accessed on 20 February 2023).

Grimsditch, G., (2004). *A Study of Potential Coral Reef Bioindicators in the Mamanucas Region, Fiji, Using Coral Cay Conservation Reef Check Data*. pp. 1–10.

Hamann, M. T., (2007). The manzamines as an example of the unique structural classes available for the discovery and optimization of infectious disease controls based on marine natural products. *Current Pharmaceutical Design, 13*, 653–660.

Harris, D. L., Rovere, A., & Casella, E., et al., (2018). Coral reef structural complexity provides important coastal protection from waves under rising sea levels. *Science Advances, 4*, eaao4350. https://doi.org/10.1126/sciadv.aao4350.

Hinrichsen, D., (1997). Coral reefs in crisis: An overview of these vanishing ecosystems, the problems that plague them, and the means for saving them. *BioScience, 47*, 554–558.

Hiroyuki, K., Manda, T., Kobayashi, K., et al., (1983). Anti-tumor activity of lemnalol isolated from the soft coral *Lemnalia tenuis* VERSEVELDT. *Chemical and Pharmaceutical Bulletin, 31*, 1086–1088.

Hoegh-Guldberg, O., Poloczanska, E. S., Skirving, W., & Dove, S., (2017). Coral reef ecosystems under climate change and ocean acidification. *Frontiers in Marine Science, 4*, 158. https://doi.org/10.3389/fmars.2017.00158.

Hu, L. C., Yen, W. H., Su, J. H., et al., (2013). Cembrane derivatives from the soft corals, *Sinularia gaweli* and *Sinularia flexibilis*. *Marine Drugs, 11*, 2154–2167.

Huang, K. J., Chen, Y. C., El-Shazly, M., Du, Y. C., Su, J. H., & Tsao, C. W., (2013). 5-episinuleptolide acetate, a norcembranoidal diterpene from the formosan soft coral *Sinularia* sp., induces leukemia cell apoptosis through hsp90 inhibition. *Molecules, 18*, 2924–2933.

Hui-Yu, F.., Chih-Chuang, L., Chih-Hua, C., et al., (2012). Bioactive pregnane-type steroids from the soft coral *Scleronephthya gracillimum*. *Tetrahedron, 68*, 9694–9700.

Jean, Y. H., Chen, W. F., Sung, C. S., et al., (2009). Capnellene, a natural marine compound derived from soft coral, attenuates chronic constriction injury-induced neuropathic pain in rats. *British Journal of Pharmacology, 158*, 713–725.

Kalaitzis, J. A., Leone, P. De. A., Harris, L., et al., (1999). Adociasulfates 1, 7, and 8: New bioactive hexaprenoid hydroquinones from the marine sponge *Adocia* sp. *The Journal of Organic Chemistry, 64*, 5571–5574.

Lee, H. P., Huang, S. Y., Lin, Y. Y., et al., (2013). Soft coral-derived lemnalol alleviates monosodium urate-induced gouty arthritis in rats by inhibiting leukocyte infiltration and iNOS, COX-2 and c-Fos protein expression. *Marine Drugs, 11*, 99–113.

Lin, Y. F., Kuo, C. Y., Wen, Z. H., et al., (2013). Flexibilisquinone, a new anti-inflammatory quinone from the cultured soft coral *Sinularia flexibilis*. *Molecules, 18*, 8160–8167.

Lin, Y. Y., Jean, Y. H., Lee, H. P., et al., (2013). A soft coral-derived compound, 11-*epi*-sinulariolide acetate suppresses inflammatory response and bone destruction in adjuvant-induced arthritis. *PLoS One, 8*, e62926. https://doi.org/10.1371/journal.pone.0062926.

Lopp, A., Pihlak, A., Paves, H., et al., (1994). The effect of 9,11-secosterol, a newly discovered compound from the soft coral *Gersemia fruticosa*, on the growth and cell cycle progression of various tumor cells in culture. *Steroids, 59*, 274–281. doi: 10.1016/0039-128x(94)90113-9.

National Ocean Service, (2020). *Are Corals Animals or Plants?* https://oceanservice.noaa.gov/facts/coral.html (accessed on 20 February 2023).

National Oceanic and Atmospheric Administration (NOAA) (2019). *Coral Reef Ecosystems.* https://www.noaa.gov/education/resource-collections/marine-life/coral-reef-ecosystems (accessed on 20 February 2023).

National Research Council (US) Committee on the Ocean's Role in Human Health. From Monsoons to Microbes: Understanding the Ocean's Role in Human Health. Washington (DC): National Academies Press (US); (1999). *4, Marine-Derived Pharmaceuticals and Related Bioactive Agents.* Available from: https://www.ncbi.nlm.nih.gov/books/NBK230700/ (accessed on 20 February 2023).

Ngoc, B. P., Mark, S. B., & Ronald, J., (2000). Quinn. isolation of psammaplin A 11'-sulfate and bisaprasin 11'-sulfate from the marine sponge *Aplysinella rhax. Journal of Natural Product, 63*, 393–395.

Novotny, L., & Rauko, P., (2009). Cytarabine conjugates with biologically active molecules and their potential anticancer activity. *Neoplasma, 56*, 177–186.

O'Malley, D. P., Li, K., Maue, M., Zografos, A. L., & Baran, P. S., (2007). Total synthesis of dimeric pyrrole-imidazole alkaloids: Sceptrin, ageliferin, nagelamide e, oxysceptrin, nakamuric acid, and the axinellamine carbon skeleton. *Journal of American Chemical Society, 129*, 4762–4775.

Qaralleh, H., (2016). Chemical and bioactive diversities of the marine sponge *Neopetrosia. Bangladesh Journal of Pharmacology, 11*, 433–452.

Radwan, M., Hanora, A., Khalifa, S., & Abou-El-Ela, S. H., (2012). Manzamines: A potential for novel cures. *Cell Cycle, 11*, 1765–1772.

Schwarzwalder, G. M., Scott, D. R., & Vanderwal, C. D., (2016). A synthesis of exiguaquinol dessulfate. *Chemistry, 22*, 17953–17957.

Shen, Y., Tzeng, G., Kuo, Y., et al., (2008). Cytotoxic activity of capnellene-8β,10α-diol derivatives from a soft coral *Capnella* sp. *Journal of the Chinese Chemical Society, 55*, 828–833.

Shi-Ying, H., Nan-Fu, C., Wu-Fu, C., et al., (2012). Sinularin from indigenous soft coral attenuates nociceptive responses and spinal neuroinflammation in carrageenan-induced inflammatory rat model. *Marine Drugs, 10*, 1899–1919.

Wang, S. K., Puu, S. Y., & Duh, C. Y., (2013). New steroids from the soft coral *Nephthea chabrolii. Marine Drugs, 11*, 571–580.

Whittaker, I., Abbiss, P., Fyfe, H., Banks, P., & Skinner, M., (2016). *AQA A-Level Geography, Hodder Education.* An Hachetee UK Company, London.

Wright, A. D., Nielson, J. L., Tapiolas, D. M., Liptrot, C. H., & Motti, C. A., (2012). A great barrier reef *Sinularia* sp. yields two new cytotoxic diterpenes. *Marine Drugs, 10*, 1619–1630.

Yamaguchi, J., Seiple, I. B., Young, I. S., O'Malley, D. P., Maue, M., & Baran, P. S., (2008). Synthesis of 1,9-dideoxy-pre-axinellamine. *Angewandte Chemie International Edition, 47*, 3578–3580.

Yang, M., Liang, L. F., Li, H., Tang, W., & Guo, Y. W., (2019). A new 5α,8α-epidioxysterol with immunosuppressive activity from the south China sea soft coral *Sinularia* sp. *Natural Product Research*, 1–6. doi: 10.1080/14786419.2018.1561683.

Zhang, H., Dong, M., Wang, H., & Crews, P., (2017). Secondary metabolites from the marine sponge genus *Phyllospongia. Marine Drugs, 15*, 12. https://doi.org/10.3390/md15010012.

PART IV

Ayurvedic and Other Sources to Boost Immunity

CHAPTER 14

IMMUNE-BOOSTING PROPERTY OF AYURVEDIC FORMULATIONS

BHARTI BAIRAGI[1] and R. S. RAGHUWANSHI[2]

[1]*Department of Pharmaceutics, Corporate Institute of Pharmacy, Bhopal, Madhya Pradesh, India*

[2]*Department of Pharmaceutical Chemistry, Corporate Institute of Pharmacy, Bhopal, Madhya Pradesh, India*

ABSTRACT

Ayurveda, known as the traditional system of medicine of India, has a big impact on prevention and cure of any disease or disorder under healthcare. Ayurveda has always been devoted to enhancing the body's overall health and prepare body's immune system to fight disease-causing agent. Ayurvedic formulations like Chyavanprash, Churna, Ghutika, Kwath, etc., are effectively works for the immune system as they prevent our body from viral infections so naturally immunity boosters are some food materials, fruits, green, and fresh vegetables or some of the supplements that help us increase the ability to of immune system. In this chapter there is description on the available herbal medicine and their marketed formulations, who act as a immunity boosters as per ayurvedic traditional medicine system.

Immune-Boosting Nutraceuticals for Better Human Health: Novel Applications. Urmila Jarouliya, PhD, Raj K. Keservani, MPharm, Rajesh K. Kesharwani, PhD, Virendra K. Patel, PhD, and Adi D. Bharti, MPharm (Eds.)

14.1 INTRODUCTION

14.1.1 IMMUNITY

'Immunity' refers to various mechanisms that works in the body as protection against environmental agent, microorganisms present in food materials, chemicals, pollen, animals that are foreign to the body. Arise from the Latin term 'immunis' is meaning 'exempt' (Allison, 2021). As the basic function of the immune system each and every part of our body is under its surveillance (Lindsay, 2016). The immune system less a thing than an organ such as the organs like heart, lungs, kidney or the liver, in aggregate, immunity consists enormous parts and resources of body, produce the large number of cells that it depends on for successful functioning of the whole-body immune system.

The raw architect of the "immune system" is defenses on several multilayered levels. The superficial barrier against infection is most obvious and primary; the skin: the first barrier against infection. Another defense is physiological factors, like the pH and temperature of the body provide inappropriate living conditions for foreign microorganisms. Once pathogens finally entered the body, they are identified by the innate or the acquired or adaptive immune response (Nicholson, 2016). Both systems consist of a number of cells and molecules that communicate in a very complex pattern to detect and eliminate pathogens causing disturbance. Detection and elimination of any pathogen depends upon bonding between chemicals: various receptors present in the surface of the immune system, some of which bind to pathogens chemically, while others bind to other immune system molecules or cells to enable the complex signaling system that regulates the immune response.

14.1.2 THE IMMUNE SYSTEM

An in-build fight mechanism of a living organism which is able to discriminate between its own "self" and "non-self" is known as "Immune System" Examples of immune systems occur in multi-cellular organisms. In fact, the life of complex multi-cellular organism would not be possible without the tendency to exclude external life from the internal system (Scott et al., 2019). Mostly an encounter with something new to the body is not harmful, but at the same time, it can be very dangerous indeed. An effective immune system must be potential enough to discriminate those differences. The ability to learn how to recognize these previously unknown structures explains that the immune system is able to recognize the problem, able to classify and recall the structures of foreign particles (Nicholson, 2016).

14.1.3 VARIOUS PARTS OF IMMUNE SYSTEM

The main parts of the immune system are demonstrated in Figure 14.1:

1. **White Blood Cells (WBC):** It is the main component in your immune system. Blood cell made in the bone marrow and found in the lymph tissue and blood. They help to fight microbial infection and other life-threatening diseases. Cells included in WBC are granulocytes (neutrophils, eosinophils, and basophils), lymphocytes (such as B-cells, T-cells, and natural killer cells) and monocytes (Dinesh et al., 2012). WBC move through throughout your body via blood and tissue, looking for foreign agents (microbes) such as bacteria, viruses, parasites, and fungi. When they find the foreign invader, they initiate an immune attack. It may be used to look for conditions such as infection, inflammation, allergies, and leukemia. Also known as leukocyte and WBC.
2. **Antibodies (Abs):** These work for the body to fight microbes and other the toxins (poisons) they produce during their lifespan. They are natural proteins produced by the body's immune system to fight disease and infections. Monoclonal antibodies are lab-made proteins, they can mimic this response.
   They do this by identifying the antigens on the surface of the microbe, or in the chemicals they produce, which mark the microbe or toxin as a foreign matter. The Abs then mark Ag to destroy them. There are many cells, proteins, and chemicals involved in this attack (Scott et al., 2019).
3. **Complement System:** It is a complex surveillance system, playing a key role in defense against pathogens and in host homeostasis. When danger signals sense generated the complement system is initiated for conformational changes in recognition molecular complexes. It works based on responses generated by antibodies.
4. **Lymphatic System:** It is a wide network of fragile tubes throughout the human body. The main aim of the lymphatic system are to maintain the fluid levels in the body, create response to bacteria, deal with cancer causing cells, deal with cell products that may be would result in disease or disorders and to absorb some of the fats in our diet from the intestine.

The lymphatic system is made up of lymph nodes, also called lymph glands, which trap microbes, lymph vessels those are tubes that carry lymph, the colorless fluid that bathes your body's tissues and contains infection-fighting WBC.

5. **Spleen:** It is an organ rich in blood supply and vascular in nature; it contains many vessels that carry and circulate fluids in your body. It works very closely with your blood and lymph, and can be affected by infection, malignancies, liver problems, parasites, and other conditions. It is a blood-filtering organ that removes microbes and destroys old or damaged red blood cells (Dinesh et al., 2012). It also makes disease-fighting components of the immune system (including antibodies and lymphocytes).
6. **Bone Marrow:** It is the spongy, soft, gelatinous tissue that fills the medullary cavities, found inside the bones. It concerns the RBCs by which our bodies need to carry oxygen, the WBC we use to defeat infection, and the platelets act in blood clot generation.
7. **Thymus:** It plays a huge role in training your body to fight infections and even cancer for the rest of your life. It's also vital to the body's chemical messaging or endocrine system. The thymus filters and monitors your blood content. It produces the WBC called T-lymphocytes (Scott et al., 2019).

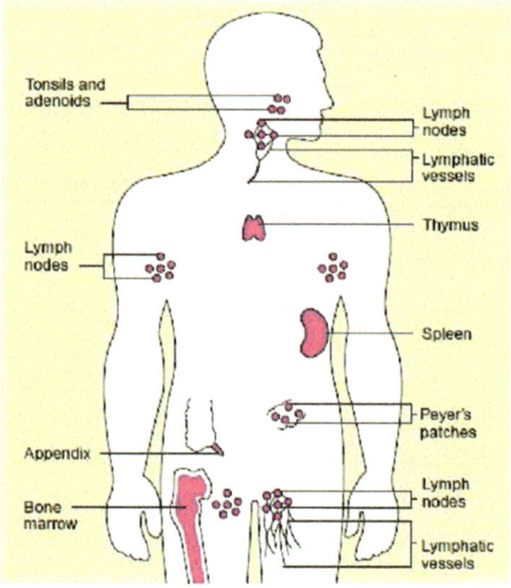

FIGURE 14.1 Various parts of the immune system.

Source: Reprinted from Mareai, 2018.

14.1.4 AYURVEDA

It is a ancient system of medicine, originated in India more than 5,000 years ago. The word "Ayurveda" was originated from two terms of the Sanskrit language – ayur (life) and veda (science or knowledge). Thus, Ayurveda translates to knowledge of life (Balasubramani et al., 2011).

Ayurveda encourages certain lifestyle interventions and natural therapies to regain a balance between the body, mind, spirit, and the environment. Ayurvedic medicine ("Ayurveda" for short) is one of the world's oldest holistic ("whole-body") healing systems. Its main goal is to promote good health, not fight disease. But treatments may be geared toward specific health problems (Kelli, 2021).

In India, Ayurveda is well known and widely practiced form of healthcare, equal to modern or Western system of medicine, traditional Chinese medicine, naturopathic medicine, and homeopathic medicine. Traditional practitioner of this ayurvedic system in India undergoes central or state-recognized training. Present times, in the United States, Ayurvedic practitioners are not licensed, and there are no standard protocols and guidelines for Ayurvedic training or certification. However, Ayurvedic schools have gained approval as educational institutions in some states (Gangadharan, 2017).

14.1.4.1 DOSHAS

As per ayurvedic system, for the proper functioning of the human body balance between "Tridoshas" is mandatory (Figure 14.2) (Pandey et al., 2013):

1. **Vata Dosha:** Air and Space, both of them are key element of Vata (also called ether). The main characteristics of Vata described as flowing, cold, dry, light, rough, and spacious. Fall season denotes vata for its cool and crisp days. In ancient system of medicine, it is said that to achieve optimal health, a person should follow a satvik routine, practice meditation for stress and anger management and other soothing practices like yoga sadhana, so a vata-dominant person can balance between a warm body temperature & cold weather outside (Bhattacharya et al., 2000).
2. **Kapha Dosha:** Earth and Water are the main element of Kapha dosha also pronounced as "kuffa." Spring season denotes kapha for its steady, stable, heavy, slow, cold, and soft key characteristics. To

maintain healthy body, a person who's kapha-dominant must follow routine of daily exercise, diet full of nutrient, maintain a warm temperature of body (e.g., by sitting in a sauna or eating warm food), and develop better sleep routine (Pandey et al., 2013).

3. **Pitta Dosha:** Fire and water are the main element of Pitta dosha which associated with a relentless personality. Summer is the season of pitta known for its main characteristic of sunny and hot days with nature like hot, sharp, liquid, light, oily, and mobile. A person with pitta-dominant dosha must balance his or her work-life, maintain stress and avoid exposure to heat environment internally as well as eternally (e.g., weather, spicy food) (Bhattacharya et al., 2000).

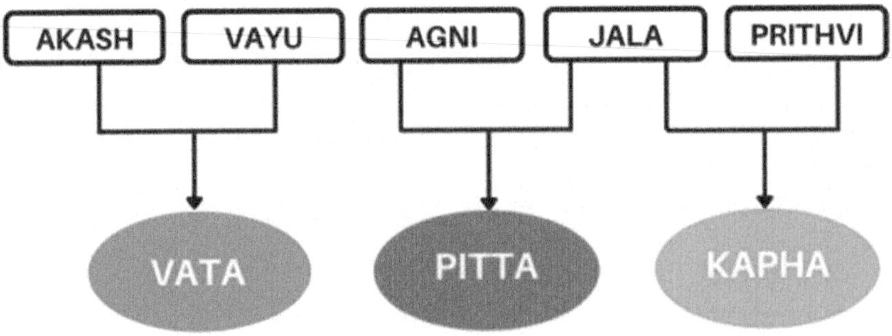

FIGURE 14.2 Doshas.

14.2 NATURALLY OCCURRING PLANT MATERIALS HAVING IMMUNE BOOSTING PROPERTY

14.2.1 BALA (SIDA CORDIFOLIA)

Sida Cordifolia, best known for its strength promoter quality. It is also known as Country mallow, Sida cordifolia Linn belongs to the family Malvaceae (Figure 14.3). It is widely used in Ayurveda and other traditional systems of medicine in India and various other countries for immunity. Ayurveda has mentioned bala as brmhaniya, a bulk promoting herb. According to Charaka maharishi this herb works for rejuvenation (rasayana) of tissue (mamsa dhatu) and muscular system (Ahmed et al., 2015).

FIGURE 14.3 Bala (*Sida cordifolia*).

Source: Photo by J.M.Garg. https://creativecommons.org/licenses/by-sa/4.0/

All species of Bala is widely distributed all over the tropical and sub-tropical plains of India and Sri Lanka. Sida cordifolia as a health tonic, mild astringent, emollient, and aphrodisiac. The immune boosting herb is also involved in preparations of various important formulations of ancient Ayurveda. Internally, bala is considered to be a excellent herb for nerve stimulation and rasayana for all kinds of vata diseases. Well known as a skin rejuvenator and cardiac stimulant.

14.2.2 DRAKSHA (VITIS VINIFERA)

Extract of Grape Seed is a notable nutraceutical or diet supplement. The extraction is performed from the skin of grape seeds and extract collected as a form of polyphenols, and OPCs. Oligomeric proanthocyanidins (OPCs) are antioxidants that act as an immunity modifier. There are several benefits of grape seed extract (Sayeed et al., 2021). It has **strong antioxidant power,** 50 times stronger antioxidant power than vitamin C & 20 times stronger than Vitamin E. Act as DNA protectant and make the immune system strong. It **benefits the cardiac health** – grape seed extract helps increase blood circulation to the cardiac system and helps enhance CVS function. It consists polyphenols, they promotes youthful skin by working as **anti-aging** ingredient and protects skin from free radical damage. Provide nourishment to the body (Figure 14.4).

FIGURE 14.4 Draksha (*Vitis vinifera*).

Source: Photo by Dragonflyir hassanahmadi. https://creativecommons.org/licenses/by-sa/3.0/

14.2.3 HARITAKI (TERMINALIA CHEBULA)

In the ancient system of medicine, haritaki is known as the 'king of herb' (Figure 14.5) that destroys all diseases, eliminates all toxins from the body, and support tissue growth and health. The main active compounds include healthy acids and metabolites because of them it works as an ideal immunity booster (Shivaprasad et al., 2008).

FIGURE 14.5 Haritaki (*Terminalia chebula*).

Source: Source: Photo by Salil Kumar Mukherjee. https://creativecommons.org/licenses/by-sa/4.0/

It acts as rejuvenating agent and helps improves overall health. When extract of *Terminalia chebula*. Administered to the body of mice, it produced an increase in humoral antibody (HA) and delayed-type hypersensitivity (DTH) in. It was concluded that the *Terminalia chebula*. extract is a promising drug with immunostimulant properties (Shivprasad et al., 2008).

14.2.4 GURUCHI (TINOSPORA CORDIFOLIA)

It acts as an immunomodulator. *T. cordifolia* (Figure 14.6) possess immune stimulatory, antimicrobial, hepatoprotective, and antioxidant properties. *Tinospora cordifolia* (*T. cordifolia*), an ephemeral shrub native to India, has been used to historically to fight acute and chronic inflammation also promote a balanced immune response.

FIGURE 14.6 Guruchi (*Tinospora cordifolia*).

Source: Photo by Tmd at Thai Wikipedia. Public domain.

A diet supplement, *T. cordifolia* formulated in dosage form by decoction method along with extracts of the Terminalia and Phyllanthus species (Charles et al., 2022).

14.2.5 SHATAVARI (ASPARAGUS RACEMOSUS)

It is a nutrient and rejuvenator. Roots of *Asparagus racemosus* (Figure 14.7) Wild has always been used in ancient medicine system of India as Rasayan for immune stimulation, as galactagogue herb, for the treatment of mouth and gastric ulcer, as primary anti-cancer. Formulations of Shatavari root extracts used as immunomodulator (Manish et al., 2008).

FIGURE 14.7 Shatavari (*Asparagus racemosus*).

Source: Photo by Neha.Vindhya. https://creativecommons.org/licenses/by-sa/3.0/

Other roles of *Asparagus racemosus* are its antioxidant, antibacterial effects and immune system stimulation.

14.2.6 GOKSHUR (TRIBULUS TERRESTRIS)

It works as nutrient and strength booster. *Tribulus terrestris* belongs to Zygophyllaceae family, commonly known as Gokshur or Gokharu (Figure 14.8). Another name of the herb is puncture vine, has been used since very long time in Indian and other traditional systems of medicine. Useful in prevention and cure of various kinds of diseases. All parts of herb contain chemical entity medicinally precious for the treatment of many sickness of the human body. These chemicals or metabolites are alkaloids, steroidal saponins, flavonoids, flavonol glycosides, etc. It consist properties like immunomodulatory, diuretic, antidiabetic, aphrodisiac, antiurolithic, hypolipidemic, cardiotonic, absorption enhancing, central nervous system, hepatoprotective, anti-inflammatory, analgesic, antispasmodic, anticancer, anthelmintic, larvicidal, antibacterial, and anticariogenic activities (Saurabh et al., 2014).

FIGURE 14.8 Gokshur (*Tribulus terrestris*).

Source: Photo by Juan Carlos Fonseca Mata. https://creativecommons.org/licenses/by-sa/4.0/

Tribulus terrestris is a tiny plant having small leaf. It cultivated worldwide including various parts of Asia, Europe, Africa, and the places of Middle East. The root and fruit of the herb have been used therapeutically in traditional system of Chinese and Indian ancient system of medicine. People have been using this medicinal plant since ages for multiple benefits, as it enhances libido, keeps the urinary functions regular and reduce water retention. Nowadays, *Tribulus terrestris* is mainly used as a health promoter, as well as diet supplements to increase level of testosterone.

14.2.7 VIDARIKAND (PUERARIA TUBEROSA)

Varahikand (*Dioscorea bulbifera*): It acts as a rejuvenator and strength promoter (Figure 14.9). It has been used in various ayurvedic formulations since ages as a health tonic, antiaging supplement and immune booster. It has been recommended for cardiovascular diseases (CVDs), hepatosplenomegaly, reproductive problems like PCOS, menopausal disturbances, sexual inability, etc. A lot of phytochemicals, bioactive compounds such as puerarin, genistein, daidzein, tuberosin, and so on have been identified in the tuber of the herb.

FIGURE 14.9 Vidarikand (*Pueraria tuberosa*).

Source: Photo by Dinesh Valke. https://creativecommons.org/licenses/by-sa/2.0/deed.en

The support data of *in vivo* and *in vitro* studies have provided knowledge that tuber of the plant have potential as spermatogenic, immune booster, aphrodisiac, anti-inflammatory, cardio tonic and brain tonic (Amal et al., 2014).

14.2.8 ASHWAGANDHA (WITHANIA SOMNIFERA)

Whole plant parts like leaves, bark, roots, fruit, and seeds are consumed for their therapeutical properties but mostly root used as medicinal agents. After a serious illness it prescribed to strengthen immune system (Bani et al., 2006). Ashwagandha (Figure 14.10) has strong properties to treat inflammatory action which prevent the body from conditions like rheumatoid arthritis, autoimmune diseases, and certain skin diseases.

FIGURE 14.10 Ashwagandha *(Withania somnifera)*.

Source: Photos by a) Piyush Kothari. https://creativecommons.org/licenses/by-sa/4.0/ b) Wowbobwow12. https://creativecommons.org/licenses/by-sa/3.0/

This herb has proven its efficacy in CNS dysfunction. It has been shown to benefit the function of brain cell, nervous activity, used in the treatment of anxiety and depression. It also helps to relieve fatigue (Ethnopharmacol, 2000).

14.2.9 MULETHI (GLYCYRRHIZA GLABRA)

Mulethi is an excellent remedy to boost the immune system, also known as Yashtimadhu, useful in the treatment of disorders like cold, cough, and other respiratory issues. According to Ayurveda, mulethi (Figure 14.11) is sweet to taste, it's slimy and heavy, and is efficacious in treating vata disorders.

FIGURE 14.11 Mulethi *(Glycyrrhiza glabra).*

Glycyrrhizin contains a saponin which makes it antimicrobial. The powdered form of root is taken with ghee and madhu (honey) to improve immunity. It is believed to be a herbal revitalizer and anti-aging agent. Some studies shown that mulethi has positive effect for brain function if used as tonic. It helps to improve strength and acts as an antioxidant (Krishan et al., 2013).

14.2.10 AMLA (EMBLICA OFFICINALIS)

The richest sources of vitamin C is said to be Amla. It is best known to improve overall immunity, act as rejuvenate and revitalize the immune systems. Amla (Figure 14.12) is cool in nature and can help maintain balance in body heat, thus often recommended in Pitta imbalance, one of Tridosha conditions. It is also helpful in in afflictions of the gastro-intestinal tract (Dasaroju et al., 2014). It is said to be that it supports regeneration of RBC in blood and help to improve hemoglobin percentage in the body. Possesses anti-inflammatory properties and it relieves joint pains.

FIGURE 14.12 Amla *(Emblica officinalis).*

Source: Photo by Salil Kumar Mukherjee. https://creativecommons.org/licenses/by-sa/4.0/

Amla is often used in powder form but is also available as tablets or liquid extracts. Amla powder can be taken orally as raw or by mixing with honey, twice a day. Chyawanprash is a versatile Ayurvedic formulation that can boost immunity also contains amla (Bhalla et al., 2017).

14.2.11 GINGER (ZINGIBER OFFICINALIS)

Ginger (Figure 14.13) is hot in potency and thus AIDS in decreasing the aggravated vata and kapha doshas. Ginger is very versatile in nature, it can be taken as raw form, oil or powder form. The dry form of this crude drug is also used as herbal medicine known as "Sonth" or dry ginger. Dry ginger powder mixed with sesame oil is used to relieve joint or muscle pains. It is also used for hot fomentation in gout, edema, arthritis or other joint pains (Shakya et al., 2015). Anti-microbial compounds in ginger helps fight infections, and boost immunity levels.

FIGURE 14.13 Ginger *(Zingiber officinalis)*.

Ginger is prescribed for many respiratory illnesses like cold and cough, lung pneumonia, acute or chronic asthma and bronchitis.

14.2.12 TULSI (OCIMUM SANCTUM)

Ocimum sanctum is the organized form of Tulsi (Figure 14.14), used for its therapeutic purposes, due to its anti-microbial properties and its use in respiratory tract infections like cough-cold, sore throat, asthma, and other respiratory problems. It helps remove excess "kapha" from lungs. This natural stimulant energizes our body, increases circulation, and has been proven to be beneficial in skin diseases and ulcers (Bhalla et al., 2017).

Fresh juice of tulsi when consumed two times a day believed to boost immune health. Mixing a few drops of ginger and honey to tulsi extract may help further to improve immunity.

14.3 AYURVEDIC FORMULATION – IMMUNITY BOOSTERS

The theory of immunity in Ayurveda, is called the "beej-bhumi theory" which means "seed and land." This explains that if the body is piled up with toxins known as "Ama" in ayurved and lacks vigor known as "Ojas" then the body serves as a very fertile place for infections easy to spread, similar as the land that is fertile to sprout seeds (Singh et al., 2012).

FIGURE 14.14 Tulsi *(Ocimum sanctum).*

As per Ayurveda, strong immunity is a result of balance between "Panchmahabhuta" – Air, Fire, Water, Space, and Earth which ultimately balance "Tridoshas" of human bosy. There are a whole lot of holistic ayurvedic formulations available that can cleanse and balance the Ama and Ojas to provide nourishment and reinstate the balance of tridoshas. This process bolsters immunity and shields the body against infections (Pukar et al., 2022).

Ayurvedic immunity boosters are available in different dosage forms like solid, liquid, and semisolid. These capsules, tablet, churna, etc., possess immune-boosting properties which can lead to promote overall health of human body.

14.3.1 CHYAWANPRASH

Chyawanprash – it is a mixture of five different tastes comprising sweet, sour, bitter, pungent, and astringent. The main ingredients are different types of herbs, amla, sesame oil, ghee, and honey (Gangadharan, 2017). It is an amazing supplement suited for all seasons as the key ingredients can be suitable for extreme climatic conditions.

Amalaki, a potent ingredient in chyawanprash is a warehouse of vitamin C and antioxidant which triggers immunity, averts the reoccurrence of infections, boosts stamina naturally and enhances the overall health of the body. Marketed product: Himalaya Chyavanaprasha, Patanjali Chawanprash, Dabur Chawanprash.

14.3.2 AMLA TABLET/CAPSULE

Amla is a wonderful antioxidant and rejuvenating agent, these properties shield the body against infections and boost up the immune system. It also purifies the circulatory system, flushes out the toxic and poisonous material and other waste from the body. Amla capsules works wonder for the skin revitalization, cureacne, reduces hair problems and alopecia, strengthen the immune system and keep body disease free (Das, 2014).

It strengthens the digestive system, relieves constipation, functions as an antioxidant, maintains healthy cholesterol levels, normalizes blood pressure, helps in proper functioning of the liver.

Marketed product: Sri Tattva Amla 500 mg Tablet 60's, Himalaya-Amalaki Immunity Wellness Capsule (Pack of 2–60 Capsules each).

14.3.3 ASHWAGANDHA CAPSULES/TABLET/CHURNA

Also known as "Indian Ginseng"; a miracle herbal drug, this rejuvenates the body. It is an excellent herbal formulation which provide vital nutrient required for human health. It increases the generation of WBC which contribute in the fight against any sort of microbial infection caused by a fungi, virus, bacteria, and other diseases causing agent (Davis et al., 2000). Previous clinical trial studies state that taking ashwagandha along with milk better the potential of the body. It also improves the production of lymphocytes (Sayeed et al., 2021).

Marketed product: Inlife Ashwagandha Capsules 60's, Add-Immune (60 Capsules), **Dabur Ashwagandha** Tablets, **Baidyanath Ashwagandha** Churna.

14.3.4 GILOY SATVA

Giloy Sat also known as Guduchi as well as Amirt Sat, it is an extract of the herbal plant Guduchi that offers a holistic healing experience. This medicine purifies the blood, treats chronic fever, jaundice, GI problems and other infections caused by microbial agents. A potent anti-bacterial and anti-inflammatory property of Guduchi cures multiple infections, GI disorder and boosts the whole immune system. In addition, it helps to deal with pitta imbalance, and it is one of the few tridosha immune-boosting herbs (PukarKhanal et al., 2022). Marketed product: Patanjali Giloy Sat Powder 5 gm, Baidyanath Giloy Satva, Dabur Giloy Satva.

14.3.5 BRAHMI TABLET

This formula is a strong neurotonic; Brahmi is prescribed in ayurveda since long time for its mind-enhancing properties. It is widely praised in the world as it supports the working of brain chemicals, Balance the functioning of matter involved in concentrating, learning, mental stability, and memory. It calms the mind and act as a potent antioxidant and also delays unnatural or age-related brain degeneration (Ashok et al., 2021).

Marketed product: Himalaya Wellness Brahmi Tablet 60's, Brahmi Tablet, 60 Tablets, Kudos Brahmi 60 Tablets, Ayurveda Redefine Brahmi Mind Wellness Tablets, Vedi Brahmi Tablets (15 Tablets).

14.3.6 IMMUNITY CAPSULES

These are daily dietary supplement consist essential nutrients, vitamins, minerals, and herbs. The utilization of oxygen in the body increases with the help of these nutrients; they improve appetite, physical performance, help in maintaining body weight, and regulate growth. It AIDS in avoiding weakness, and general debility (Pukar et al., 2022).

These are general immunity capsules that work for the betterment of mental wellness and help to avoid fatigue by improving perseverance of the body. They also correct gut health and balance body growth. Moreover, they help to utilize the essential vitamins and minerals in the body. Boosts immunity and protect against diseases, improves body weight and growth, improves physical work capacity and increase energy (Agarwal et al., 1999).

Marketed product: Vedic Amrit Immunity Capsules (60 capsules), Revite General Wellness and Immunity Capsules (100 capsules).

14.3.7 TULSI DROPS

It is prepared with a combination of five different species of Tulsi extracts also known as "Panchamrit." It helps in preventing the body from deadly diseases. It works as a natural substance to boost immunity and helps cleanse the impurities from the circulatory system of the body. It helps in managing mental stress or anxiety, also improves metabolic actions (Panda et al., 2021).

It improves immunity, keeps the stomach healthy, protects the body against infections and diseases, lowers blood sugar levels, and maintains cholesterol levels.

Marketed product: Panchamrit Tulsi Drops – A Natural Immune Booster, Healthvit Tulsi drops 30 ml, Jolly Pharma Tulsi, Dabur Tulsi Drops.

14.4 CONCLUSION

This chapter through light on Ayurveda and its co-relation with the concept of immunity. The author has collected many references of "how to enhance immunity" and keep body disease free by adopting ayurvedic principles in daily habits.

One who adopts Ayurveda principles in day today life will become free from diseases and can lead healthy and prosperous life. In modern times; there is much awareness of eating right and exercising regularly. But, despite a healthy lifestyle, diet habits and numerous hours of workout, keeping disease and stress at bay is very challenging.

Ayurveda looks at the construction of the human body in a very satvikor holistic manner. In ayurveda it is considered that all the organ systems are so self-sufficient and interdependent, this makes absolute sense when we adapt it in our daily life. Complete balance between multiple organs of the body is necessary to bring one's health to their optimum levels. Consuming immunity booster's herbal formulations as part of your diet, and daily nutrient are just a part of developing a healthy immune system which every individual must follow as required today's time to maintain a healthy and happy lifestyle.

KEYWORDS

- **ayurveda**
- **ayurvedic medicine**
- **diet**
- **health**
- **immunity**
- **immunity booster**

REFERENCES

20 Medicinal Plants OF '21 For COVID-19 Care. National Medicinal Plants Board, Ministry of AYUSH, and Government of India.

Agarwal, S. S., & Singh, V. K., (1999). Immunomodulators: A review of studies on Indian medicinal plants and synthetic peptides. Part I: Medicinal plants. *Proc. Indian Natl. Sci. Acad., 65*, 179e204.

Ahmed, G., Vijayasankar, R., & Ikhlas, A. K., (2015). *Sida cordifolia*, a traditional herb in modern perspective – A review. *Current Traditional Medicine, 1*, 5–17.

Allison, I. *Introduction to the Immune System*, 1–31. https://www.meddent.uwa.edu.au/__data/assets/pdf_file/0008/3038039/L2_Introduction-to-the-Immune-System.pdf (accessed on 17 February 2023).

Amal, K. M., Subrata, P., Pratim, B., & Debdulal, B., (2014). Pueraria tuberosa: A review on its phytochemical and therapeutic potential, national library of medicine. *Nat Prod Res., 28*(23), 2111–2127. Epub 2014 Jul 1.

Ashok, K. P., & Sarbeswar, K., (2021). Ayurvedic immuno booster: Is it myth or reality in COVID-19 pandemic. *International Journal of Current Research and Review*. Article doi: http://dx.doi.org/10.31782/IJCRR.2021.13140.

Ayurvedic Pharmacopoeia of India Published by Government of India, Ministry of Health and Family Welfare, Department of India System of Medical & Homoeopathy, New Delhi. Part 1, Volume 6, Edition 1, pp. 1–307.

Balasubramani, S. P., et al., (2011). Plant-based rasayana drugs from ayurveda. Chinese *Journal of Integrative Medicine, 17*(2), 88–94.

Bani, S., Gautam, M., Sheikh, F. A., Khan, B., Satti, N. K., Suri, K. A., et al., (2006). Selective Th1 up-regulating activity of *Withania somnifera* aqueous extraction experimental system using flow cytometry. *J. Ethnopharmacol., 107*(1), 107–115. doi: 10.1016/j.jep.2006.02.016.

Bhalla, G., Kaur, S., Kaur, J., Kaur, R., & Raina, P., (2017). Antileishmanial and immunomodulatory potential of *Ocimum sanctum* Linn. *and Cocos nucifera* Linn. in murine visceral leishmaniasis. *J. Parasit. Dis., 41*(1), 76–85. doi: 10.1007/s12639-016-0753-x.

Bhattacharya, S. K., Bhattacharya, A., & Chakrabarti, A., (2000). Adaptogenic activity of siotone, a polyherbal formulation of Ayurvedic rasayanas. *Indian J. Experimental Biol., 38*, 119e28.

Charles, R. Y., Eugene, J. B. Mary, E. D. Y., (2022). *Tinospora cordifolia*: A review of its immunomodulatory properties. *National Library of Medicine, 19*(2), 271–285.

Das, S., (2014). A review on immune modulatory effect of some traditional medicinal herbs. *Journal of Pharmaceutical, Chemical and Biological Sciences (JPCBS), 2*(1), 33–42.

Dasaroju, S., (2014). Current trends in the research of *Emblica officinalis* (Amla): A pharmacological perspective. *Int. J. Pharm. Sci. Rev. Res., 24*(2), 150–159.

Davis, L., & Kuttan, G., (2000). Immunomodulatory activity of *Withania somnifera. J. Ethnopharmacol., 71*(1, 2), 193–200.

Dinesh, K., Vikrant, A., Ranjeet, K., Zulfiqar, A. B., Vivek, K. G., & Vijender, K., (2012). A review of immunomodulators in the Indian traditional health care system. *Journal of Microbiology, Immunology and Infection, 45*(3), 165–184.

Gangadharan, G. G., (2010). Quality of ingredients used in ayurvedic herbal preparations. *Journal of Ayurveda and Integrative Medicine, 1*(1), 16, 17. PMC. Web.

https://en.wikipedia.org/wiki/Vitis_vinifera (accessed on 17 February 2023).

https://www.planetayurveda.com (accessed on 17 February 2023).

Khare, C. P., (2007). *Indian Medicinal Plants. An Illustrated Dictionary*. New York: Springer Publications.

Krishan, K., & Rakesh, K. N., (2013). "A review on immunomodulator activity of yashtimadhu (*Glycyrriza glabra*)" from the proceedings of insight Ayurveda 2013, Coimbatore. *Anc. Sci. Life, 32*(Suppl 2), S59. PA02.14.

Makare, N., Bodhankar, S., & Rangari, V., (2001). Immunomodulatory activity of alcoholic extract of *Mangifera indica* L. in mice. *J. Ethnopharmacol., 78*, 133e7.

Manish, G., Shantanu, S., Sarang, B., & Kaul, A., (2008). Immunomodulatory activity of *Asparagus racemosus* on systemic Th1/Th2 immunity: Implications for immunoadjuvant potential. *Journal of Ethnopharmacology, 121*(2), 241–7, December 2008.

Mareai, S.S. (2018). The Scopy of Immunology. Lambert Academic Publishers. https://www.researchgate.net/publication/325019907_The_Scope_of_Immunology.

McComb, S., Aude, T., Bassel, A., Lakshmi, K., & Felicity, C S., (2019). Introduction to the immune system. *Methods in Molecular Biology (Clifton, N.J.), 2024*, 1–24. doi: 10.1007/978-1-4939-9597-4_1.

Ministry of AYUSH, (2020). *Ayurveda's Immunity Boosting Measures for Self Care During COVID 19 Crisis.* Available at: https://www.ayush.gov.in/docs/123.pdf (accessed on 20 February 2023).

Mrityunjaya, M., Pavithra, V., Neelam, R., Janhavi, P., Halami, P. M., & Ravindra, P. V., (2020). Immune-boosting, antioxidant and anti-inflammatory food supplements targeting pathogenesis of COVID-19. *Review Article, Front. Immunol.* https://doi.org/10.3389/fimmu.2020.570122.

Nicholson, L. B., (2016). The immune system. *Essays Biochem., 60*(3), 275e301. https://doi.org/10.1042/EBC20160017.

Pandey, M. M., Subha, R., & Rawat, A. K. S., (2013). Review article Indian traditional ayurvedic system of medicine and nutritional supplementation. Hindawi Publishing *Corporation Evidence-Based Complementary and Alternative Medicine, 2013*, 12. Article ID 376327. http://dx.doi.org/10.1155/2013/376327.

Praveen, S., (2013). Vitamin C rich fruits can prevent heart disease. *Indian J. Clin. Biochem., 28*(3), 213–214.

Pukar, K., Taaza, D., Patil, B. M., Yadu, N. D., Ismail, P., Manish, W., Shailendra, S. G., & Arindam, M., (2022). Network pharmacology of AYUSH recommended immune-boosting medicinal plants against COVID-19. *Journal of Ayurveda and Integrative Medicine, 13*(1), 100374.

Saurabh, C., Tanuja, N., Gauresh, S., Divya, K., & Sadhana, S., (2014). Phytopharmacological overview of *Tribulus terrestris. Pharmacogn. Rev., 8*(15), 45–51.

Sayeed, A., Sultan, Z., Bushra, P., Parakh, B., Abida, P., Gaurav, Rabea, P., & Minhaj, A., (2021). Indian medicinal plants and formulations and their potential against COVID-19–preclinical and clinical research. *Review Article, Front. Pharmacol.* https://doi.org/10.3389/fphar.2020.578970.

Shakya, S. R., (2015). Medicinal uses of ginger (*Zingiber officinale* roscoe) improves growth and enhances immunity in aquaculture. *International Journal of Chemical Studies, 3*(2), 83–87.

Shastri, K. N., & Chaturvedi, G. N., (2004). "Agnivesha's Charaka Samhita, Vidyotini Commentary," In: *Chikitsasthana*, Chaukhamba Bharati Academy, Varanasi, p. 235.

Shivaprasad, H. N., Kharya, M. D., Rana, A. C., & Mohan, S., (2008). *Preliminary Immunomodulatory Activities of the Aqueous Extract of Terminalia Chebula* (Vo. 44, No.1). Taylor and Francis online.

Singh, K., et al., (2012). The concept of vyadhikshamatva (immunity) in ayurveda. *Ayurpharm Int. J. Ayur. Alli. Sci., 1*(5), 99–108.

Singh, S., Taneja, M., & Majumdar, D. K., (2007). Biological activities of *Ocimum sanctum* Linn. fixed oil e an overview. Indian *J. Exp. Biol., 45*, 403e12.

Tabar, A. I., Alvarez-Puebla, M. J., Gomez, B., et al., (2004). Diversity of asparagus allergy: Clinical and immunological features. *Clin. Exp. Allergy, 34*, 131–136.

Vaghasiya, J., Datani, M., Nandkumar, K., Malaviya, S., & Jivani, N., (2010). Comparative evaluation of alcoholic and aqueous extracts of *Ocimum sanctum* for immunomodulatory activity. *Int. J. Pharm. Biol. Res., 1*(1), 25e9.

www.ayurveda.com (accessed on 17 February 2023).

CHAPTER 15

NUTRITION, MALNUTRITION, INFECTION, AND IMMUNITY

KHEMCHAND R. SURANA,[1] EKNATH D. AHIRE,[2] SUNIL K. MAHAJAN,[3] DHANANJAY M. PATIL,[4] and SHUBHAM J. KHAIRNAR[5]

[1]*Department of Pharmaceutical Chemistry, Shreeshakti Shaikshanik Sanstha, Divine College of Pharmacy, Satana, Nashik, Maharashtra, India*

[2]*Department of Pharmaceutics, MET's Institute of Pharmacy, Affiliated to Savitribai Phule Pune University, BKC, Adgaon, Nashik, Maharashtra, India*

[3]*Department of Pharmaceutical Chemistry, MGV's Pharmacy College, Panchavati, Nashik, Maharashtra, India*

[4]*Department of Pharmaceutics, Shreeshakti Shaikshanik Sanstha, Divine College of Pharmacy, Satana, Nashik, Maharashtra, India*

[5]*Department of Pharmacology, MET's Institute of Pharmacy, Affiliated to Savitribai Phule Pune University, BKC, Adgaon, Nashik, Maharashtra, India*

ABSTRACT

Nutrition plays a main function with within the management each acute and chronic diseases, in phrases of the body's response to the infective organism. Associate array of nutrients like macro- and micro-nutrients, vitamins,

Immune-Boosting Nutraceuticals for Better Human Health: Novel Applications. Urmila Jarouliya, PhD, Raj K. Keservani, MPharm, Rajesh K. Kesharwani, PhD, Virendra K. Patel, PhD, and Adi D. Bharti, MPharm (Eds.)

etc., is expounded to boost the host's immune responses con to living thing pathogens. These nutrients have an immunomodulatory lead to dominant the infection and inflammation processes and dietary deficiency of any form, i.e., deficiency disease can even in addition result in nutritionally received immunological disorder syndrome, which considerably can increase an individual's susceptibleness to development of infection to disease. Nutrition and infectious illnesses are related to every completely different in a few aspects. First, nutrients impact the advance of human body immune system. Moreover, nutrients will have an effect on emerge of infectious diseases (e.g., channel infections), meals poisoning, intestinal illnesses, and general infectious illnesses. Overall, many elements can weaken the frame's cap potential to combat infection and reason deficiency disease. The weather contains anorexia, typical behaviors, reduction of intestinal absorption, metabolic damage, illness metabolism of lipids and carbohydrates, discount of vitamins, iron, zinc, and copper. Within the case of the association between malnutrition and infection, a giant variety of studies have illustrated bidirectional interaction of malnutrition and infection. During this aspect, to deal with severe acute deficiency disease and limit the charge of death, much preventive analysis meted out antibiotics, probiotic bacteria, and prebiotic foods. Within the future, research will be carried out on enteric or systematic immunomodulation.

15.1 INTRODUCTION

Diet nutrition, a vital factor of body homeostasis, is normally related to numerous components, systems, and processes, such as the prevalence of infectious illnesses. For this cause, nutritional reputation has critical significance at the persistence, remission, and occurrence of a huge variety of infectious illnesses and worldwide survival. Nutrition is described via way of means of the World Health Organization (WHO) because the ingestion of meals in terms of the nutritional desires of the body. Malnutrition remains common in abundant less developed areas and may be a prime reason of excessive infectious disorder morbidity and mortality. However, even in developed countries, nutritionary problems that embrace precise nutritional deficiencies, suboptimal diets, and a bit much calorie consumption still be a difficult reality. It's been aforementioned that concerning 792 million folks in growing countries and 34 million in developed countries are undernourished. Previous analysis confirmed that deficiency disease became concerning 3.1

million child deaths yearly or 45% of all child deaths in 2011 and enlarged the danger of death from infectious diseases in childhood. It's been said that 32% of the world infections is also removed via approach of means of growing nutritionary standing; however, others have reasoned that even this confirm is a real understatement of the particular burden, specifically with the emergence of human immunological disorder virus (HIV)/received immunodeficiency syndrome (AIDS). Ample proof confirms that dietary status is intently related to immunity and host resistance to contamination (Krawinkel, 2012.).

Under nutrition is considered as a vast cause for immunodeficiency across the world, inclining the malnourished to numerous forms of infections. All styles of immunodeficiency are susceptible to under nutrition, particularly opportunistic immunodeficiency that is related to altered immune fame. Under nutrition will increase the risk of the host's sicknesses, and related illnesses have a bad effect on host metabolism, exacerbating the dietary state. The complicated of malnutrition and contamination is related to vast morbidity and mortality across the world. However, comparative research on human body composition patterns in growing countries confirmed that there has been a double burden of nutrition. The populace of undernourished countries persists, whilst a developing populace is over-nourished (Cegielski & McMurray, 2004; Surana et al., 2021a).

Balanced nutrients are thought to be vital in maintaining the fitness of an individual. The body utilizes dietary supplements from food to generate energy, maintain up or fix the structures of the body, and direct or modify host metabolism. Nutrition metabolism, which starts off evolved from the ingestion of diet, is an aggregate of physical and chemical strategies that arise withinside the body to accumulate strength and put it to use to synthesize, process, transform, and dispose of materials for the premier upkeep of life. Thus, defenselessness towards the established order and improvement of the infectious disease is generally diagnosed with the aid of using the immune potential of the host (Nicolau et al., 2012; Paillaud et al., 2005; Surana et al., 2021b).

15.2 THE CYCLE BETWEEN NUTRITION AND INFECTION

Interactions of Nutrition and Infection that became revealed in 1968 by the globe Health Organization counseled that there has been a synergistic relationship between infection and malnutrition. Associate degree inadequate nutritionary consumption causes weight reduction, reduced immune ability, membrane damage, infective agent invasion, and obstructed increase in kids. It's accepted that sicknesses will adversely have an effect on the subjects' nutritional standing those effects during a lower of the body's ability to warfare infections. Thus, nutrition and infection are interacting with each other. The complex interaction among infections and under nutrition makes a threatening scenario that sustains an infinite loop that activates those benefiting from one another. Nutrition deficiency ends in expanded susceptibility to disease, and infections likewise have an effect on the dietary status and similarly make a contribution to malnutrition, which reasons an infinite loop (Figure 15.1) (Ramachandran & Gopalan, 2009).

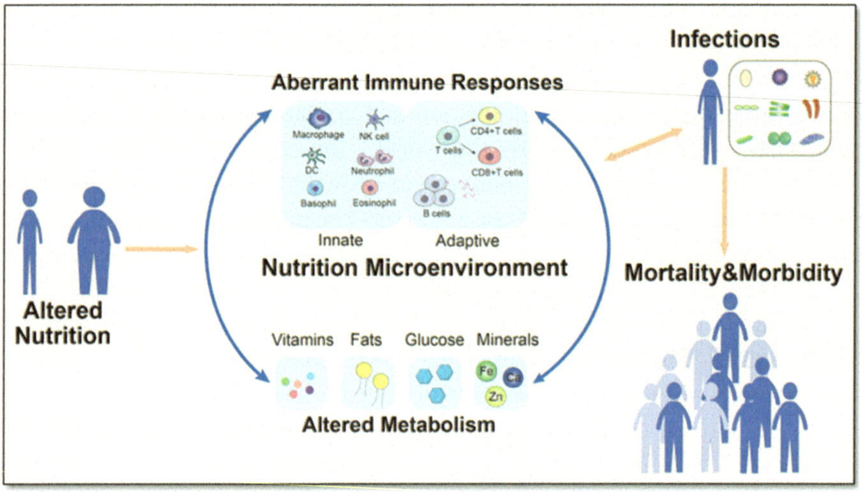

FIGURE 15.1 The cycle among nutrition and infection. Under-nutrition and over-nutrition have an effect on the immune system and make people at risk of infectious illnesses. Infections adversely have an impact on dietary status.

15.3 ABNORMAL NUTRITIONAL STATUS INCREASES THE RISKS OF INFECTION

The relationship between malnutrition and a higher risk of infections has been suggested in numerous age groups in distinctive nations, and usually in medical settings (Caulfield et al., 2004). Tuberculosis (TB) is a disease this is specially suffering from malnutrition, and it's far a prime reason of morbidity and mortality in growing countries. Roughly 33% of the populace of the world is infected with Mycobacterium TB, generally in a subclinical form. Malnutrition probably impacts the final results of TB and might cause the prevalence of drug-resistant *M. tuberculosis* strains (Cegielski & McMurray, 2004).

Another instance of infectious diseases associated with undernutrition is measles. Although there's a present effective vaccine for measles, it still reasons excessive disease and dying in kids worldwide (Hussey & Klein, 1990). Among distinctive nutrients, loss of nutrition A will increase the mortality rate. Experimental facts recommend that nutrition supplementation in kids is associated with a lower in each sickness severity and mortality rate. The unpredictability of the connection between sicknesses and undernutrition has been investigated through any other study in Angola, wherein evaluation of demographic and health survey data mentioned that nutrient deficiency and anemia are connected to malaria (Sousa-Figueiredo et al., 2012). Strikingly, immoderate caloric nutrition has additionally been determined to have an unfavorable impact on disease resistance. During the 2009 H1N1 epidemics, weight problems become diagnosed to be a risk that will increase susceptibility to the infection indicating that odd dietary repute is related to infections (Zeng et al., 2021).

15.4 EFFECTS OF INFECTION ON NUTRITIONAL STATUS

Infections may lead to reduced nutritionary status. For example, cross-sectional analysis mentioned decrease plasma concentrations of nutrients, at the side of nourishment E, in patients with infectious diseases as compared with healthful controls. In Associate in Nursingimal models of respiratory disease virus infection, slashed concentrations of nutrients, together with glutathione (GSH) and vitamins, had been discovered withinside the respiratory organ (Semba et al., 2010). Hypermetabolism and hypercatabolism are characteristics of numerous diseases, specifically infectious diseases together with T.B. and malaria. These excessive pressure things need a distended would like for dietary supplements of nutrients, named as hypermetabolism—an abnormal growth withinside the basal rate with a purpose to cause a daily power or supermolecule deficiency to satisfy the metabolic wishes of the body. Carbohydrates, lipids, and proteins are needed for providing power and form when vitamins and minerals, which modify numerous metabolic techniques love secretion homeostasis, are required for elevated metabolism. Immunological disorders caused by HIV, Plasmodium, or flagellate infection consequences in elevated host susceptibleness to completely different infections and unbalanced or degenerated metabolism and deepening of dietary deficiency during a vicious circle (Srivastava et al., 2016). Indeed, there are numerous examples displaying that malnutrition, obtained immunological disorder syndrome, TB, leishmaniasis, and helminthic sicknesses have unhealthy clinical improvement and poor response to drug treatment. Biological process standing alternations don't seem to be unusual place complications of HIV infection and exert a vital impact on morbidity and mortality. Studies on HIV have accrued our data on nutrient views withinside the discipline of hypermetabolism and hypercatabolism. Since such analysis has deeply investigated numerous events and systems, they will be allotted to different weakening infectious and inflammatory illnesses. Heaps of research have explored however co-infection (co-infection among HIV and a lot of various pathogens) takes place often and interacted with malnutrition, main to frailness and poor clinical outcomes. Intracellular parasites currently no longer best create use of this nutrient and cell power producing systems withinside the cellular, but to boot equally end in the cellular to actively offer its nutrition. In fact, proof of an association among undernutrition, leishmaniasis, and immunological disorder has been illustrated (Haanstra et al., 2016). A foul dietary standing is related to an inclined metallic element gluconate therapy response and reduced liquid body substance levels of gilded matrix metalloproteinase nine (MMP9), a protein this can be negatively related to with sickness progression. In step with ancient reports and current experimental animal research, deficiency disease is usually taken into thought to be a main motive for the event of active TB. TB is one in all the well-documented samples of primary worldwide disease burdens whereby deficiency disease ends up in worse disease outcomes. *In vitro* and animal experiments have well-tried the unfavorable outcomes of persistent supermolecule and power malnutrition at the immunity towards *M. tuberculosis*. Studies on TB sufferers to boot confirmed that the body composition cans additionally extrade as a result of altered protein, fat, and substance metabolism. Immune disorder is likewise believed to account for the robust link among polygenic disease and TB. These interactions have important ramifications in individuals with an excessive incidence of malnutrition, diabetes mellitus, and latent T.B. infection (Oliveira et al., 2013).

15.5 MALNUTRITION AND IMMUNITY

Malnutrition affects each the innate and reconciling immunity of a person rendering them vulnerable to the spread of infections. Phagocytosis and complement cascade are 2 main mechanisms attached elimination of unhealthful organisms from the body. Complement system by itself will destroy microorganisms or the complement receptors gift on the surface of phagocytes can mediate the capture of pathogens. With malnutrition, both functions get compromised—the opsonic complement issue C3 additionally because the somatic cell ability to ingest and kill pathogens are reduced considerably. In addition, the functioning of varied antigen-presenting cell varieties just like the B-lymphocytes, macrophages, and nerve fiber cells (DCs), and Kupffer cells are diminished in deficiency disease (Abe et al., 2003).

Alterations of the practical activity of the system are usually caused by protein-energy malnutrition (PEM), however usually the sort of result depends on nutrition, trace parts and different nutrient deficiency. Organic process deterioration is often aggravated by a concomitant deficiency in necessary elements resembling zinc (zinc) and other trace elements, vitamin B12, folic acid, and so forth (Shin et al., 2010). Recent analysis has incontestable that nutrition is a necessary necessity of the immune system functionality, that is mirrored even in moderate styles of PEM and in apparently healthy individuals, as a consequence of organic process deficiencies not forever realized (Badenhorst et al., 2019). Immunotoxic substances within the diet could successively impair host defenses, ultimately leading to a pseudo-ID with resulting rebellion susceptibleness to infections. Many studies highlight that infection is that the prime mover, and leads to the host's nutritional impairment, eliciting a closed cycle wherever infections have the flexibility to worsen the malnutrition induced immune alterations (Figure 15.2), that consequently augments the danger of effort infections in underfed kids (Chen et al., 2007).

15.5.1 PROTEIN ENERGY MALNUTRITION (PEM)

Malnutrition is thought to own direct results on lymphocytes. Severe PEM provokes atrophy of thymus additionally to peripheral lymphoid organs that successively reduces cell variety (leukopenia), decreases CD4/CD8 ratio, will increase numbers of CD4 and CD8 double-negative T-cells, and can increase the big variety of immature T-cells withinside the peripheral blood. An elementary lower withinside the expression of CD25 and CD27 (the molecules needed for T-cell activation and proliferation) changed into in addition demonstrated. Response of the bowel membrane is likewise full of malnutrition. Its outcomes in planate hypotrophic microvilli, small immunoglobulin A secretion, and lymphocyte counts in Peyer's patches (PP). Unnourished children have verified small producing of kind one cytokines (IL-2 and IFN-γ), which could be the principle mediators of immunity. Such changes in cell-mediated immunity end in elevated condition of a person or lady to infection. Similarly, low albumin levels (Chandrasekaran et al., 2017).

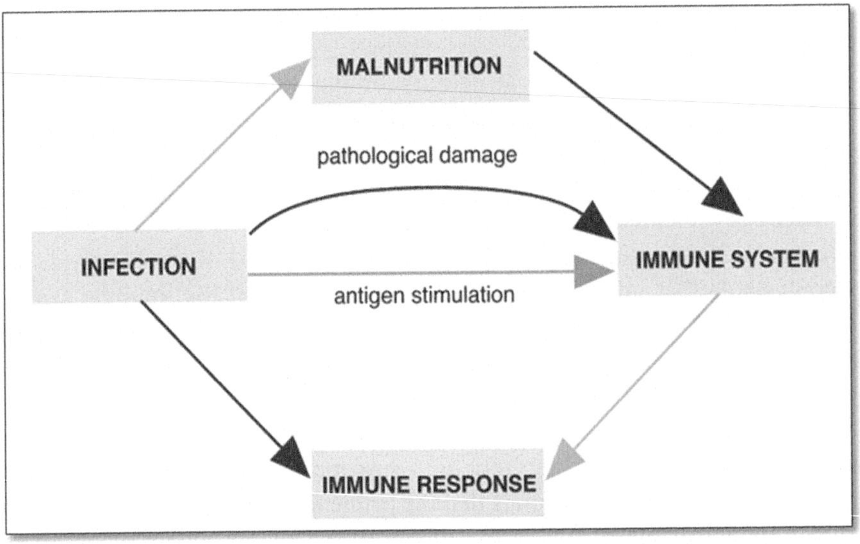

FIGURE 15.2 Interactions between nutrition, immunity, and infection.

15.5.2 ESSENTIAL FATTY ACIDS

There are instructions of vital fatty acids [n6 and n3 polyunsaturated fatty acids (PUFAs) (n6 and n3 PUFA)], which percentage the now no longer unusual place enzyme tool for their metabolism and their metabolic forestall merchandise are typically awesome in action. There are contrasting evidence to aid the outcomes of these fatty acids in M.tb survival and proliferation. One of the vital mechanisms with the resource of the usage of which the M.tb survives withinside the macrophages is with the resource of the usage of inhibiting the membrane actin assembly, this is crucial for the maturation of phagosomes (Anes & Jordao, 2008). Sphingolipids: Sphingolipids are structural lipids, which encompass sphingomyelin, ceramide, and sphingosine 1-phosphate, which is probably ample in neuronal tissue and brought into attention important bioactive molecule of inflammation. Sphingosine-1-phosphate, with results over calcium regulation, expression of growth factors and inflammatory cytokines are of essential hobby withinside the regulation of M.tb survival. There are several mechanisms proposed for the survival and proliferation of M.tb withinside the host mobileular with the resource of the usage of which the M.tb evades the safety mechanism of the host (Spiegel & Milstien, 2002).

15.5.3 VITAMINS

Many research have said nutrients as a key mediator of innate immune system with the useful resource of the usage of regulating the functions of every macrophages and DCs and immunomodulating the method of antibacterial functions, auto phagosomes formation, autophagy, and cytokine production. Multiple vitamins were tested to have a feature in immunity in the direction of M.tb infection or disease. Vitamin D has been recognized as an important modulator of every innate and adaptive immune response in the direction of TB infection, enhancing the antimicrobial homes of the phagocytes viz., monocytes, macrophages, DCs, and neutrophils (Wintergerst et al., 2007).

15.5.4 MODULATOR OF INNATE IMMUNE RESPONSE

Essentially, 1,25-dihydroxyvitamin D3 induces the differentiation of monocytes into macrophages at the site of infection and enhance the uptake of the bacterium with the useful resource of the use of promoting phagocytosis; upregulate unique markers like CD14, mannose receptor, DC signal over the floor of antigen-presenting cells to enhance the phagocytic activity, and intracellular killing. Besides, Vitamin D moreover induces the expression of several antimicrobial peptides (AMPs), mainly cathelicidin (CAMP) and β-defensin 2 (DEFB4). Vitamin D moreover induces hCAP18, which triggers the autophagy mechanism of the inflamed cells, through manner of method of mediating phagosome–lysosome fusion and plays an important feature withinside the elimination of the intracellular M.tb. In addition, vitamin D is determined to downregulate the expression of HAMP (hepcidin antibacterial protein), which AIDS withinside the intracellular shipping of iron, thereby suppressing the growth of the microorganism inner macrophages (Salahuddin et al., 2013).

15.5.5 METALS

Essentially, metals like iron, manganese, copper, and zinc are very plenty needed for the survival of M.tb withinside the macrophages. M.tb has carried out features to accumulate the ones metals from the cell milieu and employ them for survival and replication withinside the macrophages. Host immune system tackles the mycobacterial contamination by sequestering metals like copper and zinc withinside the phagosomes beyond their regular stages and tries to kill them by metallic poisoning. Alternatively, with the useful resource of dietary immunity and with the help of translocation of metallic transporter proteins, vital metals like iron and manganese will be depleted withinside the macrophages and therefore save you the replication of the bacilli (Neyrolles et al., 2015). In addition, iron homeostasis moreover differs amongst early and not on time TB-progressors, with higher ferritin and hepcidin concentrations found among early TB-progressors among their own circle of relatives' contacts (Vickers, 2017). It has been verified that every iron deficiency and overload exists in TB sufferers which have an impact on the disorder improvement and medical outcomes. Besides iron, manganese also can be depleted from the phagolysosomes thru NRAMP1 with coordinated reaction of transferrin, ferritin, and hepcidin. Thus, the poor regulatory effect of iron and IFN-γ will growth the antimicrobial function thru extended production of IL-4/IL-10/IL-5/FOXP3, which results in decreased DCs maturation, function, and differentiation. In addition, pathways related to the producing of Nos2 and TNFα mediate killing of M.tb. Innate immune response in competition to M.tb can be reinforced by pumping of copper and zinc ions withinside the phagolysosomes for bactericidal activity (Mvubu et al., 2016).

15.5.6 MINERALS

In natural conditions, upon ingestion of immunogenic overseas bodies, the macrophages get stimulated and their intracellular Ca^{2+} ion awareness will increase 10-fold than that during their nonstimulated state. The Ca^{2+} ions are very crucial for several sports associated with phagocytosis which encompass activation of the Ca^{2+}/CAM-based protein kinase II (CAMKII) and generation of reactive oxygen molecules. Selective inhibition of formation of Ca^{2+} CAM complex formation (pre translocation) or inhibition of CAMKII pathway on the surface of the phagosome (post-translocation) prevents the maturation of the phagosome. It is also been established that the effect of Ca^{2+}/CAM/CAMKII complex on phagosome maturation is associated with series of sports which is probably in particular worried withinside the phagosome–lysosome fusion (Yurie et al., 2017).

15.6 CONCLUSION

In general, poor nutrition ends up in macro- and small nutrient deficiencies, which can lead to immunodeficiency, every cell mediate and body substance immune response. This secondary immunodeficiency, in turn, influences one's capability to combat contamination succeeding in elevated susceptibleness to diseases. There seems to be a definite position for dietary or meals supplementation in assuaging the harmful consequences made through deficiency disease on immune response. There's experimental proof that a number of nutrients may be helpful but additionally, clinical analysis and irregular clinical trials could also be required for them to be counseled and to higher acknowledge the link among malnutrition. Undernutrition synergistically interacts with infections, at a similar time as infections exert a main poor impact on organic process patterns and nutritional status. Additionally to single nutrient deficiency, worldwide deficiency disease is likewise a main contributor to larger extreme or perhaps fatal infections. Interaction among infections and undernutrition is diverse, and there' a wonderful potential in preventing and dominant infections through optimizing dietary status.

ACKNOWLEDGMENTS

We, authors thank to the MET's, Institute of Pharmacy, BKC, affiliated under Savitribai Phule Pune University, Adgaon, Nashik, and Divine college of Pharmacy, Satana, Nashik. EDA, especially thankful to the Ministry of Tribal Affairs Government of India (NFST) for providing financial assistance in the form of fellowship (Award No.: 202021-NFST-MAH-01235).

KEYWORDS

- **human body immunity**
- **infection**
- **malnutrition**
- **nutrition**

REFERENCES

Abe, M., Akbar, F., Matsuura, B., Horiike, N., & Onji, M., (2003). Defective antigen-presenting capacity of murine dendritic cells during starvation. *Nutrition, 19*(3), 265–269.

Anes, E., & Jordao, L., (2008). Trick-or-treat: Dietary lipids and host resistance to infectious disease. *Mini Reviews in Medicinal Chemistry, 8*(14), 1452–1458.

Badenhorst, C. E., Black, K. E., & O'Brien, W. J., (2019). Hepcidin as a prospective individualized biomarker for individuals at risk of low energy availability. *International Journal of Sport Nutrition and Exercise Metabolism, 29*(6), 671–681.

Caulfield, L. E., Richard, S. A., & Black, R. E., (2004). Undernutrition as an underlying cause of malaria morbidity and mortality in children less than five years old. *The Intolerable Burden of Malaria II: What's New, What's Needed: Supplement of the American Journal of Tropical Medicine and Hygiene, 71*(2).

Cegielski, J. P., & McMurray, D. N., (2004). The relationship between malnutrition and tuberculosis: Evidence from studies in humans and experimental animals. *The International Journal of Tuberculosis and Lung Disease, 8*(3), 286–298.

Chandrasekaran, P., Saravanan, N., Bethunaickan, R., & Tripathy, S., (2017). Malnutrition: Modulator of immune responses in tuberculosis. *Frontiers in Immunology, 8*, 1316.

Chen, S., Sims, G. P., Chen, X. X., Gu, Y. Y., Chen, S., & Lipsky, P. E., (2007). Modulatory effects of 1, 25-dihydroxyvitamin D3 on human B cell differentiation. *The Journal of Immunology, 179*(3), 1634–1647.

Haanstra, J. R., González-Marcano, E. B., Gualdrón-López, M., & Michels, P. A., (2016). Biogenesis, maintenance and dynamics of glycosomes in trypanosomatid parasites. *Biochimica et Biophysica Acta (BBA)-Molecular Cell Research, 1863*(5), 1038–1048.

Hussey, G. D., & Klein, M., (1990). A randomized, controlled trial of vitamin A in children with severe measles. *New England Journal of Medicine, 323*(3), 160–164.

Krawinkel, M. B., (2012). Interaction of nutrition and infections globally: An overview. *Annals of Nutrition and Metabolism, 61*(Suppl. 1), 39–45.

Mvubu, N. E., Pillay, B., Gamieldien, J., Bishai, W., & Pillay, M., (2016). Canonical pathways, networks and transcriptional factor regulation by clinical strains of *Mycobacterium tuberculosis* in pulmonary alveolar epithelial cells. *Tuberculosis, 97*, 73–85.

Neyrolles, O., Wolschendorf, F., Mitra, A., & Niederweis, M., (2015). Mycobacteria, metals, and the macrophage. *Immunological Reviews, 264*(1), 249–263.

Nicolau, I., Ling, D., Tian, L., Lienhardt, C., & Pai, M. (2012). Research questions and priorities for tuberculosis: a survey of published systematic reviews and meta-analyses. *PloS one, 7*(7), e42479. https://doi.org/10.1371/journal.pone.0042479.

Oliveira, A. G. L., Brito, P. D., Schubach, A. O., Oliveira, R. V. C., Saheki, M. N., Lyra, M. R., Salgueiro, M. M., et al., (2013). Influence of the nutritional status in the clinical and therapeutical evolution in adults and elderly with American tegumentary leishmaniasis. *Acta Tropica, 128*(1), 36–40.

Paillaud, E., Herbaud, S., Caillet, P., Lejonc, J. L., Campillo, B., & Bories, P. N., (2005). Relations between undernutrition and nosocomial infections in elderly patients. *Age and Ageing, 34*(6), 619–625.

Ramachandran, P., & Gopalan, H. S., (2009). Undernutrition & risk of infections in preschool children. *Indian J. Med. Res., 130*(5), 579–583.

Salahuddin, N., Ali, F., Hasan, Z., Rao, N., Aqeel, M., & Mahmood, F., (2013). Vitamin D accelerates clinical recovery from tuberculosis: Results of the SUCCINCT study [supplementary cholecalciferol in recovery from tuberculosis]. A randomized, placebo-controlled, clinical trial of vitamin D supplementation in patients with pulmonary tuberculosis'. *BMC Infectious Diseases, 13*(1), 1–11.

Semba, R. D., Darnton-Hill, I., & De Pee, S., (2010). Addressing tuberculosis in the context of malnutrition and HIV coinfection. *Food and Nutrition Bulletin, 31*(4_suppl4), S345–S364.

Shin, D. M., Yuk, J. M., Lee, H. M., Lee, S. H., Son, J. W., Harding, C. V., Kim, J. M., et al., (2010). Mycobacterial lipoprotein activates autophagy via TLR2/1/CD14 and a functional vitamin D receptor signaling. *Cellular Microbiology, 12*(11), 1648–1665.

Sousa-Figueiredo, J. C., Gamboa, D., Pedro, J. M., Fançony, C., Langa, A. J., Soares, M. R. J., Stothard, J. R., & Nery, S. V., (2012). Epidemiology of malaria, schistosomiasis, geohelminths, anemia and malnutrition in the context of a demographic surveillance system in Northern Angola. *PloS One, 7*(4), e33189.

Spiegel, S., & Milstien, S., (2002). Sphingosine 1-phosphate, a key cell signaling molecule. *Journal of Biological Chemistry, 277*(29), 25851–25854.

Srivastava, A., Philip, N., Hughes, K. R., Georgiou, K., MacRae, J. I., Barrett, M. P., Creek, D. J., et al., (2016). Stage-specific changes in *Plasmodium* metabolism required for differentiation and adaptation to different host and vector environments. *PLoS Pathogens, 12*(12), e1006094.

Surana, K. R., Ahire, E. D., Sonawane, V. N., Talele, S. G., & Talele, G. S., (2021). Molecular modeling: Novel techniques in food and nutrition development. *Natural Food Products and Waste Recovery: Healthy Foods, Nutrition Design, and Extraction of Valuable Compounds*, 17.

Surana, K. R., Ahire, E. D., Sonawane, V. N., Talele, S. G., & Talele, G. S., (2021). Informatics and methods in nutrition design and development. *Natural Food Products and Waste Recovery: Healthy Foods, Nutrition Design, and Extraction of Valuable Compounds*, 33.

Vickers, N. J., (2017). Animal communication: When I m calling you, will you answer too? *Current Biology, 27*(14), R713–R715.

Wintergerst, E. S., Maggini, S., & Hornig, D. H., (2007). Contribution of selected vitamins and trace elements to immune function. *Annals of Nutrition and Metabolism, 51*(4), 301–323.

Yurie, H., Ikeguchi, R., Aoyama, T., Kaizawa, Y., Tajino, J., Ito, A., Ohta, S., et al., (2017). The efficacy of a scaffold-free bio 3D conduit developed from human fibroblasts on peripheral nerve regeneration in a rat sciatic nerve model. *PloS One, 12*(2), e0171448.

Zeng, C., Yang, P., Cao, T., Gu, Y., Li, N., Zhang, B., Xu, P., et al., (2021). Gut microbiota: An intermediary between metabolic syndrome and cognitive deficits in schizophrenia. *Progress in Neuro-Psychopharmacology and Biological Psychiatry, 106*, 110097.

INDEX

D

M